Algebra
DeMYSTiFieD®

DeMYSTiFieD® Series

Accounting Demystified
Advanced Calculus Demystified
Advanced Physics Demystified
Advanced Statistics Demystified
Algebra Demystified
Alternative Energy Demystified
Anatomy Demystified
asp.net 2.0 Demystified
Astronomy Demystified
Audio Demystified
Biology Demystified
Biotechnology Demystified
Business Calculus Demystified
Business Math Demystified
Business Statistics Demystified
C++ Demystified
Calculus Demystified
Chemistry Demystified
Circuit Analysis Demystified
College Algebra Demystified
Corporate Finance Demystified
Databases Demystified
Data Structures Demystified
Differential Equations Demystified
Digital Electronics Demystified
Earth Science Demystified
Electricity Demystified
Electronics Demystified
Engineering Statistics Demystified
Environmental Science Demystified
Everyday Math Demystified
Fertility Demystified
Financial Planning Demystified
Forensics Demystified
French Demystified
Genetics Demystified
Geometry Demystified
German Demystified
Home Networking Demystified
Investing Demystified
Italian Demystified
Java Demystified
JavaScript Demystified
Lean Six Sigma Demystified

Linear Algebra Demystified
Macroeconomics Demystified
Management Accounting Demystified
Math Proofs Demystified
Math Word Problems Demystified
MATLAB® Demystified
Medical Billing and Coding Demystified
Medical Terminology Demystified
Meteorology Demystified
Microbiology Demystified
Microeconomics Demystified
Nanotechnology Demystified
Nurse Management Demystified
OOP Demystified
Options Demystified
Organic Chemistry Demystified
Personal Computing Demystified
Pharmacology Demystified
Physics Demystified
Physiology Demystified
Pre-Algebra Demystified
Precalculus Demystified
Probability Demystified
Project Management Demystified
Psychology Demystified
Quality Management Demystified
Quantum Mechanics Demystified
Real Estate Math Demystified
Relativity Demystified
Robotics Demystified
Sales Management Demystified
Signals and Systems Demystified
Six Sigma Demystified
Spanish Demystified
sql Demystified
Statics and Dynamics Demystified
Statistics Demystified
Technical Analysis Demystified
Technical Math Demystified
Trigonometry Demystified
uml Demystified
Visual Basic 2005 Demystified
Visual C# 2005 Demystified
xml Demystified

Algebra DeMYSTiFieD®

Rhonda Huettenmueller

Second Edition

New York   Chicago   San Francisco   Lisbon   London   Madrid   Mexico City
Milan   New Delhi   San Juan   Seoul   Singapore   Sydney   Toronto

The McGraw·Hill Companies

Library of Congress Cataloging-in-Publication Data

Huettenmueller, Rhonda.
    Algebra demystified/Rhonda Huetenmueller.—2nd ed.
        p. cm.
    Includes index.
    ISBN 978-0-07-174361-7 (alk. paper)
  1. Algebra. I. Title.
    QA152.3.H85   2011
    512—dc22

                    2010042722

McGraw-Hill books are available at special quantity discounts to use as premiums and sales promotions, or for use in corporate training programs. To contact a representative please e-mail us at bulksales@mcgraw-hill.com.

Algebra DeMYSTiFieD®, Second Edition

3 4 5 6 7 8 9 0   DOC/DOC 1 9 8 7 6 5 4 3 2

ISBN 978-0-07-174361-7
MHID 0-07-174361-8

Sponsoring Editor	**Copy Editor**	**Art Director, Cover**
Judy Bass	Erica Orloff	Jeff Weeks
Acquisitions Coordinator	**Proofreader**	**Cover Illustration**
Michael Mulcahy	Surendra Nath Shivam	Lance Lekander
Editing Supervisor	**Production Supervisor**	
David E. Fogarty	Pamela A. Pelton	
Project Manager	**Composition**	
Vasundhara Sawhney, Glyph International	Glyph International	

About the Author

Rhonda Huettenmueller, Ph.D., has taught mathematics at the college level for more than 20 years. Popular with students for her ability to make higher math understandable and even enjoyable, she incorporates many of her teaching techniques in this book. Dr. Huettenmueller is the author of several highly successful DeMYSTiFieD titles, including *Business Calculus DeMYSTiFieD*, *Precalculus DeMYSTiFieD*, and *College Algebra DeMYSTiFieD*.

Contents

How to Use This Book

By working through this book, you will learn algebra skills that are necessary for just about any high school or college math course. This book has two features that most other math books do not. One of them is that this book presents only one idea at a time. This allows you to gradually master all of the material. The other feature is that the solutions to the problems in the examples and practice sets are complete. Usually, only one step is performed at a time so that you can clearly see how to solve the problem.

Because fractions and word problems (also called "applications") frustrate students, this book covers these topics with extra care. We begin with very simple, basic fraction operations and slowly move into more complicated fraction operations. Whenever a new algebra technique is covered, an entire section is devoted to how this technique affects fractions.

Even though two entire chapters are devoted to word problems, we develop one of the most important skills necessary for solving word problems right away. In Chapter 1 and Chapter 2, we learn how to translate English sentences into mathematical symbols. In Chapter 8, the first chapter that covers word problems, we begin with easier word problems and then we learn how to solve many of the standard problems found in an algebra course.

You will get the most from this book if you do not try to work through a section without understanding how to solve all of the problems in the previous practice set. In general, each new section extends the topic from the previous section. Once you have finished the last section in a chapter, review the chapter before taking the multiple-choice quiz. This will help you see how well you have retained the material. Once you have finished the last chapter, review all of the chapters before taking the multiple-choice Final Exam. The

xii   ALGEBRA DeMYSTiFieD

Final Exam is probably too long to take at once. Instead, you can treat it as several smaller tests. Try to improve your score with each of these "mini-tests."

If you are patient and take your time to work through all of the material, you will become comfortable with algebra and maybe even find it fun!

Rhonda Huettenmueller

chapter 1

Fractions

Being able to perform arithmetic with fractions is one of the most basic skills that we use in algebra. Though you might feel the fraction arithmetic that we learn in this chapter is not necessary (most calculators can do these computations for us), the methods that we develop in this chapter will help us when working with the kinds of fractions that frequently occur in algebra.

CHAPTER OBJECTIVES

In this chapter, you will

- Multiply and divide fractions
- Simplify fractions
- Add and subtract fractions having like denominators
- Add and subtract fractions having unlike denominators
- Convert between mixed numbers and improper fractions
- Simplify compound fractions
- Translate English sentences into mathematical symbols

Fraction Multiplication

To illustrate concepts in fraction arithmetic, we will use pie charts. For example, we represent the fraction $\frac{1}{3}$ with the shaded region in Figure 1-1. That is, $\frac{1}{3}$ is one part out of three equal parts.

Let us now develop the rule for multiplying fractions, $\frac{a}{b} \cdot \frac{c}{d} = \frac{ac}{bd}$. For example, using this rule we can compute $\frac{2}{3} \cdot \frac{1}{4}$ by multiplying the numerators, 2 and 1, and the denominators, 3 and 4. Doing so, we obtain $\frac{2}{3} \cdot \frac{1}{4} = \frac{2 \cdot 1}{3 \cdot 4} = \frac{2}{12}$ (we will concern ourselves later with simplifying fractions). Let us see how to represent the product $\frac{2}{3} \cdot \frac{1}{4}$ on the pie chart. We can think of this fraction as "two-thirds of one-fourth." We begin with one-fourth represented by a pie chart in Figure 1-2.

Let us see what happens to the representation of one-fourth if we divide the pie into twelve equal parts as in Figure 1-3.

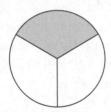

FIGURE 1-1

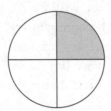

FIGURE 1-2

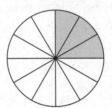

FIGURE 1-3

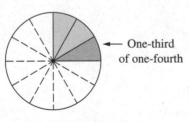

← One-third of one-fourth

FIGURE 1-4

Now we see that the fraction $\frac{1}{4}$ is the same as $\frac{3}{12}$. We can also see that when $\frac{1}{4}$ is itself divided into three equal pieces, each piece represents one-twelfth, so two-thirds of $\frac{1}{4}$ is two-twelfths. (See Figure 1-4.) This is why $\frac{2}{3} \cdot \frac{1}{4}$ is $\frac{2}{12}$.

EXAMPLE

Perform the multiplication with the rule $\frac{a}{b} \cdot \frac{c}{d} = \frac{ac}{bd}$.

$$\frac{2}{3} \cdot \frac{4}{5}$$

According to the rule, we multiply the numerators, 2 and 4, and the denominators, 3 and 5 to obtain $\frac{2}{3} \cdot \frac{4}{5} = \frac{2 \cdot 4}{3 \cdot 5} = \frac{8}{15}$.

PRACTICE

Perform the multiplication with the rule $\frac{a}{b} \cdot \frac{c}{d} = \frac{ac}{bd}$.

1. $\dfrac{7}{6} \cdot \dfrac{1}{4} =$

2. $\dfrac{8}{15} \cdot \dfrac{6}{5} =$

3. $\dfrac{5}{3} \cdot \dfrac{9}{10} =$

4. $\dfrac{40}{9} \cdot \dfrac{2}{3} =$

5. $\dfrac{3}{7} \cdot \dfrac{30}{4} =$

SOLUTIONS

1. $\dfrac{7}{6} \cdot \dfrac{1}{4} = \dfrac{7 \cdot 1}{6 \cdot 4} = \dfrac{7}{24}$

2. $\dfrac{8}{15} \cdot \dfrac{6}{5} = \dfrac{8 \cdot 6}{15 \cdot 5} = \dfrac{48}{75}$

3. $\dfrac{5}{3} \cdot \dfrac{9}{10} = \dfrac{5 \cdot 9}{3 \cdot 10} = \dfrac{45}{30}$

4. $\dfrac{40}{9} \cdot \dfrac{2}{3} = \dfrac{40 \cdot 2}{9 \cdot 3} = \dfrac{80}{27}$

5. $\dfrac{3}{7} \cdot \dfrac{30}{4} = \dfrac{3 \cdot 30}{7 \cdot 4} = \dfrac{90}{28}$

Multiplying Fractions and Whole Numbers

We now develop a rule for multiplying a whole number and a fraction. To see how we can multiply a fraction by a whole number, we use a pie chart to find the product $4 \cdot \dfrac{2}{9}$.

The shaded region in Figure 1-5 represents $\frac{2}{9}$.

We want a total of four of these shaded regions. See Figure 1-6.

As we can see, four of the $\frac{2}{9}$ regions give us a total of eight $\frac{1}{9}$'s. This is why $4 \cdot \dfrac{2}{9}$ is $4 \cdot \dfrac{2}{9} = \dfrac{4 \cdot 2}{9} = \dfrac{8}{9}$.

In general, when multiplying $W \cdot \dfrac{a}{b}$ (where W is a whole number), we have $W \cdot a$ of the $\frac{1}{b}$ fractions. This fact gives us the multiplication rule $W \cdot \dfrac{a}{b} = \dfrac{Wa}{b}$.

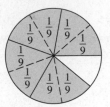

FIGURE 1-5   FIGURE 1-6

That is, the numerator of the product is the whole number times the fraction's numerator, and the denominator is the fraction's denominator.

An alternate method for finding the product of a whole number and a fraction is to treat the whole number as a fraction—the whole number over one—and then multiply as we would any two fractions. This method gives us the same rule:

$$W \cdot \frac{a}{b} = \frac{W}{1} \cdot \frac{a}{b} = \frac{W \cdot a}{1 \cdot b} = \frac{Wa}{b}$$

EXAMPLE

Find the product $5 \cdot \frac{2}{3}$ with each method outlined above.

$$5 \cdot \frac{2}{3} = \frac{5 \cdot 2}{3} = \frac{10}{3}$$

or

$$5 \cdot \frac{2}{3} = \frac{5}{1} \cdot \frac{2}{3} = \frac{5 \cdot 2}{1 \cdot 3} = \frac{10}{3}$$

PRACTICE

Find the product with either of the two methods outlined above.

1. $\frac{6}{7} \cdot 9 =$

2. $8 \cdot \frac{1}{6} =$

3. $4 \cdot \frac{2}{5} =$

4. $\frac{3}{14} \cdot 2 =$

5. $12 \cdot \frac{2}{15} =$

✔ SOLUTIONS

1. $\dfrac{6}{7} \cdot 9 = \dfrac{6 \cdot 9}{7} = \dfrac{54}{7}$ or $\dfrac{6}{7} \cdot \dfrac{9}{1} = \dfrac{6 \cdot 9}{7 \cdot 1} = \dfrac{54}{7}$

2. $8 \cdot \dfrac{1}{6} = \dfrac{8 \cdot 1}{6} = \dfrac{8}{6}$ or $\dfrac{8}{1} \cdot \dfrac{1}{6} = \dfrac{8 \cdot 1}{1 \cdot 6} = \dfrac{8}{6}$

3. $4 \cdot \dfrac{2}{5} = \dfrac{4 \cdot 2}{5} = \dfrac{8}{5}$ or $\dfrac{4}{1} \cdot \dfrac{2}{5} = \dfrac{4 \cdot 2}{1 \cdot 5} = \dfrac{8}{5}$

4. $\dfrac{3}{14} \cdot 2 = \dfrac{3 \cdot 2}{14} = \dfrac{6}{14}$ or $\dfrac{3}{14} \cdot \dfrac{2}{1} = \dfrac{3 \cdot 2}{14 \cdot 1} = \dfrac{6}{14}$

5. $12 \cdot \dfrac{2}{15} = \dfrac{12 \cdot 2}{15} = \dfrac{24}{15}$ or $\dfrac{12}{1} \cdot \dfrac{2}{15} = \dfrac{12 \cdot 2}{1 \cdot 15} = \dfrac{24}{15}$

Fraction Division

Fraction division is almost as easy as fraction multiplication. The rule for fraction division is $\dfrac{a}{b} \div \dfrac{c}{d} = \dfrac{a}{b} \cdot \dfrac{d}{c} = \dfrac{ad}{bc}$, that is, we invert (switch the numerator and denominator) the second fraction, and the fraction division problem becomes a fraction multiplication problem. To see why this might be so, consider the computation for $\dfrac{3}{1} \div \dfrac{1}{2}$. We can think of this division problem as asking the question, "How many halves go into 3?" Of course, the answer is 6, which agrees with the formula: $\dfrac{3}{1} \div \dfrac{1}{2} = 3 \cdot 2 = 6$.

EXAMPLES

Perform the division.

$$\frac{2}{3} \div \frac{4}{5} = \frac{2}{3} \cdot \frac{5}{4}\ \overset{\text{}}{=}\ \frac{10}{12}$$

$$\frac{3}{4} \div 5 = \frac{3}{4} \div \frac{5}{1} = \frac{3}{4} \cdot \frac{1}{5} = \frac{3}{20}$$

PRACTICE

Perform the division.

1. $\dfrac{7}{6} \div \dfrac{1}{4} =$

2. $\dfrac{8}{15} \div \dfrac{6}{5} =$

3. $\dfrac{5}{3} \div \dfrac{9}{10} =$

4. $\dfrac{40}{9} \div \dfrac{2}{3} =$

5. $\dfrac{3}{7} \div \dfrac{30}{4} =$

6. $4 \div \dfrac{2}{3} =$

7. $\dfrac{10}{21} \div 3 =$

SOLUTIONS

1. $\dfrac{7}{6} \div \dfrac{1}{4} = \dfrac{7}{6} \cdot \dfrac{4}{1} = \dfrac{28}{6}$

2. $\dfrac{8}{15} \div \dfrac{6}{5} = \dfrac{8}{15} \cdot \dfrac{5}{6} = \dfrac{40}{90}$

3. $\dfrac{5}{3} \div \dfrac{9}{10} = \dfrac{5}{3} \cdot \dfrac{10}{9} = \dfrac{50}{27}$

4. $\dfrac{40}{9} \div \dfrac{2}{3} = \dfrac{40}{9} \cdot \dfrac{3}{2} = \dfrac{120}{18}$

5. $\dfrac{3}{7} \div \dfrac{30}{4} = \dfrac{3}{7} \cdot \dfrac{4}{30} = \dfrac{12}{210}$

6. $4 \div \dfrac{2}{3} = \dfrac{4}{1} \div \dfrac{2}{3} = \dfrac{4}{1} \cdot \dfrac{3}{2} = \dfrac{12}{2}$

7. $\dfrac{10}{21} \div 3 = \dfrac{10}{21} \div \dfrac{3}{1} = \dfrac{10}{21} \cdot \dfrac{1}{3} = \dfrac{10}{63}$

Simplifying Fractions

When working with fractions, we are usually asked to "reduce the fraction to lowest terms" or to "write the fraction in lowest terms" or to "simplify the fraction." These phrases mean that the numerator and denominator have no common factors (other than 1). For example, $\frac{2}{3}$ is in written in lowest terms but $\frac{4}{6}$ is not because 2 is a factor of both 4 and 6. Simplifying fractions is like fraction multiplication in reverse. For now, we will use the most basic approach to simplifying fractions. In the next section, we will learn a quicker method.

First write the numerator and denominator as a product of prime numbers. (Refer to the Appendix if you need to review finding the prime factorization of a number.) Next collect the prime numbers common to both the numerator and denominator (if any) at beginning of each fraction. Split each fraction into two fractions, the first with the common prime numbers. This puts the fraction in the form of "1" times another fraction. This might seem like unnecessary work (actually, it is), but it will drive home the point that the factors that are common in the numerator and denominator form the number 1. Thinking of simplifying fractions in this way can help you avoid common fraction errors later in algebra.

 EXAMPLE

Simplify the fraction with the method outlined above.

$$\frac{6}{18}$$

 SOLUTION

We begin by factoring 6 and 18.

$$\frac{6}{18} = \frac{2 \cdot 3}{2 \cdot 3 \cdot 3} = \frac{(2 \cdot 3) \cdot 1}{(2 \cdot 3) \cdot 3}$$

We now write the common factors as a separate fraction.

$$\frac{6}{18} = \frac{2 \cdot 3}{2 \cdot 3 \cdot 3} = \frac{(2 \cdot 3) \cdot 1}{(2 \cdot 3) \cdot 3} = \frac{2 \cdot 3}{2 \cdot 3} \cdot \frac{1}{3} = \frac{6}{6} \cdot \frac{1}{3}$$

Because $\frac{6}{6}$ is 1, we see that $\frac{6}{18}$ simplifies to $\frac{1}{3}$.

$$\frac{6}{18} = \frac{2 \cdot 3}{2 \cdot 3 \cdot 3} = \frac{(2 \cdot 3) \cdot 1}{(2 \cdot 3) \cdot 3} = \frac{2 \cdot 3}{2 \cdot 3} \cdot \frac{1}{3} = \frac{6}{6} \cdot \frac{1}{3} = \frac{1}{3}$$

$$\frac{42}{49} = \frac{7 \cdot 2 \cdot 3}{7 \cdot 7} = \frac{7}{7} \cdot \frac{2 \cdot 3}{7} = 1 \cdot \frac{6}{7} = \frac{6}{7}$$

PRACTICE _____

Simplify the fraction.

1. $\dfrac{14}{42} =$

2. $\dfrac{5}{35} =$

3. $\dfrac{48}{30} =$

4. $\dfrac{22}{121} =$

5. $\dfrac{39}{123} =$

6. $\dfrac{18}{4} =$

7. $\dfrac{7}{210} =$

8. $\dfrac{240}{165} =$

9. $\dfrac{55}{33} =$

10. $\dfrac{150}{30} =$

✔**SOLUTIONS** _____

1. $\dfrac{14}{42} = \dfrac{2 \cdot 7}{2 \cdot 3 \cdot 7} = \dfrac{(2 \cdot 7) \cdot 1}{(2 \cdot 7) \cdot 3} = \dfrac{2 \cdot 7}{2 \cdot 7} \cdot \dfrac{1}{3} = \dfrac{14}{14} \cdot \dfrac{1}{3} = \dfrac{1}{3}$

2. $\dfrac{5}{35} = \dfrac{5}{5 \cdot 7} = \dfrac{5 \cdot 1}{5 \cdot 7} = \dfrac{5}{5} \cdot \dfrac{1}{7} = \dfrac{1}{7}$

3. $\dfrac{48}{30} = \dfrac{2 \cdot 2 \cdot 2 \cdot 2 \cdot 3}{2 \cdot 3 \cdot 5} = \dfrac{(2 \cdot 3) \cdot 2 \cdot 2 \cdot 2}{(2 \cdot 3) \cdot 5} = \dfrac{2 \cdot 3}{2 \cdot 3} \cdot \dfrac{2 \cdot 2 \cdot 2}{5} = \dfrac{6}{6} \cdot \dfrac{8}{5} = \dfrac{8}{5}$

4. $\dfrac{22}{121} = \dfrac{2 \cdot 11}{11 \cdot 11} = \dfrac{11}{11} \cdot \dfrac{2}{11} = \dfrac{2}{11}$

5. $\dfrac{39}{123} = \dfrac{3 \cdot 13}{3 \cdot 41} = \dfrac{3}{3} \cdot \dfrac{13}{41} = \dfrac{13}{41}$

6. $\dfrac{18}{4} = \dfrac{2 \cdot 3 \cdot 3}{2 \cdot 2} = \dfrac{2}{2} \cdot \dfrac{3 \cdot 3}{2} = \dfrac{9}{2}$

7. $\dfrac{7}{210} = \dfrac{7}{2 \cdot 3 \cdot 5 \cdot 7} = \dfrac{7 \cdot 1}{7 \cdot 2 \cdot 3 \cdot 5} = \dfrac{7}{7} \cdot \dfrac{1}{2 \cdot 3 \cdot 5} = \dfrac{1}{30}$

8. $\dfrac{240}{165} = \dfrac{2 \cdot 2 \cdot 2 \cdot 2 \cdot 3 \cdot 5}{3 \cdot 5 \cdot 11} = \dfrac{(3 \cdot 5) \cdot 2 \cdot 2 \cdot 2 \cdot 2}{(3 \cdot 5) \cdot 11} = \dfrac{3 \cdot 5}{3 \cdot 5} \cdot \dfrac{2 \cdot 2 \cdot 2 \cdot 2}{11}$

$= \dfrac{15}{15} \cdot \dfrac{16}{11} = \dfrac{16}{11}$

9. $\dfrac{55}{33} = \dfrac{5 \cdot 11}{3 \cdot 11} = \dfrac{11 \cdot 5}{11 \cdot 3} = \dfrac{11}{11} \cdot \dfrac{5}{3} = \dfrac{5}{3}$

10. $\dfrac{150}{30} = \dfrac{2 \cdot 3 \cdot 5 \cdot 5}{2 \cdot 3 \cdot 5} = \dfrac{(2 \cdot 3 \cdot 5) \cdot 5}{(2 \cdot 3 \cdot 5) \cdot 1} = \dfrac{2 \cdot 3 \cdot 5}{2 \cdot 3 \cdot 5} \cdot \dfrac{5}{1} = \dfrac{30}{30} \cdot 5 = 5$

The Greatest Common Divisor

Fortunately there is a less tedious method for writing a fraction in its lowest terms. We find the largest number that divides both the numerator and the denominator. This number is called the *greatest common divisor* (GCD). We factor the GCD from the numerator and denominator and then we rewrite the fraction in the form:

$$\dfrac{\text{GCD}}{\text{GCD}} \cdot \dfrac{\text{Other numerator factors}}{\text{Other denominator factors}}$$

In the previous practice problems, the product of the common primes was the GCD.

EXAMPLES

Identify the GCD for the numerator and denominator and write the fraction in lowest terms.

$$\dfrac{32}{48} = \dfrac{16 \cdot 2}{16 \cdot 3} = \dfrac{16}{16} \cdot \dfrac{2}{3} = 1 \cdot \dfrac{2}{3} = \dfrac{2}{3}$$

$$\dfrac{45}{60} = \dfrac{15 \cdot 3}{15 \cdot 4} = \dfrac{15}{15} \cdot \dfrac{3}{4} = 1 \cdot \dfrac{3}{4} = \dfrac{3}{4}$$

PRACTICE

Identify the GCD for the numerator and denominator and write the fraction in lowest terms.

1. $\dfrac{12}{38} =$

2. $\dfrac{12}{54} =$

3. $\dfrac{16}{52} =$

4. $\dfrac{56}{21} =$

5. $\dfrac{45}{100} =$

6. $\dfrac{48}{56} =$

7. $\dfrac{28}{18} =$

8. $\dfrac{24}{32} =$

9. $\dfrac{36}{60} =$

10. $\dfrac{12}{42} =$

SOLUTIONS

1. $\dfrac{12}{38} = \dfrac{2 \cdot 6}{2 \cdot 19} = \dfrac{2}{2} \cdot \dfrac{6}{19} = \dfrac{6}{19}$

2. $\dfrac{12}{54} = \dfrac{6 \cdot 2}{6 \cdot 9} = \dfrac{6}{6} \cdot \dfrac{2}{9} = \dfrac{2}{9}$

3. $\dfrac{16}{52} = \dfrac{4 \cdot 4}{4 \cdot 13} = \dfrac{4}{4} \cdot \dfrac{4}{13} = \dfrac{4}{13}$

4. $\dfrac{56}{21} = \dfrac{7 \cdot 8}{7 \cdot 3} = \dfrac{7}{7} \cdot \dfrac{8}{3} = \dfrac{8}{3}$

5. $\dfrac{45}{100} = \dfrac{5 \cdot 9}{5 \cdot 20} = \dfrac{5}{5} \cdot \dfrac{9}{20} = \dfrac{9}{20}$

6. $\dfrac{48}{56} = \dfrac{8 \cdot 6}{8 \cdot 7} = \dfrac{8}{8} \cdot \dfrac{6}{7} = \dfrac{6}{7}$

7. $\dfrac{28}{18} = \dfrac{2 \cdot 14}{2 \cdot 9} = \dfrac{2}{2} \cdot \dfrac{14}{9} = \dfrac{14}{9}$

8. $\dfrac{24}{32} = \dfrac{8 \cdot 3}{8 \cdot 4} = \dfrac{8}{8} \cdot \dfrac{3}{4} = \dfrac{3}{4}$

9. $\dfrac{36}{60} = \dfrac{12 \cdot 3}{12 \cdot 5} = \dfrac{12}{12} \cdot \dfrac{3}{5} = \dfrac{3}{5}$

10. $\dfrac{12}{42} = \dfrac{6 \cdot 2}{6 \cdot 7} = \dfrac{6}{6} \cdot \dfrac{2}{7} = \dfrac{2}{7}$

Sometimes the greatest common divisor is not obvious. In these cases we might want to simplify the fraction in multiple steps.

EXAMPLES

Write the fraction in lowest terms.

$$\dfrac{3990}{6762} = \dfrac{6 \cdot 665}{6 \cdot 1127} = \dfrac{665}{1127} = \dfrac{7 \cdot 95}{7 \cdot 161} = \dfrac{95}{161}$$

$$\dfrac{644}{2842} = \dfrac{2 \cdot 322}{2 \cdot 1421} = \dfrac{322}{1421} = \dfrac{7 \cdot 46}{7 \cdot 203} = \dfrac{46}{203}$$

PRACTICE

Write the fraction in lowest terms.

1. $\dfrac{600}{1280} =$

2. $\dfrac{68}{578} =$

3. $\dfrac{168}{216} =$

4. $\dfrac{72}{120} =$

5. $\dfrac{768}{288} =$

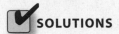 **SOLUTIONS**

1. $\dfrac{600}{1280} = \dfrac{10 \cdot 60}{10 \cdot 128} = \dfrac{60}{128} = \dfrac{4 \cdot 15}{4 \cdot 32} = \dfrac{15}{32}$

2. $\dfrac{68}{578} = \dfrac{2 \cdot 34}{2 \cdot 289} = \dfrac{34}{289} = \dfrac{17 \cdot 2}{17 \cdot 17} = \dfrac{2}{17}$

3. $\dfrac{168}{216} = \dfrac{6 \cdot 28}{6 \cdot 36} = \dfrac{28}{36} = \dfrac{4 \cdot 7}{4 \cdot 9} = \dfrac{7}{9}$

4. $\dfrac{72}{120} = \dfrac{12 \cdot 6}{12 \cdot 10} = \dfrac{6}{10} = \dfrac{2 \cdot 3}{2 \cdot 5} = \dfrac{3}{5}$

5. $\dfrac{768}{288} = \dfrac{4 \cdot 192}{4 \cdot 72} = \dfrac{192}{72} = \dfrac{2 \cdot 96}{2 \cdot 36} = \dfrac{96}{36} = \dfrac{4 \cdot 24}{4 \cdot 9} = \dfrac{24}{9} = \dfrac{3 \cdot 8}{3 \cdot 3} = \dfrac{8}{3}$

For the rest of the book, we will write fractions in lowest terms.

Adding and Subtracting Fractions with Like Denominators

If we want to add or subtract two fractions having the same denominators, we only need to add or subtract their numerators. The rule is $\dfrac{a}{b} + \dfrac{c}{b} = \dfrac{a+c}{b}$ and $\dfrac{a}{b} - \dfrac{c}{b} = \dfrac{a-c}{b}$.

Let us examine the sum $\dfrac{1}{6} + \dfrac{2}{6}$ with a pie chart.

Adding 1 one-sixth segment to 2 one-sixth segment gives us a total of 3 one-six segments, which agrees with the formula: $\dfrac{1}{6} + \dfrac{2}{6} = \dfrac{1+2}{6} = \dfrac{3}{6} = \dfrac{1}{2}$.

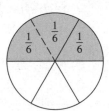

FIGURE 1-7

 EXAMPLES

Perform the addition or subtraction.

$$\frac{7}{9} - \frac{2}{9} = \frac{7-2}{9} = \frac{5}{9}$$

$$\frac{8}{15} + \frac{2}{15} = \frac{8+2}{15} = \frac{10}{15} = \frac{5 \cdot 2}{5 \cdot 3} = \frac{2}{3}$$

 PRACTICE

Perform the addition or subtraction.

1. $\dfrac{4}{7} - \dfrac{1}{7} =$

2. $\dfrac{1}{5} + \dfrac{3}{5} =$

3. $\dfrac{1}{6}+\dfrac{1}{6}=$

4. $\dfrac{5}{12}-\dfrac{1}{12}=$

5. $\dfrac{2}{11}+\dfrac{9}{11}=$

 **SOLUTIONS**

1. $\dfrac{4}{7}-\dfrac{1}{7}=\dfrac{4-1}{7}=\dfrac{3}{7}$

2. $\dfrac{1}{5}+\dfrac{3}{5}=\dfrac{1+3}{5}=\dfrac{4}{5}$

3. $\dfrac{1}{6}+\dfrac{1}{6}=\dfrac{1+1}{6}=\dfrac{2}{6}=\dfrac{1}{3}$

4. $\dfrac{5}{12}-\dfrac{1}{12}=\dfrac{5-1}{12}=\dfrac{4}{12}=\dfrac{1}{3}$

5. $\dfrac{2}{11}+\dfrac{9}{11}=\dfrac{2+9}{11}=\dfrac{11}{11}=1$

Adding and Subtracting Fractions with Unlike Denominators

If we need to find the sum or difference of two fractions having different denominators, then we must rewrite one or both fractions so that they have the *same* denominator. Let us use the pie model to find the sum $\dfrac{1}{4}+\dfrac{1}{3}$.

If we divide the pie into $3\times 4 = 12$ equal pieces, we see that $\frac{1}{4}$ is the same as $\frac{3}{12}$ and $\frac{1}{3}$ is the same as $\frac{4}{12}$.

Now that we have these fractions written so that they have the same denominator, we can add them: $\dfrac{1}{4}+\dfrac{1}{3}=\dfrac{3}{12}+\dfrac{4}{12}=\dfrac{3+4}{12}=\dfrac{7}{12}$.

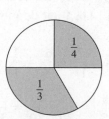

FIGURE 1-8

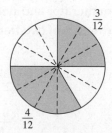

FIGURE 1-9

To compute $\frac{a}{b}+\frac{c}{d}$ or $\frac{a}{b}-\frac{c}{d}$, we can "reverse" the simplification process to rewrite the fractions so that they have the same denominator. This process is called *finding a common denominator*. Multiplying $\frac{a}{b}$ by $\frac{d}{d}$ (the second denominator over itself) and $\frac{c}{d}$ by $\frac{b}{b}$ (the first denominator over itself) gives us equivalent fractions that have the same denominator. Once this is done, we can add or subtract the numerators.

$$\frac{a}{b}+\frac{c}{d}=\frac{a}{b}\cdot\frac{d}{d}+\frac{c}{d}\cdot\frac{b}{b}= \overbrace{\frac{ad}{bd}+\frac{cb}{bd}}^{\text{Now we can add the numerators.}} = \frac{ad+cb}{bd}$$

$$\frac{a}{b}-\frac{c}{d}=\frac{a}{b}\cdot\frac{d}{d}-\frac{c}{d}\cdot\frac{b}{b}= \overbrace{\frac{ad}{bd}-\frac{cb}{bd}}^{\text{Now we can subtract the numerators.}} = \frac{ad-cb}{bd}$$

Note that this is essentially what we did with the pie chart to find $\frac{1}{4}+\frac{1}{3}$ when we divided the pie into $4\times3=12$ equal parts.

For now, we will use the formula $\frac{a}{b}\pm\frac{c}{d}=\frac{ad\pm cb}{bd}$ to add and subtract two fractions. Later, we will learn a method for finding a common denominator when the denominators have common factors.

 EXAMPLES

Find the sum or difference.

$$\frac{1}{2}+\frac{3}{7}$$

$$\frac{8}{15}-\frac{1}{2}$$

 **SOLUTIONS**

In this sum, the first denominator is 2 and the second denominator is 7. We multiply the first numerator and denominator of the first fraction, $\frac{1}{2}$, by 7 and the numerator and denominator of the second fraction, $\frac{3}{7}$, by 2. This gives us the sum of two fractions having 14 as their denominator.

$$\frac{1}{2}+\frac{3}{7}=\left(\frac{1}{2}\cdot\frac{7}{7}\right)+\left(\frac{3}{7}\cdot\frac{2}{2}\right)=\frac{7}{14}+\frac{6}{14}=\frac{13}{14}$$

$$\frac{8}{15}-\frac{1}{2}=\left(\frac{8}{15}\cdot\frac{2}{2}\right)-\left(\frac{1}{2}\cdot\frac{15}{15}\right)=\frac{16}{30}-\frac{15}{30}=\frac{1}{30}$$

PRACTICE

Find the sum or difference.

1. $\dfrac{5}{6} - \dfrac{1}{5} =$

2. $\dfrac{1}{3} + \dfrac{7}{8} =$

3. $\dfrac{5}{7} - \dfrac{1}{9} =$

4. $\dfrac{3}{14} + \dfrac{1}{2} =$

5. $\dfrac{3}{4} + \dfrac{11}{18} =$

SOLUTIONS

1. $\dfrac{5}{6} - \dfrac{1}{5} = \left(\dfrac{5}{6} \cdot \dfrac{5}{5}\right) - \left(\dfrac{1}{5} \cdot \dfrac{6}{6}\right) = \dfrac{25}{30} - \dfrac{6}{30} = \dfrac{19}{30}$

2. $\dfrac{1}{3} + \dfrac{7}{8} = \left(\dfrac{1}{3} \cdot \dfrac{8}{8}\right) + \left(\dfrac{7}{8} \cdot \dfrac{3}{3}\right) = \dfrac{8}{24} + \dfrac{21}{24} = \dfrac{29}{24}$

3. $\dfrac{5}{7} - \dfrac{1}{9} = \left(\dfrac{5}{7} \cdot \dfrac{9}{9}\right) - \left(\dfrac{1}{9} \cdot \dfrac{7}{7}\right) = \dfrac{45}{63} - \dfrac{7}{63} = \dfrac{38}{63}$

4. $\dfrac{3}{14} + \dfrac{1}{2} = \left(\dfrac{3}{14} \cdot \dfrac{2}{2}\right) + \left(\dfrac{1}{2} \cdot \dfrac{14}{14}\right) = \dfrac{6}{28} + \dfrac{14}{28} = \dfrac{20}{28} = \dfrac{5}{7}$

5. $\dfrac{3}{4} + \dfrac{11}{18} = \left(\dfrac{3}{4} \cdot \dfrac{18}{18}\right) + \left(\dfrac{11}{18} \cdot \dfrac{4}{4}\right) = \dfrac{54}{72} + \dfrac{44}{72} = \dfrac{98}{72} = \dfrac{49}{36}$

The Least Common Denominator (LCD)

Our goal is to add/subtract two fractions having the same denominator. In the previous example problems and practice problems, we found a common denominator. Now we will find the *least common denominator* (LCD). For example in $\frac{1}{3} + \frac{1}{6}$, we could compute $\frac{1}{3} + \frac{1}{6} = \left(\frac{1}{3} \cdot \frac{6}{6}\right) + \left(\frac{1}{6} \cdot \frac{3}{3}\right) = \frac{6}{18} + \frac{3}{18} = \frac{9}{18} = \frac{1}{2}$.

But we really only need to rewrite $\frac{1}{3}$: $\frac{1}{3} + \frac{1}{6} = \left(\frac{1}{3} \cdot \frac{2}{2}\right) + \frac{1}{6} = \frac{2}{6} + \frac{1}{6} = \frac{3}{6} = \frac{1}{2}$.

While 18 is *a* common denominator in the above example, 6 is the *smallest* common denominator. When denominators get more complicated, either by being large or by having variables in them, it usually easier to use the LCD to add or subtract fractions. The solution requires less simplifying, too.

In the following practice problems one of the denominators will be the LCD; you only need to rewrite one fraction before computing the sum or difference.

 PRACTICE

Find the sum or difference.

1. $\dfrac{1}{8} + \dfrac{1}{2} =$

2. $\dfrac{2}{3} - \dfrac{5}{12} =$

3. $\dfrac{4}{5} + \dfrac{1}{20} =$

4. $\dfrac{7}{30} - \dfrac{2}{15} =$

5. $\dfrac{5}{24} + \dfrac{5}{6} =$

 SOLUTIONS

1. $\dfrac{1}{8} + \dfrac{1}{2} = \dfrac{1}{8} + \left(\dfrac{1}{2} \cdot \dfrac{4}{4}\right) = \dfrac{1}{8} + \dfrac{4}{8} = \dfrac{5}{8}$

2. $\dfrac{2}{3} - \dfrac{5}{12} = \left(\dfrac{2}{3} \cdot \dfrac{4}{4}\right) - \dfrac{5}{12} = \dfrac{8}{12} - \dfrac{5}{12} = \dfrac{3}{12} = \dfrac{1}{4}$

3. $\dfrac{4}{5} + \dfrac{1}{20} = \left(\dfrac{4}{5} \cdot \dfrac{4}{4}\right) + \dfrac{1}{20} = \dfrac{16}{20} + \dfrac{1}{20} = \dfrac{17}{20}$

4. $\dfrac{7}{30} - \dfrac{2}{15} = \dfrac{7}{30} - \left(\dfrac{2}{15} \cdot \dfrac{2}{2}\right) = \dfrac{7}{30} - \dfrac{4}{30} = \dfrac{3}{30} = \dfrac{1}{10}$

5. $\dfrac{5}{24} + \dfrac{5}{6} = \dfrac{5}{24} + \left(\dfrac{5}{6} \cdot \dfrac{4}{4}\right) = \dfrac{5}{24} + \dfrac{20}{24} = \dfrac{25}{24}$

Finding the LCD

We have a couple of ways for finding the LCD. Take, for example, $\dfrac{1}{12} + \dfrac{9}{14}$. We could list the multiples of 12 and 14—the first number that appears on each list is the LCD: 12, 24, 36, 48, 60, 72, **84** and 14, 28, 42, 56, 70, **84**.

Because 84 is the first number on each list, **84** is the LCD for $\frac{1}{12}$ and $\frac{9}{14}$. This method works fine as long as the lists aren't too long. But what if the denominators are 6 and 291 for example? The LCD for these denominators (which is 582) occurs 97th on the list of multiples of 6.

We can use the prime factors of the denominators to find the LCD more efficiently. The LCD consists of every prime factor in each denominator (at its most frequent occurrence). To find the LCD for $\frac{1}{12}$ and $\frac{9}{14}$, we factor 12 and 14 into their prime factorizations: $12 = 2 \cdot 2 \cdot 3$ and $14 = 2 \cdot 7$. There are two 2's and one 3 in the prime factorization of 12, so the LCD will have two 2's and one 3. There is one 2 in the prime factorization of 14, but this 2 is covered by the 2's from 12. There is one 7 in the prime factorization of 14, so the LCD will also have a 7 as a factor. Once we have computed the LCD, we divide the LCD by each denominator and then multiply the fractions by these numbers over themselves.

$$LCD = 2 \cdot 2 \cdot 3 \cdot 7 = 84$$

$$84 \div 12 = 7\text{: multiply } \tfrac{1}{12} \text{ by } \tfrac{7}{7} \qquad 84 \div 14 = 6\text{: multiply } \tfrac{9}{14} \text{ by } \tfrac{6}{6}$$

$$\frac{1}{12} + \frac{9}{14} = \left(\frac{1}{12} \cdot \frac{7}{7} \right) + \left(\frac{9}{14} \cdot \frac{6}{6} \right) = \frac{7}{84} + \frac{54}{84} = \frac{61}{84}$$

EXAMPLE

Find the sum or difference after computing the LCD.

$$\frac{5}{6} + \frac{4}{15}$$

 **SOLUTION**

We begin by factoring the denominators: $6 = 2 \cdot 3$ and $15 = 3 \cdot 5$. The LCD is $2 \cdot 3 \cdot 5 = 30$. Dividing 30 by each denominator gives us $30 \div 6 = 5$ and $30 \div 15 = 2$. Once we multiply $\frac{5}{6}$ by $\frac{5}{5}$ and $\frac{4}{15}$ by $\frac{2}{2}$, we can add the fractions.

$$\frac{5}{6} + \frac{4}{15} = \left(\frac{5}{6} \cdot \frac{5}{5} \right) + \left(\frac{4}{15} \cdot \frac{2}{2} \right) = \frac{25}{30} + \frac{8}{30} = \frac{33}{30} = \frac{11}{10}$$

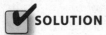

EXAMPLE

Find the sum or difference after computing the LCD.

$$\frac{17}{24}+\frac{5}{36}$$

SOLUTION

$$24 = 2 \cdot 2 \cdot 2 \cdot 3 \quad \text{and} \quad 36 = 2 \cdot 2 \cdot 3 \cdot 3$$

The LCD = $2 \cdot 2 \cdot 2 \cdot 3 \cdot 3 = 72$; $72 \div 24 = 3$; and $72 \div 36 = 2$. We multiply $\frac{17}{24}$ by $\frac{3}{3}$ and $\frac{5}{36}$ by $\frac{2}{2}$.

$$\frac{17}{24}+\frac{5}{36}=\left(\frac{17}{24}\cdot\frac{3}{3}\right)+\left(\frac{5}{36}\cdot\frac{2}{2}\right)=\frac{51}{72}+\frac{10}{72}=\frac{61}{72}$$

PRACTICE

Find the sum or difference after computing the LCD.

1. $\dfrac{11}{12}-\dfrac{5}{18}=$

2. $\dfrac{7}{15}+\dfrac{9}{20}=$

3. $\dfrac{23}{24}+\dfrac{7}{16}=$

4. $\dfrac{3}{8}+\dfrac{7}{20}=$

5. $\dfrac{1}{6}+\dfrac{4}{15}=$

6. $\dfrac{8}{75}+\dfrac{3}{10}=$

7. $\dfrac{35}{54}-\dfrac{7}{48}=$

8. $\dfrac{15}{88}+\dfrac{3}{28}=$

9. $\dfrac{119}{180}+\dfrac{17}{210}=$

✔ **SOLUTIONS**

1. $\dfrac{11}{12}-\dfrac{5}{18}=\left(\dfrac{11}{12}\cdot\dfrac{3}{3}\right)-\left(\dfrac{5}{18}\cdot\dfrac{2}{2}\right)=\dfrac{33}{36}-\dfrac{10}{36}=\dfrac{23}{36}$

2. $\dfrac{7}{15}+\dfrac{9}{20}=\left(\dfrac{7}{15}\cdot\dfrac{4}{4}\right)+\left(\dfrac{9}{20}\cdot\dfrac{3}{3}\right)=\dfrac{28}{60}+\dfrac{27}{60}=\dfrac{55}{60}=\dfrac{11}{12}$

3. $\dfrac{23}{24}+\dfrac{7}{16}=\left(\dfrac{23}{24}\cdot\dfrac{2}{2}\right)+\left(\dfrac{7}{16}\cdot\dfrac{3}{3}\right)=\dfrac{46}{48}+\dfrac{21}{48}=\dfrac{67}{48}$

4. $\dfrac{3}{8}+\dfrac{7}{20}=\left(\dfrac{3}{8}\cdot\dfrac{5}{5}\right)+\left(\dfrac{7}{20}\cdot\dfrac{2}{2}\right)=\dfrac{15}{40}+\dfrac{14}{40}=\dfrac{29}{40}$

5. $\dfrac{1}{6}+\dfrac{4}{15}=\left(\dfrac{1}{6}\cdot\dfrac{5}{5}\right)+\left(\dfrac{4}{15}\cdot\dfrac{2}{2}\right)=\dfrac{5}{30}+\dfrac{8}{30}=\dfrac{13}{30}$

6. $\dfrac{8}{75}+\dfrac{3}{10}=\left(\dfrac{8}{75}\cdot\dfrac{2}{2}\right)+\left(\dfrac{3}{10}\cdot\dfrac{15}{15}\right)=\dfrac{16}{150}+\dfrac{45}{150}=\dfrac{61}{150}$

7. $\dfrac{35}{54}-\dfrac{7}{48}=\left(\dfrac{35}{54}\cdot\dfrac{8}{8}\right)-\left(\dfrac{7}{48}\cdot\dfrac{9}{9}\right)=\dfrac{280}{432}-\dfrac{63}{432}=\dfrac{217}{432}$

8. $\dfrac{15}{88}+\dfrac{3}{28}=\left(\dfrac{15}{88}\cdot\dfrac{7}{7}\right)+\left(\dfrac{3}{28}\cdot\dfrac{22}{22}\right)=\dfrac{105}{616}+\dfrac{66}{616}=\dfrac{171}{616}$

9. $\dfrac{119}{180}+\dfrac{17}{210}=\left(\dfrac{119}{180}\cdot\dfrac{7}{7}\right)+\left(\dfrac{17}{210}\cdot\dfrac{6}{6}\right)=\dfrac{833}{1260}+\dfrac{102}{1260}=\dfrac{935}{1260}=\dfrac{187}{252}$

Adding More than Two Fractions

Finding the LCD for three or more fractions is pretty much the same as finding the LCD for two fractions. One way to approach the problem is to work with two fractions at a time. For instance, in the sum $\dfrac{5}{6}+\dfrac{3}{4}+\dfrac{1}{10}$, we can begin with $\frac{5}{6}$ and $\frac{3}{4}$. The LCD for these fractions is 12.

$$\dfrac{5}{6}=\dfrac{5}{6}\cdot\dfrac{2}{2}=\dfrac{10}{12} \qquad \text{and} \qquad \dfrac{3}{4}=\dfrac{3}{4}\cdot\dfrac{3}{3}=\dfrac{9}{12}$$

The sum $\dfrac{5}{6}+\dfrac{3}{4}+\dfrac{1}{10}$ can be condensed to the sum of two fractions.

$$\dfrac{5}{6}+\dfrac{3}{4}+\dfrac{1}{10}=\left(\dfrac{5}{6}+\dfrac{3}{4}\right)+\dfrac{1}{10}=\left(\dfrac{10}{12}+\dfrac{9}{12}\right)+\dfrac{1}{10}=\dfrac{19}{12}+\dfrac{1}{10}$$

We can now work with $\frac{19}{12}+\frac{1}{10}$. The LCD for these fractions is 60.

$$\frac{19}{12}+\frac{1}{10}=\frac{19}{12}\cdot\frac{5}{5}+\frac{1}{10}\cdot\frac{6}{6}=\frac{95}{60}+\frac{6}{60}=\frac{101}{60}$$

To work with all three fractions at the same time, factor each denominator into its prime factors and list the primes that appear in each. As before, the LCD includes any prime number that appears in a denominator. If a prime number appears in more than one denominator, the highest power is a factor in the LCD.

 EXAMPLE

Find the sum.

$$\frac{4}{5}+\frac{7}{15}+\frac{9}{20}$$

 **SOLUTION**

Prime factorization of the denominators:

$$5 = 5$$
$$15 = 3 \cdot 5$$
$$20 = 2 \cdot 2 \cdot 5$$

The LCD = 2 · 2 · 3 · 5 = 60

$$\frac{4}{5}+\frac{7}{15}+\frac{9}{20}=\left(\frac{4}{5}\cdot\frac{12}{12}\right)+\left(\frac{7}{15}\cdot\frac{4}{4}\right)+\left(\frac{9}{20}\cdot\frac{3}{3}\right)=\frac{48}{60}+\frac{28}{60}+\frac{27}{60}=\frac{103}{60}$$

EXAMPLE

Find the sum.

$$\frac{3}{10}+\frac{5}{12}+\frac{1}{18}$$

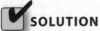 **SOLUTION**

Prime factorization of the denominators:

$$10 = 2 \cdot 5$$
$$12 = 2 \cdot 2 \cdot 3$$
$$18 = 2 \cdot 3 \cdot 3$$

$$LCD = 2 \cdot 2 \cdot 3 \cdot 3 \cdot 5 = 180$$

$$\frac{3}{10}+\frac{5}{12}+\frac{1}{18}=\left(\frac{3}{10}\cdot\frac{18}{18}\right)+\left(\frac{5}{12}\cdot\frac{15}{15}\right)+\left(\frac{1}{18}\cdot\frac{10}{10}\right)=\frac{54}{180}+\frac{75}{180}+\frac{10}{180}=\frac{139}{180}$$

PRACTICE

Find the sum.

1. $\dfrac{5}{36}+\dfrac{4}{9}+\dfrac{7}{12}=$

2. $\dfrac{11}{24}+\dfrac{3}{10}+\dfrac{1}{8}=$

3. $\dfrac{1}{4}+\dfrac{5}{6}+\dfrac{9}{20}=$

4. $\dfrac{3}{35}+\dfrac{9}{14}+\dfrac{7}{10}=$

5. $\dfrac{5}{48}+\dfrac{3}{16}+\dfrac{1}{6}+\dfrac{7}{9}=$

✔ SOLUTIONS

1. $\dfrac{5}{36}+\dfrac{4}{9}+\dfrac{7}{12}=\dfrac{5}{36}+\left(\dfrac{4}{9}\cdot\dfrac{4}{4}\right)+\left(\dfrac{7}{12}\cdot\dfrac{3}{3}\right)=\dfrac{5}{36}+\dfrac{16}{36}+\dfrac{21}{36}=\dfrac{42}{36}=\dfrac{7}{6}$

2. $\dfrac{11}{24}+\dfrac{3}{10}+\dfrac{1}{8}=\left(\dfrac{11}{24}\cdot\dfrac{5}{5}\right)+\left(\dfrac{3}{10}\cdot\dfrac{12}{12}\right)+\left(\dfrac{1}{8}\cdot\dfrac{15}{15}\right)=\dfrac{55}{120}+\dfrac{36}{120}+\dfrac{15}{120}=\dfrac{106}{120}=\dfrac{53}{60}$

3. $\dfrac{1}{4}+\dfrac{5}{6}+\dfrac{9}{20}=\left(\dfrac{1}{4}\cdot\dfrac{15}{15}\right)+\left(\dfrac{5}{6}\cdot\dfrac{10}{10}\right)+\left(\dfrac{9}{20}\cdot\dfrac{3}{3}\right)=\dfrac{15}{60}+\dfrac{50}{60}+\dfrac{27}{60}=\dfrac{92}{60}=\dfrac{23}{15}$

4. $\dfrac{3}{35}+\dfrac{9}{14}+\dfrac{7}{10}=\left(\dfrac{3}{35}\cdot\dfrac{2}{2}\right)+\left(\dfrac{9}{14}\cdot\dfrac{5}{5}\right)+\left(\dfrac{7}{10}\cdot\dfrac{7}{7}\right)=\dfrac{6}{70}+\dfrac{45}{70}+\dfrac{49}{70}=\dfrac{100}{70}=\dfrac{10}{7}$

5. $\dfrac{5}{48}+\dfrac{3}{16}+\dfrac{1}{6}+\dfrac{7}{9}=\left(\dfrac{5}{48}\cdot\dfrac{3}{3}\right)+\left(\dfrac{3}{16}\cdot\dfrac{9}{9}\right)+\left(\dfrac{1}{6}\cdot\dfrac{24}{24}\right)+\left(\dfrac{7}{9}\cdot\dfrac{16}{16}\right)$

   $=\dfrac{15}{144}+\dfrac{27}{144}+\dfrac{24}{144}+\dfrac{112}{144}=\dfrac{178}{144}=\dfrac{89}{72}$

Whole Number–Fraction Arithmetic

A whole number can be written as a fraction whose denominator is 1. With this in mind, we can see that addition and subtraction of whole numbers and fractions are nothing new. To add a whole number to a fraction, we multiply the whole number by the fraction's denominator and add this product to the fraction's numerator. The sum is the new numerator.

$$W + \frac{a}{b} = \frac{W}{1} + \frac{a}{b} = \frac{W}{1} \cdot \frac{b}{b} + \frac{a}{b} = \frac{Wb}{b} + \frac{a}{b} = \frac{Wb+a}{b}$$

 EXAMPLE

Add the fractions with the rule $W + \dfrac{a}{b} = \dfrac{Wb+a}{b}$

 **SOLUTION**

$$3 + \frac{7}{8} = \frac{(3 \cdot 8)+7}{8} = \frac{24+7}{8} = \frac{31}{8}$$

PRACTICE

Find the sum.

1. $4 + \dfrac{1}{3} =$

2. $5 + \dfrac{2}{11} =$

3. $1 + \dfrac{8}{9} =$

4. $2 + \dfrac{2}{5} =$

5. $3 + \dfrac{6}{7} =$

SOLUTIONS

1. $4 + \dfrac{1}{3} = \dfrac{(4 \cdot 3)+1}{3} = \dfrac{12+1}{3} = \dfrac{13}{3}$

2. $5 + \dfrac{2}{11} = \dfrac{(5 \cdot 11)+2}{11} = \dfrac{55+2}{11} = \dfrac{57}{11}$

3. $1 + \dfrac{8}{9} = \dfrac{(1 \cdot 9)+8}{9} = \dfrac{17}{9}$

4. $2 + \dfrac{2}{5} = \dfrac{(2 \cdot 5)+2}{5} = \dfrac{10+2}{5} = \dfrac{12}{5}$

5. $3 + \dfrac{6}{7} = \dfrac{(3 \cdot 7)+6}{7} = \dfrac{21+6}{7} = \dfrac{27}{7}$

To subtract a fraction from a whole number, we multiply the whole number by the fraction's denominator and then subtract the fraction's numerator from this product. The difference will be the new numerator:

$$W - \frac{a}{b} = \frac{Wb - a}{b}$$

 EXAMPLE

$$2 - \frac{5}{7} = \frac{(2 \cdot 7) - 5}{7} = \frac{14 - 5}{7} = \frac{9}{7}$$

 **PRACTICE**
Find the difference.

1. $1 - \frac{1}{4} =$

2. $2 - \frac{3}{8} =$

3. $5 - \frac{6}{11} =$

4. $2 - \frac{4}{5} =$

✓ **SOLUTIONS**

1. $1 - \frac{1}{4} = \frac{(1 \cdot 4) - 1}{4} = \frac{3}{4}$

2. $2 - \frac{3}{8} = \frac{(2 \cdot 8) - 3}{8} = \frac{16 - 3}{8} = \frac{13}{8}$

3. $5 - \frac{6}{11} = \frac{(5 \cdot 11) - 6}{11} = \frac{55 - 6}{11} = \frac{49}{11}$

4. $2 - \frac{4}{5} = \frac{(2 \cdot 5) - 4}{5} = \frac{10 - 4}{5} = \frac{6}{5}$

To subtract a whole number from the fraction, we again multiply the whole number by the fraction's denominator and then subtract this product from the fraction's numerator. This difference will be the new numerator. The rule is:

$$\frac{a}{b} - W = \frac{a - Wb}{b}$$

EXAMPLE

$$\frac{8}{3} - 2 = \frac{8 - (2 \cdot 3)}{3} = \frac{8 - 6}{3} = \frac{2}{3}$$

PRACTICE
Find the difference.

1. $\dfrac{12}{5} - 1 =$

2. $\dfrac{14}{3} - 2 =$

3. $\dfrac{19}{4} - 2 =$

4. $\dfrac{18}{7} - 1 =$

SOLUTIONS

1. $\dfrac{12}{5} - 1 = \dfrac{12 - (1 \cdot 5)}{5} = \dfrac{7}{5}$

2. $\dfrac{14}{3} - 2 = \dfrac{14 - (2 \cdot 3)}{3} = \dfrac{14 - 6}{3} = \dfrac{8}{3}$

3. $\dfrac{19}{4} - 2 = \dfrac{19 - (2 \cdot 4)}{4} = \dfrac{19 - 8}{4} = \dfrac{11}{4}$

4. $\dfrac{18}{7} - 1 = \dfrac{18 - (1 \cdot 7)}{7} = \dfrac{11}{7}$

Compound Fractions

Remember what a fraction is—the division of the numerator by the denominator. For example, $\frac{15}{3}$ another way of saying "15 ÷ 3." A compound fraction, a fraction where the numerator or denominator or both are fractions, is merely a fraction division problem. For this reason, this section is almost the same as the section on fraction division. We use one of three rules, depending on whether there is a fraction in the numerator, denominator, or both.

1. If the fraction is in the numerator: $\dfrac{\frac{a}{b}}{W} = \dfrac{a}{b} \div W = \dfrac{a}{b} \div \dfrac{W}{1} = \dfrac{a}{b} \cdot \dfrac{1}{W} = \dfrac{a}{bW}$.

2. If the fraction is in the denominator: $\dfrac{W}{\frac{a}{b}} = W \div \dfrac{a}{b} = \dfrac{W}{1} \cdot \dfrac{b}{a} = \dfrac{Wb}{a}$.

3. If fractions are in both numerator and denominator: $\dfrac{\frac{a}{b}}{\frac{c}{d}} = \dfrac{a}{b} \div \dfrac{c}{d} = \dfrac{a}{b} \cdot \dfrac{d}{c} = \dfrac{ad}{bc}$.

  EXAMPLES
Simplify the compound fraction.

$$\frac{\frac{2}{3}}{\frac{1}{6}} = \frac{2}{3} \div \frac{1}{6} = \frac{2}{3} \cdot \frac{6}{1} = \frac{12}{3} = 4$$

$$\frac{1}{\frac{2}{3}} = 1 \div \frac{2}{3} = 1 \cdot \frac{3}{2} = \frac{3}{2}$$

$$\frac{\frac{8}{9}}{5} = \frac{8}{9} \div 5 = \frac{8}{9} \cdot \frac{1}{5} = \frac{8}{45}$$

PRACTICE
Simplify the compound fraction.

1. $\dfrac{3}{\frac{5}{9}} =$

2. $\dfrac{8}{\frac{4}{3}} =$

3. $\dfrac{\frac{5}{7}}{2} =$

4. $\dfrac{\frac{4}{11}}{2} =$

5. $\dfrac{\frac{10}{27}}{\frac{4}{7}} =$

✔ **SOLUTIONS**

1. $\dfrac{3}{\frac{5}{9}} = 3 \div \dfrac{5}{9} = 3 \cdot \dfrac{9}{5} = \dfrac{27}{5}$

2. $\dfrac{8}{\frac{4}{3}} = 8 \div \dfrac{4}{3} = 8 \cdot \dfrac{3}{4} = \dfrac{24}{4} = 6$

3. $\dfrac{\frac{5}{7}}{2} = \dfrac{5}{7} \div 2 = \dfrac{5}{7} \cdot \dfrac{1}{2} = \dfrac{5}{14}$

4. $\dfrac{\frac{4}{11}}{2} = \dfrac{4}{11} \div 2 = \dfrac{4}{11} \cdot \dfrac{1}{2} = \dfrac{4}{22} = \dfrac{2}{11}$

5. $\dfrac{\frac{10}{27}}{\frac{4}{7}} = \dfrac{10}{27} \div \dfrac{4}{7} = \dfrac{10}{27} \cdot \dfrac{7}{4} = \dfrac{70}{108} = \dfrac{35}{54}$

Mixed Numbers and Improper Fractions

An improper fraction is a fraction whose numerator is larger than its denominator. For example, $\frac{6}{5}$ is an improper fraction. A mixed number consists of the sum of a whole number and a fraction. For example, $1\frac{1}{5}$ (which is really $1 + \frac{1}{5}$) is a mixed number. For now, we will practice writing a mixed number as an improper fraction.

To convert a mixed number into an improper fraction, we multiply the whole number by the fraction's denominator and then add this to the numerator. The sum is the new numerator. The rule is:

$$a\frac{b}{c} = \frac{ac + b}{c}$$

■ **EXAMPLES**

Write the mixed number as an improper fraction.

$$2\tfrac{6}{25} = \frac{(2 \cdot 25) + 6}{25} = \frac{50 + 6}{25} = \frac{56}{25}$$

$$1\tfrac{2}{9} = \frac{(1 \cdot 9) + 2}{9} = \frac{11}{9}$$

$$4\tfrac{1}{6} = \frac{(4 \cdot 6) + 1}{6} = \frac{24 + 1}{6} = \frac{25}{6}$$

PRACTICE

Write the mixed number as an improper fraction.

1. $1\frac{7}{8} =$

2. $5\frac{1}{3} =$

3. $2\frac{4}{7} =$

4. $9\frac{6}{11} =$

5. $8\frac{5}{8} =$

SOLUTIONS

1. $1\frac{7}{8} = \dfrac{(1 \cdot 8) + 7}{8} = \dfrac{8 + 7}{8} = \dfrac{15}{8}$

2. $5\frac{1}{3} = \dfrac{(5 \cdot 3) + 1}{3} = \dfrac{15 + 1}{3} = \dfrac{16}{3}$

3. $2\frac{4}{7} = \dfrac{(2 \cdot 7) + 4}{7} = \dfrac{14 + 4}{7} = \dfrac{18}{7}$

4. $9\frac{6}{11} = \dfrac{(9 \cdot 11) + 6}{11} = \dfrac{99 + 6}{11} = \dfrac{105}{11}$

5. $8\frac{5}{8} = \dfrac{(8 \cdot 8) + 5}{8} = \dfrac{64 + 5}{8} = \dfrac{69}{8}$

Fractions and Division of Whole Numbers

There is a close relationship between improper fractions and division of whole numbers. First let us review the parts of a division problem.

$$\text{divisor} \overline{)\text{dividend}}^{\text{quotient}} \qquad \frac{\text{dividend}}{\text{divisor}} = \text{quotient} + \frac{\text{remainder}}{\text{divisor}}$$

$$\underline{\quad \dots \quad}$$
$$\text{remainder}$$

In an improper fraction, the numerator is the dividend and the divisor is the denominator. In a mixed number, the quotient is the whole number, the remainder is the new numerator, and the denominator is the divisor.

$$\frac{dividend}{divisor} \xrightarrow{\text{becomes}} divisor\overline{)dividend} \begin{array}{c} quotient \\ \\ \overline{remainder} \end{array}$$

For example, $\frac{22}{7} = 3\frac{1}{7}$:

$$\frac{22}{7} \xrightarrow{\text{becomes}} \begin{array}{r} 3 \\ 7\overline{)22} \\ \underline{-21} \\ 1 \end{array} \text{ so, } \frac{22}{7} = 3\frac{1}{7}.$$

To convert an improper fraction to a mixed number, divide the numerator into the denominator. The remainder will be the new numerator and the quotient will be the whole number.

EXAMPLES

Write the improper fraction to a mixed number.

$$\frac{14}{5} \qquad \begin{array}{r} 2 \\ 5\overline{)14} \\ \underline{-10} \\ 4 \end{array} \text{ new numerator}$$

$$\frac{14}{5} = 2\frac{4}{5}$$

$$\frac{21}{5} \qquad \begin{array}{r} 4 \\ 5\overline{)21} \\ \underline{-20} \\ 1 \end{array} \text{ new numerator}$$

$$\frac{21}{5} = 4\frac{1}{5}$$

PRACTICE

Write the improper fraction to a mixed number.

1. $\dfrac{13}{4}$

2. $\dfrac{19}{3}$

3. $\dfrac{39}{14}$

4. $\dfrac{24}{5}$

5. $\dfrac{26}{7}$

SOLUTIONS

1. $\dfrac{13}{4}$

$$4\overline{)13} \;\; \begin{array}{r} 3 \\ \hline \end{array}$$
$$\underline{-12}$$
$$1$$

$\dfrac{13}{4} = 3\frac{1}{4}$

2. $\dfrac{19}{3}$

$$3\overline{)19} \;\; \begin{array}{r} 6 \\ \hline \end{array}$$
$$\underline{-18}$$
$$1$$

$\dfrac{19}{3} = 6\frac{1}{3}$

3. $\dfrac{39}{14}$

$$14\overline{)39} \;\; \begin{array}{r} 2 \\ \hline \end{array}$$
$$\underline{-28}$$
$$11$$

$\dfrac{39}{14} = 2\frac{11}{14}$

4. $\dfrac{24}{5}$

$$5\overline{)24}$$
$$\underline{-20}$$
$$4$$

$$\dfrac{24}{5}=4\tfrac{4}{5}$$

5. $\dfrac{26}{7}$

$$7\overline{)26}$$
$$\underline{-21}$$
$$5$$

$$\dfrac{26}{7}=3\tfrac{5}{7}$$

Mixed Number Arithmetic

We can add (or subtract) two mixed numbers in one of two ways. We will compute $4\tfrac{2}{3}+3\tfrac{1}{2}$ both ways. One strategy is to add the whole numbers and then to add the fractions.

$$4\tfrac{2}{3}+3\tfrac{1}{2}=\left(4+\dfrac{2}{3}\right)+\left(3+\dfrac{1}{2}\right)=(4+3)+\left(\dfrac{2}{3}+\dfrac{1}{2}\right)=7+\left(\dfrac{4}{6}+\dfrac{3}{6}\right)=7+\dfrac{7}{6}=7+1+\dfrac{1}{6}=8\tfrac{1}{6}$$

The other strategy is to convert the mixed numbers in improper fractions before adding them.

$$4\tfrac{2}{3}+3\tfrac{1}{2}=\dfrac{14}{3}+\dfrac{7}{2}=\dfrac{28}{6}+\dfrac{21}{6}=\dfrac{49}{6}=8\tfrac{1}{6}$$

PRACTICE

Find the sum or difference.

1. $2\tfrac{3}{7}+1\tfrac{1}{2}$

2. $2\tfrac{5}{16}+1\tfrac{11}{12}$

3. $4\tfrac{5}{6}+1\tfrac{2}{3}$

4. $3\frac{4}{9} - 1\frac{1}{6}$

5. $2\frac{3}{4} + \frac{5}{6}$

6. $4\frac{2}{3} + 2\frac{1}{5}$

7. $5\frac{1}{12} - 3\frac{3}{8}$

✔ SOLUTIONS

1. $2\frac{3}{7} + 1\frac{1}{2} = \frac{17}{7} + \frac{3}{2} = \frac{34}{14} + \frac{21}{14} = \frac{55}{14} = 3\frac{13}{14}$

2. $2\frac{5}{16} + 1\frac{11}{12} = \frac{37}{16} + \frac{23}{12} = \frac{111}{48} + \frac{92}{48} = \frac{203}{48} = 4\frac{11}{48}$

3. $4\frac{5}{6} + 1\frac{2}{3} = \frac{29}{6} + \frac{5}{3} = \frac{29}{6} + \frac{10}{6} = \frac{39}{6} = \frac{13}{2} = 6\frac{1}{2}$

4. $3\frac{4}{9} - 1\frac{1}{6} = \frac{31}{9} - \frac{7}{6} = \frac{62}{18} - \frac{21}{18} = \frac{41}{18} = 2\frac{5}{18}$

5. $2\frac{3}{4} + \frac{5}{6} = \frac{11}{4} + \frac{5}{6} = \frac{33}{12} + \frac{10}{12} = \frac{43}{12} = 3\frac{7}{12}$

6. $4\frac{2}{3} + 2\frac{1}{5} = \frac{14}{3} + \frac{11}{5} = \frac{70}{15} + \frac{33}{15} = \frac{103}{15} = 6\frac{13}{15}$

7. $5\frac{1}{12} - 3\frac{3}{8} = \frac{61}{12} - \frac{27}{8} = \frac{122}{24} - \frac{81}{24} = \frac{41}{24} = 1\frac{17}{24}$

Multiplying Mixed Numbers

When multiplying mixed numbers we first convert them to improper fractions before multiplying. Simply multiplying the whole numbers and the fractions is incorrect because there are really two operations involved—addition and multiplication:

$$1\frac{1}{2} \cdot 4\frac{1}{3} = \left(1 + \frac{1}{2}\right) \cdot \left(4 + \frac{1}{3}\right)$$

Instead, let us rewrite the mixed numbers as improper fractions before multiplying:

$$1\tfrac{1}{2} \cdot 4\tfrac{1}{3} = \frac{3}{2} \cdot \frac{13}{3} = \frac{39}{6} = \frac{13}{2} = 6\tfrac{1}{2}$$

PRACTICE _____

Convert the mixed numbers to improper fractions, and then find the product.

1. $1\tfrac{3}{4} \cdot 2\tfrac{1}{12} =$

2. $2\tfrac{2}{25} \cdot \dfrac{3}{7} =$

3. $2\tfrac{1}{8} \cdot 2\tfrac{1}{5} =$

4. $7\tfrac{1}{2} \cdot 1\tfrac{1}{3} =$

5. $\dfrac{3}{4} \cdot 2\tfrac{1}{4} =$

SOLUTIONS _____

1. $1\tfrac{3}{4} \cdot 2\tfrac{1}{12} = \dfrac{7}{4} \cdot \dfrac{25}{12} = \dfrac{175}{48} = 3\tfrac{31}{48}$

2. $2\tfrac{2}{25} \cdot \dfrac{3}{7} = \dfrac{52}{25} \cdot \dfrac{3}{7} = \dfrac{156}{175}$

3. $2\tfrac{1}{8} \cdot 2\tfrac{1}{5} = \dfrac{17}{8} \cdot \dfrac{11}{5} = \dfrac{187}{40} = 4\tfrac{27}{40}$

4. $7\tfrac{1}{2} \cdot 1\tfrac{1}{3} = \dfrac{15}{2} \cdot \dfrac{4}{3} = \dfrac{60}{6} = 10$

5. $\dfrac{3}{4} \cdot 2\tfrac{1}{4} = \dfrac{3}{4} \cdot \dfrac{9}{4} = \dfrac{27}{16} = 1\tfrac{11}{16}$

Dividing Mixed Numbers

Division of mixed numbers is similar to multiplication in that we first convert the mixed numbers into improper fractions. Recall that we perform division with fractions by rewriting the problem as a multiplication problem. That is, we use the following rule:

$$\frac{\frac{a}{b}}{\frac{c}{d}} = \frac{a}{b} \div \frac{c}{d} = \frac{a}{b} \cdot \frac{d}{c} = \frac{ad}{bc}$$

EXAMPLE _____

$$\frac{3\frac{1}{3}}{1\frac{2}{5}} = \frac{\frac{10}{3}}{\frac{7}{5}} = \frac{10}{3} \div \frac{7}{5} = \frac{10}{3} \cdot \frac{5}{7} = \frac{50}{21} = 2\frac{8}{21}$$

PRACTICE _____

Convert the mixed number to an improper fraction and then simplify the compound fraction.

1. $\dfrac{1\frac{3}{8}}{2\frac{1}{3}} =$

2. $\dfrac{\frac{7}{16}}{1\frac{2}{3}} =$

3. $\dfrac{1\frac{4}{15}}{1\frac{1}{4}} =$

4. $\dfrac{5\frac{1}{2}}{3} =$

5. $\dfrac{2\frac{1}{2}}{\frac{1}{2}} =$

SOLUTIONS _____

1. $\dfrac{1\frac{3}{8}}{2\frac{1}{3}} = \dfrac{\frac{11}{8}}{\frac{7}{3}} = \frac{11}{8} \div \frac{7}{3} = \frac{11}{8} \cdot \frac{3}{7} = \frac{33}{56}$

2. $\dfrac{\frac{7}{16}}{1\frac{2}{3}} = \dfrac{\frac{7}{16}}{\frac{5}{3}} = \frac{7}{16} \div \frac{5}{3} = \frac{7}{16} \cdot \frac{3}{5} = \frac{21}{80}$

3. $\dfrac{1\frac{4}{15}}{1\frac{1}{4}} = \dfrac{\frac{19}{15}}{\frac{5}{4}} = \dfrac{19}{15} \div \dfrac{5}{4} = \dfrac{19}{15} \cdot \dfrac{4}{5} = \dfrac{76}{75} = 1\frac{1}{75}$

4. $\dfrac{5\frac{1}{2}}{3} = \dfrac{\frac{11}{2}}{3} = \dfrac{11}{2} \div 3 = \dfrac{11}{2} \cdot \dfrac{1}{3} = \dfrac{11}{6} = 1\frac{5}{6}$

5. $\dfrac{2\frac{1}{2}}{\frac{1}{2}} = \dfrac{\frac{5}{2}}{\frac{1}{2}} = \dfrac{5}{2} \div \dfrac{1}{2} = \dfrac{5}{2} \cdot \dfrac{2}{1} = \dfrac{10}{2} = 5$

Recognizing Quantities and Relationships in Word Problems

Success in solving word problems depends on the mastery of three skills—"translating" English into mathematics, setting the variable equal to an appropriate unknown quantity, and using knowledge of mathematics to solve the equation or inequality. This book will help you develop these skills.

TABLE 1-1	
English	**Mathematical Symbol**
"Is," "are," "will be" (any form of the verb "to be") mean "equal"	=
"More than," "increased by," "sum of" mean "add"	+
"Less than," "decreased by," "difference of" mean "subtract"	−
"Of" means "multiply"	·
"Per" means "divide"	÷
"More than" and "greater than" both mean the relation "greater than" although "more than" can mean "add"	>
"Less than," means the relation "less than" although it can also mean "subtract"	<
"At least," and "no less than," mean the relation "greater than or equal to"	≥
"No more than," and "at most," mean the relation "less than or equal to"	≤

We will ease into the topic of word problems by translating English sentences into mathematical sentences. We will not solve word problems until later in the book.

EXAMPLES

- **Five is two more than three.**

  5 = 2   +   3

- **Ten less six is four.**

  10 − 6 = 4

- **One half of twelve is six.**

  $\dfrac{1}{2}$ ·   12 = 6

- **Eggs cost $1.15 per dozen.**

  1.15 /   12 (this gives the price per egg)

- **The difference of sixteen and five is eleven.**

  16 − 5 = 11

- **Fourteen decreased by six is eight.**

  14   −   6 = 8

- **Seven increased by six is thirteen.**

  7   +   6 = 13

- **Eight is less than eleven.**

  8   <   11

- **Eight is at most eleven.**

  8   ≤   11

- **Eleven is more than eight.**

  11   >   8

- **Eleven is at least eight.**

  11   ≥   8

- **One hundred is twice fifty.**

  100   = 2 · 50

- **Five more than eight is thirteen.**

  5   +   8 = 13

PRACTICE

Translate the English sentence into a mathematical sentence.

1. Fifteen less four is eleven.

2. Seven decreased by two is five.

3. Six increased by one is seven.

4. The sum of two and three is five.

5. Nine more than four is thirteen.

6. One-third of twelve is four.

7. One-third of twelve is greater than two.

8. Half of sixteen is eight.

9. The car gets 350 miles per eleven gallons.

10. Ten is less than twelve.

11. Ten is no more than twelve.

12. Three-fourths of sixteen is twelve.

13. Twice fifteen is thirty.

14. The difference of fourteen and five is nine.

15. Nine is more than six.

16. Nine is at least six.

SOLUTIONS

1. $15 - 4 = 11$

2. $7 - 2 = 5$

3. $6 + 1 = 7$

4. $2 + 3 = 5$

5. $9 + 4 = 13$

6. $\frac{1}{3} \cdot 12 = 4$

7. $\frac{1}{3} \cdot 12 > 2$

8. $\frac{1}{2} \cdot 16 = 8$

9. $350 \div 11$ (miles per gallon)

10. $10 < 12$

11. $10 \leq 12$

12. $\dfrac{3}{4} \cdot 16 = 12$

13. $2 \cdot 15 = 30$

14. $14 - 5 = 9$

15. $9 > 6$

16. $9 \geq 6$

Summary

In this chapter, we learned how to:

- *Perform multiplication and division with fractions.* Multiply fractions by multiplying the numerators together and the denominators together. To divide two fractions, write the problem as a multiplication problem by inverting (switching the numerator and denominator) the second fraction.

- *Write a fraction in lowest terms.* A fraction is in lowest terms if the numerator and denominator have no common factors. This involves factoring the numerator and denominator and dividing out (canceling) the common factors. Writing a fraction in lowest terms is also called simplifying the fraction and reducing the fraction.

- *Perform addition and subtraction with fractions having the same denominator.* Add (subtract) two fractions having the same denominator by adding (subtracting) their numerators.

- *Find the least common denominator (LCD) of two fractions.* Begin by finding the smallest number that is a multiple of each denominator. Divide this multiple by the first denominator. Multiply the numerator and denominator by this number. This gets an equivalent fraction having the LCD as its denominator. Do the same thing to the second fraction.

- *Perform addition and subtraction with fractions having different denominators.* Write each fraction so that its denominator is the LCD and then add/subtract the numerators.

- *Perform arithmetic with fractions and whole numbers.* Write the whole number in fraction form (its denominator is 1) and then follow one of the procedures outlined above.

- *Simplify compound fractions.* Rewrite the compound fraction as a division problem and then as a multiplication problem. After this is done, find the product.

- *Write a mixed number as an improper fraction and an improper fraction as a mixed number.* Write a mixed number as an improper fraction by multiplying the whole number by the denominator and then adding the numerator of the fraction. This gives us the numerator of the improper fraction. The improper fraction has the same denominator as the original fraction. Write an improper fraction as a mixed number by dividing the numerator by the denominator. The quotient is the whole number part of the mixed number, and the numerator of the fraction part is the remainder while the denominator is the same as the original denominator.

- *Perform arithmetic with mixed numbers.* One method is to convert the mixed number to an improper fraction and then use one of the strategies outlined above to perform the arithmetic.

- *Translate certain English sentences and phrases into mathematical symbols.*

In Chapter 2, we will extend what we learned in Chapter 1 to include fractions having variables in them and English sentences that have unknown quantities in them.

QUIZ

1. Write $5\frac{2}{3}$ as an improper fraction.

   A. $\dfrac{10}{3}$

   B. $\dfrac{18}{3}$

   C. $\dfrac{7}{3}$

   D. $\dfrac{17}{3}$

2. $\dfrac{11}{18} - \dfrac{1}{6} =$

   A. $\dfrac{5}{3}$

   B. $\dfrac{2}{3}$

   C. $\dfrac{4}{9}$

   D. $\dfrac{5}{6}$

3. $\dfrac{2\frac{1}{3}}{4\frac{2}{5}} =$

   A. $\dfrac{5}{12}$

   B. $\dfrac{5}{36}$

   C. $\dfrac{35}{66}$

   D. $\dfrac{56}{15}$

4. $\dfrac{2}{9} + \dfrac{5}{6} =$

    A. $\dfrac{7}{15}$

    B. $\dfrac{19}{18}$

    C. $\dfrac{1}{2}$

    D. $\dfrac{10}{9}$

5. Write $\frac{16}{3}$ as a mixed number.

    A. $5\frac{1}{3}$

    B. $5\frac{2}{3}$

    C. $6\frac{1}{3}$

    D. $6\frac{2}{3}$

6. $\dfrac{4}{3} \cdot \dfrac{2}{5} =$

    A. $\dfrac{8}{15}$

    B. $\dfrac{10}{3}$

    C. $\dfrac{3}{4}$

    D. $\dfrac{2}{5}$

7. $\dfrac{1}{15} + \dfrac{1}{3} =$

    A. $\dfrac{7}{15}$

    B. $\dfrac{1}{5}$

    C. $\dfrac{2}{5}$

    D. $\dfrac{8}{15}$

8. $1\frac{3}{5} - \dfrac{1}{2} =$

    A. $\dfrac{7}{5}$

    B. $\dfrac{9}{10}$

    C. $1\frac{1}{5}$

    D. $1\frac{1}{10}$

9. $\dfrac{4}{\frac{2}{3}} =$

    A. $\dfrac{8}{3}$

    B. $\dfrac{1}{6}$

    C. 6

    D. 4

10. $\dfrac{2}{3} + \dfrac{7}{12} + \dfrac{1}{18} =$

 A. $\dfrac{5}{4}$

 B. $\dfrac{47}{36}$

 C. $\dfrac{4}{3}$

 D. $\dfrac{43}{36}$

11. $3\frac{3}{4} + 1\frac{5}{6} =$

 A. $\dfrac{65}{12}$

 B. $\dfrac{17}{3}$

 C. $\dfrac{11}{2}$

 D. $\dfrac{67}{12}$

12. $2\frac{3}{5} \cdot 3\frac{1}{4} =$

 A. $\dfrac{10}{3}$

 B. $\dfrac{117}{20}$

 C. $\dfrac{123}{20}$

 D. $\dfrac{169}{20}$

13. $\dfrac{7}{2} \div \dfrac{1}{4} =$

 A. 14

 B. $\dfrac{4}{3}$

 C. $\dfrac{15}{2}$

 D. $\dfrac{7}{8}$

14. **Write in mathematical symbols: Eight is three more than five.**
    A. $8 > 3 + 5$
    B. $3 > 5 = 8$
    C. $8 = 3 + 5$
    D. $8 \geq 3 + 5$

15. **Write in mathematical symbols: Twice eight is more than twelve.**
    A. $2(8) > 12$
    B. $2(18) \geq 12$
    C. $2 + 18 > 12$
    D. $2 + 18 \geq 12$

chapter 2

Introduction to Variables

Variables lie at the heart of courses from algebra to calculus and beyond. In this introduction, we will learn how to perform arithmetic on fractions containing one or more variables, and we will learn how to use variables to represent quantities in word problems. In later chapters, we will learn how to work with variables so that we can rewrite expressions. These skills are useful for solving equations and applied problems.

CHAPTER OBJECTIVES

In this chapter, you will

- Simplify fractions containing variables
- Perform arithmetic on fractions containing variables
- Rewrite a fraction as a product of a number and a variable expression
- Use variables to represent quantities in applied problems

A variable is a symbol for a number whose value is unknown. A variable might represent quantities at different times. For example if you are paid by the hour for your job and you earn $10 per hour, letting x represent the number of hours worked would allow you to write your earnings as "$10x$." The value of your earnings *varies* depending on the number of hours worked. If an equation has one variable, we can use algebra to determine what value the variable is representing.

Variables are treated like numbers because they are numbers. For instance $2 + x$ means two plus the quantity x and $2x$ means the quantity two times x (when no operation sign is given, the operation is assumed to be multiplication). The expression $3x + 4$ means three times x plus four. This is not the same as $3x + 4x$ which is three x's plus four x's for a total of seven x's: $3x + 4x = 7x$.

Simplifying Fractions Containing Variables

We can simplify fractions containing variables with the same techniques that we used to simplify ordinary fractions. If a variable appears as a factor in the numerator and denominator, we can simplify the fraction. We simplify fractions containing variables with the same strategy that we used in Chapter 1; that is, we divide common factors from the numerator and denominator.

 EXAMPLES

Write the fraction in lowest terms.

$$\frac{2x}{x} = \frac{2}{1} \cdot \frac{x}{x} = 2$$

$$\frac{6x}{9x} = \frac{6}{9} \cdot \frac{x}{x} = \frac{6}{9} = \frac{2}{3}$$

$$\frac{7xy}{5x} = \frac{7y}{5} \cdot \frac{x}{x} = \frac{7y}{5}$$

Still Struggling

When you see a plus or minus sign in a fraction, be *very careful* when simplifying. For example, we cannot remove x from the numerator and denominator in the expression $\frac{2+x}{x}$; x cannot be canceled. The only quantities that can be divided are factors. Many students mistakenly "cancel" the x and conclude that $\frac{2+x}{x} = \frac{2+1}{1} = 3$.

To see why we cannot "cancel" x, let us consider a similar fraction that does not contain a variable.

$$\frac{7}{6} = \frac{4+3}{6} = \frac{4+\cancel{3}^{1}}{\cancel{6}_{2}} = \frac{4+1}{2} = \frac{5}{2}$$

"Canceling" 3 from the numerator and denominator leaves us with the false equation $\frac{7}{6} = \frac{5}{2}$. The reason that the 3 can't be canceled from the fraction is that 3 is a *term* in the numerator, not a *factor*. (A term is a quantity separated from others by a plus or minus sign.)

If we wish to divide the numerator and denominator of $\frac{2+x}{x}$ by x we can rewrite the fraction as the sum of two separate fractions before removing the common factor.

$$\frac{2+x}{x} = \frac{2}{x} + \frac{\cancel{x}^{1}}{\cancel{x}_{1}} = \frac{2}{x} + \frac{1}{1} = \frac{2}{x} + 1$$

Simply because a plus or minus sign appears in a fraction does not automatically mean that dividing it out is not appropriate. For instance $\frac{3+x}{3+x} = 1$ because any nonzero number divided by itself is one.

EXAMPLE

Simplify the fraction:

$$\frac{(2+3x)(x-1)}{2+3x}$$

 **SOLUTION**

The factor $2 + 3x$ is in the numerator and denominator, so it can be divided out (or canceled).

$$\frac{(2+3x)(x-1)}{2+3x} = \frac{2+3x}{2+3x} \cdot \frac{x-1}{1} = x-1$$

$$\frac{2(x+7)(3x+1)}{2} = \frac{2}{2} \cdot \frac{(x+7)(3x+1)}{1} = (x+7)(3x+1)$$

$$\frac{15(x+6)(x-2)}{3(x-2)} = \frac{3 \cdot 5(x+6)(x-2)}{3(x-2)} = \frac{3(x-2)}{3(x-2)} \cdot \frac{5(x+6)}{1} = 5(x+6)$$

PRACTICE

Simplify the fraction.

1. $\dfrac{3xy}{2x} =$

2. $\dfrac{8x}{4} =$

3. $\dfrac{30xy}{16y} =$

4. $\dfrac{72x}{18xy} =$

5. $\dfrac{x(x-6)}{2x} =$

6. $\dfrac{6xy(2x-1)}{3x} =$

7. $\dfrac{(5x+16)(2x+7)}{6(2x+7)} =$

8. $\dfrac{24x(y+8)(x+1)}{15x(y+8)} =$

9. $\dfrac{150xy(2x+17)(8x-3)}{48y} =$

✔ **SOLUTIONS**

1. $\dfrac{3xy}{2x} = \dfrac{3y}{2} \cdot \dfrac{x}{x} = \dfrac{3y}{2}$

2. $\dfrac{8x}{4} = \dfrac{2x}{1} \cdot \dfrac{4}{4} = 2x$

3. $\dfrac{30xy}{16y} = \dfrac{15x}{8} \cdot \dfrac{2y}{2y} = \dfrac{15x}{8}$

4. $\dfrac{72x}{18xy} = \dfrac{4}{y} \cdot \dfrac{18x}{18x} = \dfrac{4}{y}$

5. $\dfrac{x(x-6)}{2x} = \dfrac{x-6}{2} \cdot \dfrac{x}{x} = \dfrac{x-6}{2}$

6. $\dfrac{6xy(2x-1)}{3x} = \dfrac{2y(2x-1)}{1} \cdot \dfrac{3x}{3x} = 2y(2x-1)$

7. $\dfrac{(5x+16)(2x+7)}{6(2x+7)} = \dfrac{5x+16}{6} \cdot \dfrac{2x+7}{2x+7} = \dfrac{5x+16}{6}$

8. $\dfrac{24x(y+8)(x+1)}{15x(y+8)} = \dfrac{8(x+1)}{5} \cdot \dfrac{3x(y+8)}{3x(y+8)} = \dfrac{8(x+1)}{5}$

9. $\dfrac{150xy(2x+17)(8x-3)}{48y} = \dfrac{25x(2x+17)(8x-3)}{8} \cdot \dfrac{6y}{6y}$

$$= \dfrac{25x(2x+17)(8x-3)}{8}$$

Operations on Fractions Containing Variables

Multiplication of fractions with variables is done in exactly the same way as multiplication of fractions without variables—we multiply the numerators and multiply the denominators and divide out common factors.

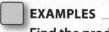 **EXAMPLES**

Find the product.

$$\frac{7}{10} \cdot \frac{3x}{4} = \frac{7 \cdot 3x}{10 \cdot 4} = \frac{21x}{40}$$

$$\frac{24}{5y} \cdot \frac{3}{14} = \frac{24 \cdot 3}{5y \cdot 14} = \frac{12 \cdot 3}{5y \cdot 7} = \frac{36}{35y}$$

$$\frac{34x}{15} \cdot \frac{3}{16} = \frac{34x \cdot 3}{15 \cdot 16} = \frac{17x \cdot 1}{5 \cdot 8} = \frac{17x}{40}$$

PRACTICE

Find the product.

1. $\dfrac{4x}{9} \cdot \dfrac{2}{7} =$

2. $\dfrac{2}{3x} \cdot \dfrac{6y}{5} =$

3. $\dfrac{18x}{19} \cdot \dfrac{2}{11x} =$

4. $\dfrac{5x}{9} \cdot \dfrac{4y}{3} =$

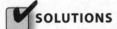 **SOLUTIONS**

1. $\dfrac{4x}{9} \cdot \dfrac{2}{7} = \dfrac{8x}{63}$

2. $\dfrac{2}{3x} \cdot \dfrac{6y}{5} = \dfrac{2 \cdot 6y}{3x \cdot 5} = \dfrac{2 \cdot 2y}{x \cdot 5} = \dfrac{4y}{5x}$

3. $\dfrac{18x}{19} \cdot \dfrac{2}{11x} = \dfrac{18x \cdot 2}{19 \cdot 11x} = \dfrac{18 \cdot 2}{19 \cdot 11} = \dfrac{36}{209}$

4. $\dfrac{5x}{9} \cdot \dfrac{4y}{3} = \dfrac{20xy}{27}$

At times, especially in calculus, students need to separate a variable from the rest of the fraction. This involves writing the fraction as a product of two fractions, or of one fraction and a whole number, or of one fraction and a variable. The steps we follow are the same as in multiplying fractions—only in reverse. For the following practice problems, we will use one of the following properties to separate the variable from the rest of the fraction.

$$\frac{a}{b} = a \cdot \frac{1}{b}$$

$$\frac{a}{b} = \frac{1}{b} \cdot a$$

$$\frac{ab}{c} = a \cdot \frac{b}{c} = \frac{a}{c} \cdot b$$

$$\frac{a}{bc} = a \cdot \frac{1}{bc} = \frac{a}{b} \cdot \frac{1}{c}$$

EXAMPLES

Write the expression as the product of a whole number or fraction and a variable expression. The appropriate rule is written next to the problem.

$$\frac{x}{3} = \frac{1 \cdot x}{3 \cdot 1} = \frac{1}{3} \cdot \frac{x}{1} = \frac{1}{3}x \qquad \frac{a}{b} = \frac{1}{b} \cdot a$$

$$\frac{3}{x} = \frac{3 \cdot 1}{1 \cdot x} = \frac{3}{1} \cdot \frac{1}{x} = 3 \cdot \frac{1}{x} \qquad \frac{a}{b} = a \cdot \frac{1}{b}$$

$$\frac{7}{8x} = \frac{7 \cdot 1}{8 \cdot x} = \frac{7}{8} \cdot \frac{1}{x} \qquad \frac{a}{bc} = \frac{a}{b} \cdot \frac{1}{c}$$

$$\frac{7x}{8} = \frac{7 \cdot x}{8 \cdot 1} = \frac{7}{8}x \qquad \frac{ab}{c} = \frac{a}{c} \cdot b$$

$$\frac{x+1}{2} = \frac{1 \cdot (x+1)}{2 \cdot 1} = \frac{1}{2}(x+1) \qquad \frac{a}{b} = \frac{1}{b} \cdot a$$

$$\frac{2}{x+1} = \frac{2 \cdot 1}{1 \cdot (x+1)} = \frac{2}{1} \cdot \frac{1}{x+1} = 2\left(\frac{1}{x+1}\right) \qquad \frac{a}{b} = a \cdot \frac{1}{b}$$

PRACTICE

Separate the factor having a variable from the rest of the fraction.

1. $\dfrac{4x}{5} =$

2. $\dfrac{7y}{15} =$

3. $\dfrac{3}{4t} =$

4. $\dfrac{11}{12x} =$

5. $\dfrac{x-3}{4} =$

6. $\dfrac{5}{7(4x-1)} =$

7. $\dfrac{10}{2x+1} =$

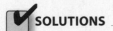 **SOLUTIONS**

1. $\dfrac{4x}{5} = \dfrac{4 \cdot x}{5 \cdot 1} = \dfrac{4}{5} \cdot \dfrac{x}{1} = \dfrac{4}{5}x$

2. $\dfrac{7y}{15} = \dfrac{7 \cdot y}{15 \cdot 1} = \dfrac{7}{15} \cdot \dfrac{y}{1} = \dfrac{7}{15}y$

3. $\dfrac{3}{4t} = \dfrac{3 \cdot 1}{4 \cdot t} = \dfrac{3}{4} \cdot \dfrac{1}{t}$

4. $\dfrac{11}{12x} = \dfrac{11 \cdot 1}{12 \cdot x} = \dfrac{11}{12} \cdot \dfrac{1}{x}$

5. $\dfrac{x-3}{4} = \dfrac{1 \cdot (x-3)}{4 \cdot 1} = \dfrac{1}{4}(x-3)$

6. $\dfrac{5}{7(4x-1)} = \dfrac{5 \cdot 1}{7 \cdot (4x-1)} = \dfrac{5}{7} \cdot \dfrac{1}{4x-1}$

7. $\dfrac{10}{2x+1} = \dfrac{10 \cdot 1}{1 \cdot (2x+1)} = 10 \cdot \dfrac{1}{2x+1}$

Fraction Division and Compound Fractions

Division of fractions with variables can become multiplication of fractions by inverting the second fraction. Because compound fractions are really only fraction division problems, we can rewrite the compound fraction as fraction division and then as fraction multiplication.

 EXAMPLE

Simplify the compound fraction.

$$\frac{\frac{4}{5}}{\frac{x}{3}} = \frac{4}{5} \div \frac{x}{3} = \frac{4}{5} \cdot \frac{3}{x} = \frac{12}{5x}$$

PRACTICE

Simplify the compound fraction.

1. $\dfrac{\frac{15}{32}}{\frac{x}{6}} =$

2. $\dfrac{\frac{9x}{14}}{\frac{2x}{17}} =$

3. $\dfrac{\frac{4}{3}}{\frac{21x}{8}} =$

4. $\dfrac{\frac{3}{8y}}{\frac{1}{y}} =$

5. $\dfrac{\frac{2}{x}}{\frac{7}{11y}} =$

6. $\dfrac{\frac{12x}{7}}{\frac{23}{6y}} =$

7. $\dfrac{3}{\frac{2}{9y}} =$

8. $\dfrac{\frac{x}{2}}{5} =$

✔ SOLUTIONS

1. $\dfrac{\frac{15}{32}}{\frac{x}{6}} = \dfrac{15}{32} \div \dfrac{x}{6} = \dfrac{15}{32} \cdot \dfrac{6}{x} = \dfrac{15 \cdot 6}{32 \cdot x} = \dfrac{15 \cdot 3}{16 \cdot x} = \dfrac{45}{16x}$

2. $\dfrac{\frac{9x}{14}}{\frac{2x}{17}} = \dfrac{9x}{14} \div \dfrac{2x}{17} = \dfrac{9x}{14} \cdot \dfrac{17}{2x} = \dfrac{9x \cdot 17}{14 \cdot 2x} = \dfrac{9 \cdot 17}{14 \cdot 2} = \dfrac{153}{28}$

3. $\dfrac{\frac{4}{3}}{\frac{21x}{8}} = \dfrac{4}{3} \div \dfrac{21x}{8} = \dfrac{4}{3} \cdot \dfrac{8}{21x} = \dfrac{32}{63x}$

4. $\dfrac{\frac{3}{8y}}{\frac{1}{y}} = \dfrac{3}{8y} \div \dfrac{1}{y} = \dfrac{3}{8y} \cdot \dfrac{y}{1} = \dfrac{3 \cdot y}{8y \cdot 1} = \dfrac{3}{8}$

5. $\dfrac{\frac{2}{x}}{\frac{7}{11y}} = \dfrac{2}{x} \div \dfrac{7}{11y} = \dfrac{2}{x} \cdot \dfrac{11y}{7} = \dfrac{22y}{7x}$

6. $\dfrac{\frac{12x}{7}}{\frac{23}{6y}} = \dfrac{12x}{7} \div \dfrac{23}{6y} = \dfrac{12x}{7} \cdot \dfrac{6y}{23} = \dfrac{72xy}{161}$

7. $\dfrac{3}{\frac{2}{9y}} = \dfrac{3}{1} \div \dfrac{2}{9y} = \dfrac{3}{1} \cdot \dfrac{9y}{2} = \dfrac{27y}{2}$

8. $\dfrac{\frac{x}{2}}{5} = \dfrac{x}{2} \div \dfrac{5}{1} = \dfrac{x}{2} \cdot \dfrac{1}{5} = \dfrac{x}{10}$

Adding and Subtracting Fractions Containing Variables

When adding or subtracting fractions with variables, we treat the variables as if they were prime numbers. If a variable appears in the denominator, the LCD has the variable, too.

EXAMPLES

Find the sum.

$$\dfrac{6}{25} + \dfrac{y}{4} = \dfrac{6}{25} \cdot \dfrac{4}{4} + \dfrac{y}{4} \cdot \dfrac{25}{25} = \dfrac{24}{100} + \dfrac{25y}{100} = \dfrac{24+25y}{100}$$

$$\dfrac{2t}{15} + \dfrac{14}{33} = \dfrac{2t}{15} \cdot \dfrac{11}{11} + \dfrac{14}{33} \cdot \dfrac{5}{5} = \dfrac{22t}{165} + \dfrac{70}{165} = \dfrac{22t+70}{165}$$

$$\dfrac{18}{x} - \dfrac{3}{4} = \dfrac{18}{x} \cdot \dfrac{4}{4} - \dfrac{3}{4} \cdot \dfrac{x}{x} = \dfrac{72}{4x} - \dfrac{3x}{4x} = \dfrac{72-3x}{4x}$$

$$\dfrac{9}{16t} + \dfrac{5}{12}$$

$16t = 2 \cdot 2 \cdot 2 \cdot 2 \cdot t$ and $12 = 2 \cdot 2 \cdot 3$ so the LCD $= 2 \cdot 2 \cdot 2 \cdot 2 \cdot 3 \cdot t = 48t$

$$48t \div 16t = 3 \text{ and } 48t \div 12 = 4t \left(\dfrac{48t}{16t} = 3 \text{ and } \dfrac{48t}{12} = 4t \right)$$

$$\dfrac{9}{16t} + \dfrac{5}{12} = \dfrac{9}{16t} \cdot \dfrac{3}{3} + \dfrac{5}{12} \cdot \dfrac{4t}{4t} = \dfrac{27}{48t} + \dfrac{20t}{48t} = \dfrac{27+20t}{48t}$$

$$\dfrac{71}{84} - \dfrac{13}{30x}$$

$84 = 2 \cdot 2 \cdot 3 \cdot 7$ and $30x = 2 \cdot 3 \cdot 5 \cdot x$

$$\text{LCD} = 2 \cdot 2 \cdot 3 \cdot 5 \cdot 7 \cdot x = 420x \text{ and } 420x \div 84 = 5x \text{ and } 420x \div 30x = 14$$

$$\left(\frac{420x}{84} = 5x \text{ and } \frac{420x}{30x} = 14 \right)$$

$$\frac{71}{84} - \frac{13}{30x} = \frac{71}{84} \cdot \frac{5x}{5x} - \frac{13}{30x} \cdot \frac{14}{14} = \frac{355x}{420x} - \frac{182}{420x} = \frac{355x - 182}{420x}$$

PRACTICE

Find the sum. Do not try to simplify the solutions. We will learn how to simplify this type of fraction in a later chapter.

1. $\dfrac{4}{15} + \dfrac{2x}{33} =$

2. $\dfrac{x}{48} - \dfrac{7}{30} =$

3. $\dfrac{3}{x} + \dfrac{4}{25} =$

4. $\dfrac{2}{45x} + \dfrac{6}{35} =$

5. $\dfrac{11}{150x} + \dfrac{7}{36} =$

6. $\dfrac{2}{12x} + \dfrac{3}{98x} =$

7. $\dfrac{1}{12x} + \dfrac{5}{9y} =$

8. $\dfrac{19}{51y} - \dfrac{1}{6x} =$

9. $\dfrac{7x}{24y} + \dfrac{2}{15y} =$

10. $\dfrac{3}{14y} + \dfrac{2}{35x} =$

SOLUTIONS

1. $\dfrac{4}{15} + \dfrac{2x}{33} = \dfrac{4}{15} \cdot \dfrac{11}{11} + \dfrac{2x}{33} \cdot \dfrac{5}{5} = \dfrac{44}{165} + \dfrac{10x}{165} = \dfrac{44 + 10x}{165}$

2. $\dfrac{x}{48} - \dfrac{7}{30} = \dfrac{x}{48} \cdot \dfrac{5}{5} - \dfrac{7}{30} \cdot \dfrac{8}{8} = \dfrac{5x}{240} - \dfrac{56}{240} = \dfrac{5x - 56}{240}$

3. $\dfrac{3}{x} + \dfrac{4}{25} = \dfrac{3}{x} \cdot \dfrac{25}{25} + \dfrac{4}{25} \cdot \dfrac{x}{x} = \dfrac{75}{25x} + \dfrac{4x}{25x} = \dfrac{75 + 4x}{25x}$

4. $\dfrac{2}{45x} + \dfrac{6}{35} = \dfrac{2}{45x} \cdot \dfrac{7}{7} + \dfrac{6}{35} \cdot \dfrac{9x}{9x} = \dfrac{14}{315x} + \dfrac{54x}{315x} = \dfrac{14 + 54x}{315x}$

5. $\dfrac{11}{150x} + \dfrac{7}{36} = \dfrac{11}{150x} \cdot \dfrac{6}{6} + \dfrac{7}{36} \cdot \dfrac{25x}{25x} = \dfrac{66}{900x} + \dfrac{175x}{900x} = \dfrac{66 + 175x}{900x}$

6. $\dfrac{2}{21x} + \dfrac{3}{98x} = \dfrac{2}{21x} \cdot \dfrac{14}{14} + \dfrac{3}{98x} \cdot \dfrac{3}{3} = \dfrac{28}{294x} + \dfrac{9}{294x} = \dfrac{37}{294x}$

7. $\dfrac{1}{12x} + \dfrac{5}{9y} = \dfrac{1}{12x} \cdot \dfrac{3y}{3y} + \dfrac{5}{9y} \cdot \dfrac{4x}{4x} = \dfrac{3y}{36xy} + \dfrac{20x}{36xy} = \dfrac{3y + 20x}{36xy}$

8. $\dfrac{19}{51y} - \dfrac{1}{6x} = \dfrac{19}{51y} \cdot \dfrac{2x}{2x} - \dfrac{1}{6x} \cdot \dfrac{17y}{17y} = \dfrac{38x}{102xy} - \dfrac{17y}{102xy} = \dfrac{38x - 17y}{102xy}$

9. $\dfrac{7x}{24y} + \dfrac{2}{15y} = \dfrac{7x}{24y} \cdot \dfrac{5}{5} + \dfrac{2}{15y} \cdot \dfrac{8}{8} = \dfrac{35x}{120y} + \dfrac{16}{120y} = \dfrac{35x + 16}{120y}$

10. $\dfrac{3}{14y} + \dfrac{2}{35x} = \dfrac{3}{14y} \cdot \dfrac{5x}{5x} + \dfrac{2}{35x} \cdot \dfrac{2y}{2y} = \dfrac{15x}{70xy} + \dfrac{4y}{70xy} = \dfrac{15x + 4y}{70xy}$

Variables in Word Problems

Often the equations used to solve word problems should have only one variable, so other unknowns must be written in terms of a single variable. The goal of this section is to acquainte you with setting the variable equal to an appropriate unknown quantity and writing other unknown quantities in terms of the variable.

EXAMPLES

Andrea is twice as old as Sarah.

Because Andrea's age is being compared to Sarah's, the easiest thing to do is to let x represent Sarah's age:

Let x = Sarah's age.

Andrea is twice as old as Sarah, so Andrea's age = $2x$. We could have let x represent Andrea's age, but we would have to re-think the statement as "Sarah is *half* as old as Andrea." This would mean Sarah's age would be represented by $\frac{1}{2}x$.

John has eight more nickels than Larry has.

The number of John's nickels is being compared to the number of Larry's nickels, so it is easier to let x represent the number of nickels Larry has.

Let x = the number of nickels Larry has.

$x + 8$ = the number of nickels John has.

A used car costs $5000 less than a new car.

Let x = the price of the new car.

$x - 5000$ = the price of the used car.

A box's length is three times its width.

Let x = width (in the given units).

$3x$ = length (in the given units).

Jack is two-thirds as tall as Jill.

Let x = Jill's height (in the given units).

$\frac{2}{3}x$ = Jack's height (in the given units).

From 6 P.M. to 6 A.M. the temperature dropped 30 degrees.

Let x = temperature (in degrees) at 6 P.M.

$x - 30$ = temperature (in degrees) at 6 A.M.

One-eighth of an employee's time is spent cleaning his work station.

Let x = the number of hours he is on the job.

$\frac{1}{8}x$ = the number of hours he spends cleaning his work station.

$10,000 was deposited between two savings accounts, Account A and Account B.

Let x = amount deposited in Account A.

How much is left to represent the amount invested in Account B? If x dollars is taken from $10,000, then it must be that $10,000 - x$ dollars is left to be deposited in Account B.

Or if x represents the amount deposited in Account B, then $10,000 - x$ is left to be deposited in Account A.

A wire is cut into three pieces of unequal length. The shortest piece is $\frac{1}{4}$ the length of the longest piece, and the middle piece is $\frac{1}{3}$ the length of the longest piece.

Let x = length of the longest piece.

$\frac{1}{3}x$ = length of the middle piece.

$\frac{1}{4}x$ = length of the shortest piece.

A store is having a one-third off sale on a certain model of air conditioner. Let x = regular price of the air conditioner. Then $\frac{2}{3}x$ = sale price of the air conditioner.

We can't say that the sale price is $x - \frac{1}{3}$ because $\frac{1}{3}$ is not "one-third off the price of the air conditioner"—it is simply "one-third." "One-third the price of the air conditioner" is represented by $\frac{1}{3}x$. "One-third off the price of the air conditioner" is represented by subtracting one-third of the price from the price:

$$\overbrace{x - \frac{1}{3}x}^{\text{Price − one-third the price}} = \frac{x}{1} - \frac{x}{3} = \frac{3x}{3} - \frac{x}{3} = \frac{2x}{3} = \frac{2}{3}x$$

PRACTICE

Fill in the blanks.

1. Tony is three years older than Marie.

    Marie's age = _____.

    Tony's age = _____.

2. Sandie is three-fourths as tall as Mona.

    Mona's height (in the given unit of measure) = _____.

    Sandie's height (in the given unit of measure) = _____.

3. Sam takes two hours longer than Lisa to compute his taxes.

    Number of hours Lisa takes to compute her taxes = _____.

    Number of hours Sam takes to compute his taxes = _____.

4. Three-fifths of a couple's net income is spent on rent.

    Net income = _____.

    Amount spent on rent = _____.

5. A rectangle's length is four times its width.

    Width (in the given unit of measure) = _____.

    Length (in the given unit of measure) = _____.

6. Candice paid $5000 last year in federal and state income taxes.

    Amount paid in federal income taxes = _____.

    Amount paid in state income taxes = _____.

7. Nikki has $8000 in her bank, some in a checking account, some in a certificate of deposit (CD).

   Amount in checking account = _____.

   Amount in CD = _____.

8. A total of 450 tickets were sold, some adult tickets, some children's tickets.

   Number of adult tickets sold = _____.

   Number of children's tickets sold = _____.

9. A boutique is selling a sweater for three-fourths off retail.

   Regular selling price = _____.

   Sale price = _____.

10. A string is cut into three pieces of unequal length. The shortest piece is $\frac{1}{5}$ as long as the longest piece. The mid-length piece is $\frac{1}{2}$ the length of the longest piece.

    Length of the longest piece (in the given units) = _____.

    Length of the shortest piece (in the given units) = _____.

    Length of the middle piece (in the given units) = _____.

SOLUTIONS

1. Marie's age = __x__.
   Tony's age = __x + 3__.

2. Mona's height (in the given unit of measure) = __x__.
   Sandie's height (in the given unit of measure) = __$\frac{3}{4}x$__.

3. Number of hours Lisa takes to compute her taxes = __x__.
   Number of hours Sam takes to compute his taxes = __x + 2__.

4. Net income = __x__.
   Amount spent on rent = __$\frac{3}{5}x$__.

5. Width (in the given unit of measure) = __x__.
   Length (in the given unit of measure) = __4x__.

6. Amount paid in federal income taxes = __x__.
   Amount paid in state income taxes = __5000 − x__.
   (Or x = amount paid in state income taxes and 5000 − x = amount paid in federal taxes).

7. Amount in checking account = __x__.
   Amount in CD = __8000 − x__.
   (Or x = amount in CD and 8000 − x = amount in checking account)

8. Number of adult tickets sold = __x__ .
    Number of children's tickets sold = __450 − x__ .
    (Or x = number of children's tickets and $450 − x$ = number of adult tickets)

9. Retail selling price = __x__ .
    Sale price = $\dfrac{1}{4}x$ or $\dfrac{x}{4}\left(x - \dfrac{3}{4}x = \dfrac{4x}{4} - \dfrac{3x}{4} = \dfrac{4x - 3x}{4} = \dfrac{x}{4}\right)$.

10. Length of the longest piece (in the given units) = __x__ .
    Length of the shortest piece (in the given units) = __$\frac{1}{5}x$__ .
    Length of the mid-length piece (in the given units) = __$\frac{1}{2}x$__ .

Summary

In this chapter, we learned how to:

- *Simplify fractions containing variables.* Simplify a fraction whose numerator and/or denominator contain variables by dividing out common factors. Treat variables as if they were prime numbers. Only factors can be divided out (canceled), not terms.

- *Perform arithmetic on fractions that contain variables.* The steps for adding, subtracting, multiplying, and dividing fractions containing variables are the same we used in Chapter 1 to perform these operations on fractions consisting of whole numbers.

- *Simplify compound fractions containing variables.* The steps for simplifying compound fractions containing variables are the same as simplifying fractions that don't contain variables. Rewrite the fraction as a division problem and then as a multiplication problem.

- *Use variables to represent unknown quantities in word problems.* If a word problem has only one unknown, we use x (or some other variable) to represent the unknown quantity. If there are multiple unknowns, we usually let x represent one of the unknowns and then represent the other unknowns in terms of x.

After covering the distributive property, we will work with a larger family of fractions. Later in the book, we will use what we learned about fractions with variables in them to solve equations, inequalities, and word problems.

QUIZ

1. Write $\dfrac{10y}{21}$ as a product of a number and a variable.

    A. $\dfrac{10}{21}y$

    B. $\dfrac{10}{21}+y$

    C. $\dfrac{21}{10}+y$

    D. $\dfrac{10}{21}\cdot\dfrac{1}{y}$

2. Write $\dfrac{8}{12x}$ in lowest terms.

    A. $\dfrac{2x}{3}$

    B. $\dfrac{2}{3}$

    C. $\dfrac{1}{2x}$

    D. $\dfrac{2}{3x}$

3. The width of a room is three-fourths its length. If x represent the length, in feet, what is Its width?

    A. $\dfrac{3}{4}$ feet

    B. $\dfrac{3}{4}-x$ feet

    C. $\dfrac{3}{4}x$ feet

    D. $x-\dfrac{3}{4}$ feet

4. $\dfrac{9}{5}\cdot\dfrac{10x}{21}=$

    A. $\dfrac{6x}{15}$

    B. $\dfrac{x}{16}$

    C. $\dfrac{3x}{7}$

    D. $\dfrac{6x}{7}$

5. $\dfrac{6x}{5} + \dfrac{1}{2} =$

    A. $\dfrac{6x+1}{7}$

    B. $\dfrac{12x+5}{10}$

    C. $\dfrac{6x+1}{10}$

    D. $\dfrac{12x+1}{10}$

6. A cable company advertises an introductory monthly charge that is one-fourth off its regular monthly charge. If x represents its regular monthly fee, what is the introductory monthly fee?

    A. $x - \dfrac{1}{4}$

    B. $x - 25\%$

    C. $\dfrac{3}{4}x$

    D. $\dfrac{1}{4} - x$

7. Separate the factor having the variable from the rest of the fraction: $\dfrac{16}{5x}$.

    A. $\dfrac{16}{5} \cdot \dfrac{1}{x}$

    B. $\dfrac{16}{5} \cdot x$

    C. $\dfrac{5}{16} \cdot \dfrac{1}{x}$

    D. $\dfrac{5}{16} \cdot x$

8. $\dfrac{\frac{5}{3}}{15x} =$

    A. $\dfrac{25x}{3}$

    B. $\dfrac{1}{9x}$

    C. $9x$

    D. $\dfrac{1}{25x}$

9. Suppose $20,000 is invested in two mutual funds, Fund A and Fund B. If x represents the amount invested in Fund A, how much is invested in Fund B?

  A. $x - 20{,}000$

  B. $\dfrac{20{,}000}{x}$

  C. $20{,}000 - x$

  D. There is not enough information to answer this question.

10. Write $\dfrac{12(x+9)(x-1)}{20(x-1)}$ in lowest terms.

  A. $\dfrac{3}{5(x+9)}$

  B. $\dfrac{3(x+9)}{5}$

  C. $\dfrac{5(x+9)}{3(x+1)}$

  D. $\dfrac{5(x+9)}{3}$

11. $\dfrac{4x}{9} + \dfrac{5}{6} =$

  A. $\dfrac{4x+15}{18}$

  B. $\dfrac{8x+15}{18}$

  C. $\dfrac{8x+5}{18}$

  D. $\dfrac{24x+54}{15}$

12. $\dfrac{\frac{4x}{3}}{\frac{16}{15}} =$

  A. $\dfrac{5x}{4}$

  B. $\dfrac{64x}{45}$

  C. $\dfrac{4}{5x}$

  D. $\dfrac{4x}{5}$

13. $\dfrac{4(x+6)(x+3)}{18(x+3)(x+5)} =$

    A. $\dfrac{12}{45}$

    B. $\dfrac{2(x+2)}{3(x+5)}$

    C. $\dfrac{2(x+6)}{9(x+5)}$

    D. $\dfrac{2(x+6)}{9(x+3)}$

14. $\dfrac{9x}{4y} \cdot \dfrac{8y}{15} =$

    A. $\dfrac{6x}{5}$

    B. $\dfrac{3x}{10}$

    C. $\dfrac{2x}{5}$

    D. $\dfrac{3x}{20}$

15. Separate the factor having the variable from the rest of the fraction: $\dfrac{8}{3(2x+1)}$.

    A. $\dfrac{8}{3}(2x+1)$

    B. $\dfrac{8}{3} \cdot \dfrac{1}{2x+1}$

    C. $\dfrac{4}{3} \cdot \dfrac{1}{x+1}$

    D. The variable cannot be separated from the fraction.

Decimals

We now use what we learned in Chapter 1 to develop rules for arithmetic with decimal numbers. Although calculators can do the arithmetic for you, you might find yourself without a calculator or in a class that does not permit calculators. Some of what we cover in this chapter will help us later when we solve equations and word problems.

CHAPTER OBJECTIVES

In this chapter, you will

- Write a decimal number as a fraction
- Perform arithmetic with decimal numbers
- Simplify fractions containing decimal numbers
- Rewrite division problems containing decimal numbers

A decimal number is convenient notation for the sum of fractions. For example the decimal number 0.183 is the sum of three fractions.

$$\frac{1}{10}+\frac{8}{100}+\frac{3}{1000}=\frac{1}{10}\cdot\frac{100}{100}+\frac{8}{100}\cdot\frac{10}{10}+\frac{3}{1000}=\frac{100+80+3}{1000}=\frac{183}{1000}$$

As you can see, the denominators of these fractions are powers of 10. Here are the powers of 10 that we will use most in this chapter.

$$10^1 = 10$$
$$10^2 = 100$$
$$10^3 = 1000$$
$$10^4 = 10,000$$

For now, we will learn how to convert a decimal number to its fraction form. Later, we will learn rules for adding, subtracting, multiplying, and dividing with decimal numbers.

The number in front of the decimal point is the whole number, and the number behind the decimal point is the numerator of a fraction whose denominator is a power of 10. The denominator consists of 1 followed by one or more zeros. The number of zeros is the same as the number of digits behind the decimal point.

 EXAMPLES

Rewrite the decimal number as a fraction.

$$2.8 = 2\tfrac{8}{10} \text{ (one decimal place—one zero)}$$

$$0.7695 = \frac{7695}{10,000} \text{ (four decimal places—four zeros)}$$

PRACTICE

Rewrite the decimal number as a fraction. If the decimal number is more than 1, rewrite the number both as a mixed number and an improper fraction. You do not need to simplify the answer.

1. $1.71 =$

2. $34.598 =$

3. $0.6 =$

4. $0.289421 =$

✓ SOLUTIONS

1. $1.71 = 1\frac{71}{100} = \dfrac{171}{100}$

2. $34.598 = 34\frac{598}{1000} = \dfrac{34{,}598}{1000}$

3. $0.6 = \dfrac{6}{10}$

4. $0.289421 = \dfrac{289{,}421}{1{,}000{,}000}$

There are two types of decimal numbers—*terminating* and *nonterminating*. The numbers in the above problems are terminating decimal numbers. A nonterminating decimal number has infinitely many nonzero digits following the decimal point. For example, 0.333333333... is a nonterminating decimal number. Some nonterminating decimal numbers represent fractions, such as 0.333333333... = $\frac{1}{3}$. But some nonterminating decimals, such as π = 3.1415926654... and $\sqrt{2}$ = 1.414213562..., do not represent fractions. We will be concerned mostly with terminating decimal numbers in this book.

We can add as many zeros at the end of a terminating decimal number as we want because the extra zeros can be divided out.

$$0.7 = \frac{7}{10}$$

$$0.70 = \frac{70}{100} = \frac{7 \cdot 10}{10 \cdot 10} = \frac{7}{10}$$

$$0.700 = \frac{700}{1000} = \frac{7 \cdot 100}{10 \cdot 100} = \frac{7}{10}$$

Adding and Subtracting Decimal Numbers

In order to add or subtract decimal numbers, each number needs to have the same number of digits behind the decimal point. Writing the problem vertically helps us avoid the common problem of adding the numbers incorrectly. For instance, 1.2 + 3.41 is *not* 4.43. The "2" needs to be added to the "4," not to the "1."

$$\begin{array}{r} 1.20 \text{ (Add as many zeros at the end as needed.)} \\ + \underline{4.43} \\ 5.63 \end{array}$$

510.3 – 422.887   becomes   510.300
                                  – 422.887
                                  ──────────
                                   87.413

 PRACTICE

Rewrite as a vertical problem and find the sum or difference.

1. 7.26 + 18.1
2. 5 – 2.76
3. 15.01 – 6.328
4. 968.323 – 13.08
5. 28.56 – 16.7342
6. 0.446 + 1.2
7. 2.99 + 3

 SOLUTIONS

1. 7.26 + 18.1

      7.26
   + 18.10
   ──────
    25.36

2. 5 – 2.76

    5.00
   – 2.76
   ──────
    2.24

3. 15.01 – 6.328

    15.010
   – 6.328
   ──────
    8.682

4. 968.323 – 13.08

    968.323
   – 13.080
   ────────
    955.243

5. 28.56 – 16.7342

    28.5600
   – 16.7342
   ────────
    11.8258

6. 0.446 + 1.2

$$\begin{array}{r} 0.446 \\ + 1.200 \\ \hline 1.646 \end{array}$$

7. 2.99 + 3

$$\begin{array}{r} 2.99 \\ + 3.00 \\ \hline 5.99 \end{array}$$

Multiplying Decimal Numbers

To multiply decimal numbers, we perform multiplication, ignoring the decimal point, and then we decide where to put the decimal point. We count the number of digits that follow the decimal points in the factors. This total will be the number of digits that follow the decimal point in the product.

 EXAMPLES

Find the product.

12.<u>83</u> · 7.<u>91</u> The product has four digits following the decimal point.

1283 · 791 = 1,014,853, so 12.83 · 7.91 = 101.4853

3.<u>782</u> · 19.<u>41</u> The product has five digits following the decimal point.

3782 · 1941 = 7340862, so 3.782 · 19.41 = 73.40862

14 · 3.<u>55</u> The product has two digits following the decimal point.

14 · 355 = 4970, so 14 · 3.55 = 49.70

To see why this strategy works, let us compute 1.4×5.12 *after* writing these numbers as fractions.

$$1.4 \cdot 5.12 = \frac{14}{10} \cdot \frac{512}{1000} = \frac{14(512)}{10(100)} = \frac{7168}{1000} = 7.168$$

Because the denominator of the final fraction is 1000 (1 followed by three 0s), we move the decimal place *three* places.

 PRACTICE

Find the product.

1. 3.2 · 1.6 =

2. 4.11 · 2.84 =

   3. $8 \cdot 2.5 =$
   4. $0.153 \cdot 6.8 =$
   5. $0.0351 \cdot 5.6 =$

 **SOLUTIONS** _____

   1. $3.2 \cdot 1.6 = 5.12$
   2. $4.11 \cdot 2.84 = 11.6724$
   3. $8 \cdot 2.5 = 20.0 = 20$
   4. $0.153 \cdot 6.8 = 1.0404$
   5. $0.0351 \cdot 5.6 = 0.19656$

Fractions Containing Decimals

Fractions having a decimal number in their numerator and/or denominator can be rewritten as fractions without decimal points. We multiply the numerator and denominator by a power of 10—the same power of 10—large enough so that the decimal point becomes unnecessary (more about this in a moment).

$$\frac{1.3}{3.9} = \frac{1.3 \cdot 10}{3.9 \cdot 10} = \frac{13}{39} = \frac{1}{3}$$

To determine what power of 10 we need, we count the number of digits behind each decimal point.

$$\underline{1.28} \leftarrow \text{Two digits behind the decimal point}$$
$$4.6 \leftarrow \text{One digit behind the decimal point}$$

We need to multiply the numerator and denominator by $10^2 = 100$ in order to eliminate the need for decimal points.

$$\frac{1.28 \cdot 100}{4.6 \cdot 100} = \frac{128}{460} = \frac{32}{115}$$

In general, if there is only one digit behind the decimal point, we multiply the fraction by $\frac{10}{10}$, if there are at most two digits behind the decimal point, we multiply the fraction by $\frac{100}{100}$, and if there are at most three digits behind the decimal place, we multiply the fraction by $\frac{1000}{1000}$. This process is called *clearing the decimal*.

EXAMPLES

Clear the decimal from the fraction.

$$\frac{7.1}{2.285}$$ There are 3 digits behind the decimal point.

$$\frac{7.1}{2.285} = \frac{7.1 \cdot 1000}{2.285 \cdot 1000} = \frac{7100}{2285} = \frac{1420}{457}$$

$$\frac{6}{3.14}$$ There are 2 digits behind the decimal point.

$$\frac{6}{3.14} = \frac{6 \cdot 100}{3.14 \cdot 100} = \frac{600}{314} = \frac{300}{157}$$

PRACTICE

Clear the decimal from the fraction.

1. $\dfrac{4.58}{2.15} =$

2. $\dfrac{3.6}{18.11} =$

3. $\dfrac{2.123}{5.6} =$

4. $\dfrac{8}{2.4} =$

5. $\dfrac{6.25}{5} =$

6. $\dfrac{0.31}{1.2} =$

7. $\dfrac{0.423}{0.6} =$

✔ SOLUTIONS

1. $\dfrac{4.58}{2.15} = \dfrac{4.58 \cdot 100}{2.15 \cdot 100} = \dfrac{458}{215}$

2. $\dfrac{3.6}{18.11} = \dfrac{3.6 \cdot 100}{18.11 \cdot 100} = \dfrac{360}{1811}$

3. $\dfrac{2.123}{5.6} = \dfrac{2.123 \cdot 1000}{5.6 \cdot 1000} = \dfrac{2123}{5600}$

4. $\dfrac{8}{2.4} = \dfrac{8 \cdot 10}{2.4 \cdot 10} = \dfrac{80}{24} = \dfrac{10}{3}$

5. $\dfrac{6.25}{5} = \dfrac{6.25 \cdot 100}{5 \cdot 100} = \dfrac{625}{500} = \dfrac{5}{4}$

6. $\dfrac{0.13}{1.2} = \dfrac{0.13 \cdot 100}{1.2 \cdot 100} = \dfrac{13}{120}$

7. $\dfrac{0.423}{0.6} = \dfrac{0.423 \cdot 1000}{0.6 \cdot 1000} = \dfrac{423}{600} = \dfrac{141}{200}$

Division with Decimals

We can use the method in the previous section to rewrite decimal division problems as whole-number division problems. We rewrite the division problem as a fraction, clear the decimal, and then rewrite the fraction as a division problem without decimal points.

 EXAMPLES

Rewrite as a division problems without decimal points.

$1.2\overline{)6.03}$

$1.2\overline{)6.03}$ is another way of writing $\dfrac{6.03}{1.2}$.

$\dfrac{6.03}{1.2} = \dfrac{6.03 \cdot 100}{1.2 \cdot 100} = \dfrac{603}{120}$ and $\dfrac{603}{120}$ becomes $120\overline{)603}$.

$0.51\overline{)3.7}$

$0.51\overline{)3.7}$ becomes $\dfrac{3.7}{0.51} = \dfrac{3.7 \cdot 100}{0.51 \cdot 100} = \dfrac{370}{51}$ which becomes $51\overline{)370}$.

$8\overline{)12.8}$

$8\overline{)12.8}$ becomes $\dfrac{12.8}{8} = \dfrac{12.8 \cdot 10}{8 \cdot 10} = \dfrac{128}{80}$ which becomes $80\overline{)128}$. We could reduce the fraction and get an even simpler division problem.

$\dfrac{128}{80} = \dfrac{8}{5}$ which becomes $5\overline{)8}$.

PRACTICE

Rewrite as a division problem without decimal points.

1. $6.85\overline{)15.11}$

2. $0.9\overline{)8.413}$

3. $4\overline{)8.8}$

4. $19.76\overline{)60.4}$

5. $3.413\overline{)7}$

SOLUTIONS

1. $6.85\overline{)15.11}$ $\dfrac{15.11}{6.85} = \dfrac{15.11 \cdot 100}{6.85 \cdot 100} = \dfrac{1511}{685}$ becomes $685\overline{)1511}$

2. $0.9\overline{)8.413}$ $\dfrac{8.413}{0.9} = \dfrac{8.413 \cdot 1000}{0.9 \cdot 1000} = \dfrac{8413}{900}$ becomes $900\overline{)8413}$

3. $4\overline{)8.8}$ $\dfrac{8.8}{4} = \dfrac{8.8 \cdot 10}{4 \cdot 10} = \dfrac{88}{40}$ becomes $40\overline{)88}$

   Reduce to get a simpler division problem.

   $\dfrac{88}{40} = \dfrac{11}{5}$ becomes $5\overline{)11}$.

4. $19.76\overline{)60.4}$ $\dfrac{60.4}{19.76} = \dfrac{60.4 \cdot 100}{19.76 \cdot 100} = \dfrac{6040}{1976}$

   Reduce to get a simpler division problem: $\dfrac{6040}{1976} = \dfrac{755}{247}$, which becomes $247\overline{)755}$.

5. $3.413\overline{)7}$ $\dfrac{7}{3.413} = \dfrac{7 \cdot 1000}{3.413 \cdot 1000} = \dfrac{7000}{3413}$ becomes $3413\overline{)7000}$.

Summary

In this chapter, we learned how to:

- *Write a terminating decimal number as a fraction.* The numerator of the fraction is the number, with the decimal point removed, and the denominator is a power of 10. Determine the power of 10 by counting the number of digits behind the decimal point.

- *Perform addition and subtraction with decimal numbers.* Add (or subtract) two decimal numbers (or a whole number to a decimal number), by lining up the decimal points, adding 0s behind the last digit if necessary. Once this is done, perform the addition (or subtraction).

- *Perform multiplication with decimal numbers.* Multiply two decimal numbers (or a whole number and a decimal number), by carrying out the multiplication as if the decimal points were not there. Once this is done, count the total number of digits behind the decimal points and move the decimal point to the left in the product number of places.

- *Rewrite division problems containing decimal points as division problems without decimal points.* Divide two decimal numbers by rewriting the quotient as a fraction and then clearing the decimal—that is, multiplying the numerator and denominator by the same power of 10, large enough to eliminate any decimal points.

In Chapter 7, some equations will contain decimal numbers. We will use what we learned in this chapter to clear the equation of decimal points before solving the equation.

QUIZ

1. 3.64 =

   A. $\dfrac{67}{100}$

   B. $\dfrac{48}{25}$

   C. $\dfrac{91}{25}$

   D. $\dfrac{72}{25}$

2. $3 \cdot 2.4 =$

   A. 6.2
   B. 6.8
   C. 7.2
   D. 7.6

3. $1.76 + 3.2 =$

   A. 4.96
   B. 4.78
   C. 2.08
   D. 4.52

4. $\dfrac{2.4}{3.2} =$

   A. $\dfrac{3}{4}$

   B. $\dfrac{3}{40}$

   C. $\dfrac{30}{4}$

   D. $\dfrac{11}{13}$

5. Rewrite $2.65\overline{)8.4}$ without decimal numbers.

   A. $2650\overline{)840}$

   B. $265\overline{)8400}$

   C. $265\overline{)84}$

   D. $265\overline{)840}$

6. **4.8 · 2.15 =**
   A. 8.23
   B. 8.95
   C. 10.32
   D. 9.2

7. **1.024 =**

   A. $\dfrac{128}{1250}$

   B. $\dfrac{128}{125}$

   C. $\dfrac{1280}{125}$

   D. $\dfrac{1024}{125}$

8. **4.6 − 3.12 =**
   A. 1.48
   B. 1.72
   C. 1.52
   D. 0.52

9. $\dfrac{\mathbf{2.85}}{\mathbf{1.75}} =$

   A. $\dfrac{5700}{350}$

   B. $\dfrac{570}{35}$

   C. $\dfrac{57}{350}$

   D. $\dfrac{57}{35}$

10. **5 · 3.8 =**
    A. 15.4
    B. 19
    C. 10.9
    D. 6.5

chapter 4

Negative Numbers

Negative numbers arise all around us. They represent the number necessary to bring a quantity *up* to 0. In this chapter, we will learn rules for performing arithmetic with negative numbers. We will also learn how to rewrite a subtraction problem as an addition problem, a technique that is useful for solving equations.

CHAPTER OBJECTIVES

In this chapter, you will

- To perform arithmetic with negative numbers
- To rewrite subtraction problems as addition problems
- To write the negative of a number or variable

To understand why some of the rules for negative number arithmetic are true, we will consider the readings on a thermometer. A reading of −10° means the temperature is 10° *below* 0°, and that the temperature would need to increase 10° to reach 0°. A reading of 10° means the temperature would need to decrease 10° to reach 0°, so the sign in front of the number tells us on which side of 0 the number lies.

Suppose the temperature is −15°. If the temperature rises by 10°, the temperature, −5°, is *still* 5° below 0°.

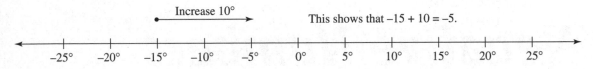

FIGURE 4-1

If the temperature increases by 20°, then the temperature is 5° above 0°.

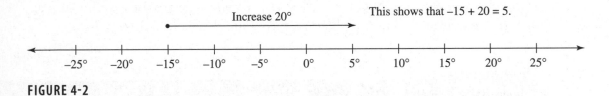

FIGURE 4-2

The Sum of a Positive Number and a Negative Number

From these examples, we see that the sum of a negative number and a positive number might be positive or might be negative, depending on whether the positive number is large enough. When adding a positive number to a negative number (or a negative number to a positive number), we take the *difference* of the numbers. The sum is *positive* if the "larger" number is positive. The sum is *negative* if the sign of the "larger" number is negative.

EXAMPLES

Find the sum.

$-82 + 30 = $ ____.

The difference of 82 and 30 is 52. Because 82 is larger than 30, we use the sign on −82 for the sum: $-82 + 30 = -52$.

$-125 + 75 = $ ____.

The difference of 125 and 75 is 50. Because 125 is larger than 75, we use the sign on –125 for the sum: $-125 + 75 = -50$.

$-10 + 48 =$ ____.

The difference of 48 and 10 is 38. Because 48 is larger than 10, we use the sign on 48 for the sum: $-10 + 48 = 38$.

PRACTICE

Find the sum.

1. $-10 + 8 =$
2. $-65 + 40 =$
3. $13 + (-20) =$
4. $-5 + 9 =$
5. $-24 + 54 =$
6. $-6 + 19 =$
7. $-71 + 11 =$
8. $40 + (-10) =$
9. $12 + (-18) =$

SOLUTIONS

1. $-10 + 8 = -2$
2. $-65 + 40 = -25$
3. $13 + (-20) = -7$
4. $-5 + 9 = 4$
5. $-24 + 54 = 30$
6. $-6 + 19 = 13$
7. $-71 + 11 = -60$
8. $40 + (-10) = 30$
9. $12 + (-18) = -6$

Subtracting a Larger Number from a Smaller Number

To illustrate the next arithmetic rule, let us consider the temperature being above 0° and the temperature dropping. If it drops only a little, the temperature remains above 0°, but if it drops enough, the temperature will be below 0°. Suppose the temperature is 20°. If it decreases by 10°, then it is still above 0°.

This shows that 20 − 10 = 10.

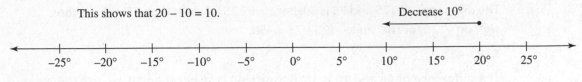

FIGURE 4-3

If the temperature decreases 30°, then the temperature drops *below* 0°.

This shows that 20 − 30 = −10.

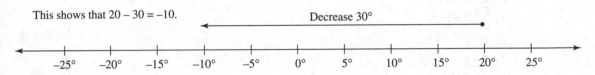

FIGURE 4-4

In general, subtracting a larger positive number from a smaller positive number gives us a negative number.

 EXAMPLES
Find the difference.

$$410 - 500 = -90$$
$$10 - 72 = -62$$

CAUTION *Be careful what you call these signs; a negative sign in of a number indicates that the number is smaller than zero. A minus sign between two numbers indicates the operation of subtraction. In the equation 3 − 5 = −2, the sign in front of 5 is a minus sign and the sign in front of 2 is a negative sign. A minus sign requires two quantities and a negative sign requires one quantity.*

 PRACTICE
Find the difference.
1. 28 − 30 =
2. 88 − 100 =
3. 25 − 110 =
4. 4 − 75 =
5. 5 − 90 =

 **SOLUTIONS**
1. $28 - 30 = -2$
2. $88 - 100 = -12$
3. $25 - 110 = -85$
4. $4 - 75 = -71$
5. $5 - 90 = -85$

Subtracting a Positive Number from a Negative Number

Let us see the effect of subtracting a positive number from a negative number. Returning to the temperature example, suppose the temperature is $-10°$ and the temperature drops another $5°$, then the temperature is even further below $0°$.

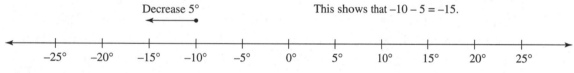

FIGURE 4-5

When subtracting a positive number from a negative number, add the numbers (ignoring the negative sign). The difference is negative.

 EXAMPLES
Find the difference.

$$-30 - 15 = -45$$
$$-18 - 7 = -25$$
$$-500 - 81 = -581$$

 PRACTICE
Find the difference.
1. $-16 - 4 =$
2. $-70 - 19 =$
3. $-35 - 5 =$
4. $-100 - 8 =$
5. $-99 - 1 =$

SOLUTIONS

1. $-16 - 4 = -20$
2. $-70 - 19 = -89$
3. $-35 - 5 = -40$
4. $-100 - 8 = -108$
5. $-99 - 1 = -100$

Double Negatives

A negative sign in front of a quantity can be interpreted to mean "opposite." For instance -3 can be called "the opposite of 3." Viewed in this way, we can see that $-(-4)$ means "the opposite of -4." But the opposite of -4 is $+4$, so $-(-4) = +4$.

 EXAMPLES

$$-(-25) = 25$$
$$-(-x) = x$$
$$-(-3y) = 3y$$

Rewriting a Subtraction Problem as an Addition Problem

Sometimes in algebra it is easier to think of a subtraction problem as an addition problem. One advantage to this is that we can rearrange the terms in an addition problem but not a subtraction problem: $3 + 4 = 4 + 3$ but $4 - 3 \neq 3 - 4$. The minus sign can be replaced with a plus sign if we change the sign of the number following it: $4 - 3 = 4 + (-3)$. The parentheses are used to show that the sign in front of the number is a negative sign and not a minus sign. The rule to rewrite a subtraction problem as an addition problem is $a - b = a + (-b)$.

EXAMPLES
Rewrite as an addition problem.

$$-82 - 14 = -82 + (-14)$$
$$20 - (-6) = 20 + 6$$
$$x - y = x + (-y)$$

PRACTICE

Rewrite as an addition problem.

1. $8 - 5 =$

2. $-29 - 4 =$

3. $-6 - (-10) =$

4. $15 - x =$

5. $40 - 85 =$

6. $y - 37 =$

7. $-x - (-14) =$

8. $-x - 9 =$

9. $\dfrac{-4}{5} + \dfrac{2}{3} =$

Calculate. Remember to convert a mixed number to an improper fraction before subtracting.

10. $2\frac{1}{8} - 3\frac{1}{4} =$

11. $-4\frac{2}{9} - 1\frac{1}{2} =$

12. $\dfrac{5}{36} - 2 =$

13. $\dfrac{6}{25} + \dfrac{2}{3} - \dfrac{14}{15} =$

14. $\dfrac{-4}{3} + \dfrac{5}{6} - \dfrac{8}{21} =$

15. $1\frac{4}{5} - 3\frac{1}{2} - 1\frac{6}{7} =$

SOLUTIONS

1. $8 - 5 = 8 + (-5)$

2. $-29 - 4 = -29 + (-4)$

3. $-6 - (-10) = -6 + 10$

4. $15 - x = 15 + (-x)$

5. $40 - 85 = 40 + (-85)$

6. $y - 37 = y + (-37)$

7. $-x - (-14) = -x + 14$

8. $-x - 9 = -x + (-9)$

9. $\dfrac{-4}{5} + \dfrac{2}{3} = \dfrac{-4}{5} \cdot \dfrac{3}{3} + \dfrac{2}{3} \cdot \dfrac{5}{5} = \dfrac{-12}{15} + \dfrac{10}{15} = \dfrac{-12 + 10}{15} = \dfrac{-2}{15}$

10. $2\frac{1}{8} - 3\frac{1}{4} = \dfrac{17}{8} - \dfrac{13}{4} = \dfrac{17}{8} - \dfrac{13}{4} \cdot \dfrac{2}{2} = \dfrac{17}{8} - \dfrac{26}{8} = \dfrac{17 - 26}{8} = \dfrac{-9}{8}$

11. $-4\frac{2}{9} - 1\frac{1}{2} = \dfrac{-38}{9} - \dfrac{3}{2} = \dfrac{-38}{9} \cdot \dfrac{2}{2} - \dfrac{3}{2} \cdot \dfrac{9}{9} = \dfrac{-76}{18} - \dfrac{27}{18} = \dfrac{-76 - 27}{18} = \dfrac{-103}{18}$

12. $\dfrac{5}{36} - 2 = \dfrac{5}{36} - \dfrac{2}{1} \cdot \dfrac{36}{36} = \dfrac{5}{36} - \dfrac{72}{36} = \dfrac{5 - 72}{36} = \dfrac{-67}{36}$

13. $\dfrac{6}{25} + \dfrac{2}{3} - \dfrac{14}{15} = \dfrac{6}{25} \cdot \dfrac{3}{3} + \dfrac{2}{3} \cdot \dfrac{25}{25} - \dfrac{14}{15} \cdot \dfrac{5}{5} = \dfrac{18}{75} + \dfrac{50}{75} - \dfrac{70}{75} = \dfrac{18 + 50 - 70}{75} = \dfrac{-2}{75}$

14. $\dfrac{-4}{3} + \dfrac{5}{6} - \dfrac{8}{21} = \dfrac{-4}{3} \cdot \dfrac{14}{14} + \dfrac{5}{6} \cdot \dfrac{7}{7} - \dfrac{8}{21} \cdot \dfrac{2}{2} = \dfrac{-56}{42} + \dfrac{35}{42} - \dfrac{16}{42} = \dfrac{-56 + 35 - 16}{42} = \dfrac{-37}{42}$

15. $1\frac{4}{5} - 3\frac{1}{2} - 1\frac{6}{7} = \dfrac{9}{5} - \dfrac{7}{2} - \dfrac{13}{7} = \dfrac{9}{5} \cdot \dfrac{14}{14} - \dfrac{7}{2} \cdot \dfrac{35}{35} - \dfrac{13}{7} \cdot \dfrac{10}{10} = \dfrac{126 - 245 - 130}{70} = \dfrac{-249}{70}$

Multiplication and Division with Negative Numbers

When taking the product of two or more quantities when one or more of them is negative, we take the product as if the negative signs were not there. An even number of negative factors gives us a positive product and an odd number of negative factors gives us a negative product. Similarly, for a quotient (or fraction), two negative numbers give us a positive quotient, and one negative number and one positive number gives us a negative quotient. The rules for multiplying or dividing negative numbers are below.

$$(-a)(-b) = ab$$
$$a(-b) = (-a)b = -ab$$
$$a \div (-b) = -a \div b$$
$$(-a) \div (-b) = a \div b$$

EXAMPLES

Find the product or quotient.

$(4)(-3)(2) = -24$

$(-16)(-2) = 32$

$8 \div (-2) = -4$

$\dfrac{-55}{5} = -11$

$\dfrac{-2}{-3} = \dfrac{2}{3}$

$(-5)(-6)(-1)(3) = -90$

PRACTICE

Find the product or quotient.

1. $(15)(-2) =$

2. $-32 \div (-8) =$

3. $3(-3)(4) =$

4. $62 \div (-2) =$

5. $\dfrac{1}{2}\left(\dfrac{-3}{7}\right) =$

6. $(-4)(6)(-3) =$

7. $\dfrac{\frac{4}{3}}{-\frac{1}{2}} =$

8. $(-2)(5)(-6)(-8) =$

9. $\dfrac{-\frac{3}{5}}{-\frac{6}{5}} =$

10. $-28 \div (-4) =$

✔ SOLUTIONS

1. $(15)(-2) = -30$

2. $-32 \div (-8) = 4$

3. $3(-3)(4) = -36$

4. $62 \div (-2) = -31$

5. $\dfrac{1}{2}\left(\dfrac{-3}{7}\right) = \dfrac{-3}{14}$

6. $(-4)(6)(-3) = 72$

7. $\dfrac{\frac{4}{3}}{-\frac{1}{2}} = \dfrac{4}{3} \div \dfrac{-1}{2} = \dfrac{4}{3} \cdot \dfrac{2}{-1} = \dfrac{8}{-3} = -\dfrac{8}{3}$

8. $(-2)(5)(-6)(-8) = -480$

9. $\dfrac{\frac{-3}{5}}{\frac{-6}{5}} = \dfrac{-3}{5} \div \dfrac{-6}{5} = \dfrac{-3}{5} \cdot \dfrac{5}{-6} = \dfrac{1}{2}$

10. $-28 \div (-4) = 7$

Negating Variables

Negating a variable does not automatically mean that the quantity is negative: $-x$ means "the opposite" of x. We can't conclude that $-x$ is a negative number unless we have reason to believe x itself is a positive number. If x is a negative number, $-x$ is a positive number. (Although in practice we verbally say "negative x" for "$-x$" when we really mean "the opposite of x.")

The same rules above apply when multiplying "negative" variables.

 EXAMPLES

Use the rules for negating numbers in a product to rewrite the product.

$-3(5x) = -15x$ $5(-x) = -5x$

$-12(-4x) = 48x$ $-x(-y) = xy$

$-2x(3y) = -6xy$ $x(-y) = -xy$

$-16x(-4y) = 64xy$ $4(-1.83x)(2.36y) = -17.2752xy$

$-3(-x) = 3x$

PRACTICE

Use the rules for negating numbers in a product to rewrite the product.

1. $18(-3x) =$

2. $-4(2x)(-9y) =$

3. $28(-3x) =$

4. $-5x(-7y) =$

5. $-1(-6)(-7x) =$

6. $1.1x(2.5y) =$

7. $-8.3(4.62x) =$

8. $-2.6(-13.14)(-6x) =$

9. $0.36(-8.1x)(-1.6y) =$

10. $4(-7)(2.1x)y =$

SOLUTIONS

1. $18(-3x) = -54x$

2. $-4(2x)(-9y) = 72xy$

3. $28(-3x) = -84x$

4. $-5x(-7y) = 35xy$

5. $-1(-6)(-7x) = -42x$

6. $1.1x(2.5y) = 2.75xy$

7. $-8.3(4.62x) = -38.346x$

8. $-2.6(-13.14)(-6x) = -204.984x$

9. $0.36(-8.1x)(-1.6y) = 4.6656xy$

10. $4(-7)(2.1x)y = -58.8xy$

Fractions and Negative Signs

Because a negative number divided by a positive number is negative, and a positive number divided by a negative number is a negative number, a negative sign in a fraction can go wherever we want to put it.

$$\frac{negative}{positive} = \frac{positive}{negative} = -\frac{positive}{positive}$$

The fraction rules are below.

$$\frac{-a}{-b} = \frac{a}{b}$$

$$\frac{-a}{b} = \frac{a}{-b} = -\frac{a}{b}$$

EXAMPLES

Rewrite the fraction two different ways.

$$\frac{-2}{3} = \frac{2}{-3} = -\frac{2}{3} \qquad \frac{-x}{4} = \frac{x}{-4} = -\frac{x}{4}$$

PRACTICE

Rewrite the fraction two different ways.

1. $\dfrac{-3}{5} =$

2. $\dfrac{-2x}{19} =$

3. $\dfrac{4}{-3x} =$

4. $-\dfrac{5}{9y} =$

SOLUTIONS

1. $\dfrac{-3}{5} = \dfrac{3}{-5} = -\dfrac{3}{5}$

2. $\dfrac{-2x}{19} = \dfrac{2x}{-19} = -\dfrac{2x}{19}$

3. $\dfrac{4}{-3x} = \dfrac{-4}{3x} = -\dfrac{4}{3x}$

4. $-\dfrac{5}{9y} = \dfrac{-5}{9y} = \dfrac{5}{-9y}$

Summary

Table 4-1 summarizes the rules we learned for arithmetic with negative numbers.

TABLE 4-1 Arithmetic with negative numbers	
Let a and b be positive numbers, so $-a$ and $-b$ are negative numbers.	
Rule	**Example**
$-a - b = -(a + b)$	$-5 - 3 = -(5 + 3) = -8$
When adding a negative to a positive take the difference. The sum has the same sign as the "larger number."	$24 + (-10) = 14$ $6 + (-15) = -9$
$a - b = a + (-b)$	$3 - 7 = 3 + (-7) = -4$; $10 + (-2) = 10 - 2 = 8$
$-(-a) = a$	$-(-4) = 4$
$(-a)(-b) = ab$	$(-2)(-3) = (2)(3) = 6$
$\dfrac{-a}{b} = \dfrac{a}{-b} = -\dfrac{a}{b}$	$\dfrac{-2}{5} = \dfrac{2}{-5} = -\dfrac{2}{5}$
$\dfrac{-a}{-b} = \dfrac{a}{b}$	$\dfrac{-2}{-15} = \dfrac{2}{15}$

In this chapter, we also learned how to

- *Add a negative number to a positive number.* Take the difference of the numbers. The sign on the sum is the same as the sign on the "larger" number.

- *Subtract a larger positive number from a smaller positive number.* Momentarily disregarding the negative sign. Find the difference of the numbers. The sign on the difference is negative.

- *Subtract a positive number from a negative number.* Momentarily disregarding the negative sign, add the two numbers. The sign on the sum is negative.

- *Interpret a double negative.* Negating a quantity means using the opposite, so the opposite of the opposite is the original number.

- *Rewrite a subtraction problem as an addition problem.* Rewrite a subtraction problem as an addition problem by changing the sign of the second term.

- *Perform multiplication and division with negative numbers.* Momentarily disregarding the negative sign(s), find the product or quotient of the numbers. The sign of the product is positive if there are an even number of negative factors and negative if there are an odd number of negative factors. The same holds true for quotients.

QUIZ

1. $(-6)(-3)(-1) =$
   A. 10
   B. −18
   C. −10
   D. 18

2. $-30 + 12 =$
   A. 18
   B. −18
   C. −42
   D. 42

3. $4 - 9 =$
   A. −13
   B. 13
   C. −5
   D. 5

4. $-5\frac{3}{4} + 2\frac{1}{3} =$
   A. $3\frac{5}{12}$
   B. $-2\frac{1}{12}$
   C. $-4\frac{1}{12}$
   D. $-3\frac{5}{12}$

5. $-3 - 10 =$
   A. −13
   B. 7
   C. −7
   D. 30

6. $\dfrac{-\frac{3}{5}}{\frac{6}{25}} =$

   A. $-\dfrac{5}{2}$

   B. $-\dfrac{2}{5}$

   C. $\dfrac{5}{2}$

   D. $\dfrac{2}{5}$

7. $-\dfrac{3}{4}$ is *not* equal to which of the following?

   A. $\dfrac{-3}{4}$

   B. $\dfrac{-3}{-4}$

   C. $\dfrac{3}{-4}$

   D. $-\dfrac{3}{4}$ is equal to all these numbers.

8. $\dfrac{-5(-x)}{-4} =$

   A. $\dfrac{-5x}{-4}$

   B. $-\dfrac{5x}{4}$

   C. $-\dfrac{5}{4}(-x)$

   D. $\dfrac{5x}{4}$

9. $-(-4x) =$
   A. $4(-x)$
   B. $4x$
   C. $-4x$
   D. None of the above

10. **Rewrite $8 - 3x$ as an addition problem.**

   A. $-(8 + 3x)$

   B. $8 + (-3x)$

   C. $-8 + 3x$

   D. None of the above

chapter **5**

Exponents and Roots

Expressions involving roots and exponents occur frequently in algebra and calculus. By mastering the properties in this chapter, you will find working with such expressions much easier. Here, we learn exponent and root properties, how to use them to rewrite some expressions, and how to simplify others. We will use these properties later to solve equations.

CHAPTER OBJECTIVES

In this chapter, you will

- Use exponent properties to rewrite expressions
- Use exponent properties to perform arithmetic
- Use exponent properties to simplify fractions
- Write a fraction as a product
- Use root properties to rewrite expressions

The expression $4x$ is shorthand for the sum $x + x + x + x$, that is, x added to itself four times. Likewise x^4 is shorthand for the product $x \cdot x \cdot x \cdot x$, that is, x multiplied by itself four times. In the expression x^4, x is called the *base* and 4 is the *power* or *exponent*. We say "x raised to the fourth power" or simply "x to the fourth." Exponents have many useful properties.

Property 1	$a^n a^m = a^{m+n}$

When multiplying two numbers whose bases are the same, we add their exponents.

EXAMPLES

$$2^3 \cdot 2^4 = (2 \cdot 2 \cdot 2)(2 \cdot 2 \cdot 2 \cdot 2) = 2^7$$

$$x^9 \cdot x^3 = x^{9+3} = x^{12}$$

Property 2	$\dfrac{a^m}{a^n} = a^{m-n}$

(For the rest of the chapter, we will assume that a is not zero.)

When dividing two numbers whose bases are the same, we subtract their exponents.

EXAMPLES

$$\frac{3^4}{3^2} = \frac{3 \cdot 3 \cdot \cancel{3}^{\,1} \cdot \cancel{3}^{\,1}}{\cancel{3}_{\,1} \cdot \cancel{3}_{\,1}} = 3^2$$

$$\frac{y^7}{y^3} = y^{7-3} = y^4$$

Property 3	$(a^n)^m = a^{mn}$

If a number is raised to a power which is itself is raised to another power, we multiply the exponents.

EXAMPLE

Rewrite using a single exponent.

$$(5^3)^2 = (5 \cdot 5 \cdot 5)^2 = (5 \cdot 5 \cdot 5)(5 \cdot 5 \cdot 5) = 5^6$$

$$(x^6)^7 = x^{(6)(7)} = x^{42}$$

Still Struggling

Be careful, Properties 1 and 2 are easily confused.

$$4^3 \cdot 4^2 = 4^{3+2} = 4^5 \text{ and } (4^3)^2 = 4^{3(2)} = 4^6$$

Property 4	$a^0 = 1$

Any nonzero number raised to the zero power is one. We will see that this is true by Property 2 and the fact that any nonzero number over itself is one.

$$1 = \frac{16}{16} = \frac{4^2}{4^2} = 4^{2-2} = 4^0 \qquad \text{From this we see that } 4^0 \text{ must be 1.}$$

 PRACTICE

Rewrite using a single exponent.

1. $\dfrac{x^8}{x^2} =$

2. $(x^3)^2 =$

3. $y' y^3 =$

4. $x^{10} x =$

5. $\dfrac{x^3}{x^2} =$

6. $(y^5)^5 =$

 SOLUTIONS

1. $\dfrac{x^8}{x^2} = x^{8-2} = x^6$

2. $(x^3)^2 = x^{(3)(2)} = x^6$

3. $y^7 y^3 = y^{7+3} = y^{10}$

4. $x^{10} x = x^{10} x^1 = x^{10+1} = x^{11}$

5. $\dfrac{x^3}{x^2} = x^{3-2} = x^1 = x$

6. $(y^5)^5 = y^{(5)(5)} = y^{25}$

Exponent Properties and Algebraic Expressions

Exponent properties also work with algebraic expressions.

 EXAMPLES

Rewrite using a single exponent.

$$(x+3)^2(x+3)^4 = (x+3)^{2+4} = (x+3)^6$$

$$[(x+11)^3]^5 = (x+11)^{3(5)} = (x+11)^{15}$$

$$\frac{(3x-4)^7}{(3x-4)^5} = (3x-7)^{7-5} = (3x-4)^2$$

 ## Still Struggling

Be careful not to write $(3x-4)^2$ as $(3x)^2 - 4^2$. We will see later that $(3x-4)^2$ is $9x^2 - 24x + 16$.

 PRACTICE

Rewrite using a single exponent.

1. $\dfrac{(5x^2 + x + 1)^3}{5x^2 + x + 1} =$

2. $\dfrac{(7x)^9}{(7x)^3} =$

3. $(2x - 5)^0 =$

4. $(x + 1)^{11}(x + 1)^6 =$

5. $(x^2 - 1)(x^2 - 1)^3 =$

6. $((16x - 4)^5)^2 =$

✔ SOLUTIONS

1. $\dfrac{(5x^2 + x + 1)^3}{5x^2 + x + 1} = \dfrac{(5x^2 + x + 1)^3}{(5x^2 + x + 1)^1} = (5x^2 + x + 1)^{3-1} = (5x^2 + x + 1)^2$

2. $\dfrac{(7x)^9}{(7x)^3} = (7x)^{9-3} = (7x)^6$

3. $(2x - 5)^0 = 1$

4. $(x + 1)^{11}(x + 1)^6 = (x + 1)^{11+6} = (x + 1)^{17}$

5. $(x^2 - 1)(x^2 - 1)^3 = (x^2 - 1)^1(x^2 - 1)^3 = (x^2 - 1)^{1+3} = (x^2 - 1)^4$

6. $((16x - 4)^5)^2 = (16x - 4)^{(5)(2)} = (16x - 4)^{10}$

Adding/Subtracting Fractions

When adding fractions with variables in one or more denominators, the LCD has each variable (or algebraic expression) to its highest power as a factor. For example, the LCD for $\dfrac{1}{x^2} + \dfrac{1}{x} + \dfrac{1}{y^3} + \dfrac{1}{y^2}$ is x^2y^3 because the highest power on x is 2, and the highest power on y is 3.

EXAMPLES

Identify the LCD and then find the sum or difference.

$$\frac{4}{x^2} - \frac{3}{x}$$

The LCD is x^2 because the highest power on x is 2.

$$\frac{4}{x^2} - \frac{3}{x} = \frac{4}{x^2} - \frac{3}{x} \cdot \frac{x}{x} = \frac{4}{x^2} - \frac{3x}{x^2} = \frac{4 - 3x}{x^2}$$

$$\frac{13}{xy^2} - \frac{6}{yz}$$

The LCD includes xy, and z. Because 2 is the highest power on y, the LCD also includes y^2. The LCD is xy^2z. We multiply the first fraction by $\dfrac{z}{z}$ and the second fraction by $\dfrac{xy}{xy}$.

$$\frac{13}{xy^2} - \frac{6}{yz} = \frac{13}{xy^2} \cdot \frac{z}{z} - \frac{6}{yz} \cdot \frac{xy}{xy} = \frac{13z}{xy^2z} - \frac{6xy}{xy^2z} = \frac{13z - 6xy}{xy^2z}$$

$$\frac{2x}{(x + 1)^2(4x + 5)} + \frac{1}{x + 1}$$

The LCD is $(x+1)^2(4x+5)$.

$$\frac{2x}{(x+1)^2(4x+5)}+\frac{1}{x+1}=\frac{2x}{(x+1)^2(4x+5)}+\frac{1}{x+1}\cdot\frac{(x+1)(4x+5)}{(x+1)(4x+5)}$$

$$=\frac{2x+(x+1)(4x+5)}{(x+1)^2(4x+5)}$$

$$\frac{1}{x^2(x+y)}+\frac{1}{x}+\frac{1}{(x+y)^3}$$

The LCD is $x^2(x+y)^3$.

$$\frac{1}{x^2(x+y)}+\frac{1}{x}+\frac{1}{(x+y)^3}$$

$$=\frac{1}{x^2(x+y)}\cdot\frac{(x+y)^2}{(x+y)^2}+\frac{1}{x}\cdot\frac{x(x+y)^3}{x(x+y)^3}+\frac{1}{(x+y)^3}\cdot\frac{x^2}{x^2}$$

$$=\frac{(x+y)^2}{x^2(x+y)^3}+\frac{x(x+y)^3}{x^2(x+y)^3}+\frac{x^2}{x^2(x+y)^3}$$

$$=\frac{(x+y)^2+x(x+y)^3+x^2}{x^2(x+y)^3}$$

$$\frac{2}{xy}+\frac{1}{x^3y^2}+\frac{2}{xy^4}$$

The LCD is x^3y^4.

$$\frac{2}{xy}+\frac{1}{x^3y^2}+\frac{2}{xy^4}=\frac{2}{xy}\cdot\frac{x^2y^3}{x^2y^3}+\frac{1}{x^3y^2}\cdot\frac{y^2}{y^2}+\frac{2}{xy^4}\cdot\frac{x^2}{x^2}$$

$$=\frac{2x^2y^3}{x^3y^4}+\frac{y^2}{x^3y^4}+\frac{2x^2}{x^3y^4}=\frac{2x^2y^3+y^2+2x^2}{x^3y^4}$$

PRACTICE

Identify the LCD and then find the sum or difference.

1. $\dfrac{6}{x}+\dfrac{2}{xy}=$

2. $\dfrac{3}{xy^2}-\dfrac{1}{y}=$

3. $\dfrac{1}{2x}+\dfrac{3}{10x^2}=$

4. $\dfrac{1}{x}+\dfrac{1}{xyz^2}+\dfrac{1}{x^2yz}=$

5. $2 + \dfrac{x-1}{(x+4)^2} =$

6. $\dfrac{6}{2(x-1)(x+1)} + \dfrac{1}{6(x-1)^2} =$

7. $\dfrac{4}{3xy^2} + \dfrac{9}{2x^5y} - \dfrac{1}{x^3y^3} =$

SOLUTIONS

1. $\dfrac{6}{x} + \dfrac{2}{xy} = \dfrac{6}{x} \cdot \dfrac{y}{y} + \dfrac{2}{xy} = \dfrac{6y}{xy} + \dfrac{2}{xy} = \dfrac{6y+2}{xy}$

2. $\dfrac{3}{xy^2} - \dfrac{1}{y} = \dfrac{3}{xy^2} - \dfrac{1}{y} \cdot \dfrac{xy}{xy} = \dfrac{3}{xy^2} - \dfrac{xy}{xy^2} = \dfrac{3-xy}{xy^2}$

3. $\dfrac{1}{2x} + \dfrac{3}{10x^2} = \dfrac{1}{2x} \cdot \dfrac{5x}{5x} + \dfrac{3}{10x^2} = \dfrac{5x}{10x^2} + \dfrac{3}{10x^2} = \dfrac{5x+3}{10x^2}$

4. $\dfrac{1}{x} + \dfrac{1}{xyz^2} + \dfrac{1}{x^2yz} = \dfrac{1}{x} \dfrac{xyz^2}{xyz^2} + \dfrac{1}{xyz^2} \cdot \dfrac{x}{x} + \dfrac{1}{x^2yz} \cdot \dfrac{z}{z}$

$$= \dfrac{xyz^2}{x^2yz^2} + \dfrac{x}{x^2yz^2} + \dfrac{z}{x^2yz^2} = \dfrac{xyz^2 + x + z}{x^2yz^2}$$

5. $2 + \dfrac{x-1}{(x+4)^2} = \dfrac{2}{1} + \dfrac{x-1}{(x+4)^2} = \dfrac{2}{1} \cdot \dfrac{(x+4)^2}{(x+4)^2} + \dfrac{x-1}{(x+4)^2}$

$$= \dfrac{2(x+4)^2 + x - 1}{(x+4)^2}$$

6. $\dfrac{6}{2(x-1)(x+1)} + \dfrac{1}{6(x-1)^2} = \dfrac{6}{2(x-1)(x+1)} \cdot \dfrac{3(x-1)}{3(x-1)} + \dfrac{1}{6(x-1)^2} \cdot \dfrac{x+1}{x+1}$

$$= \dfrac{18(x-1)}{6(x-1)^2(x+1)} + \dfrac{x+1}{6(x-1)^2(x+1)}$$

$$= \dfrac{18(x-1) + x + 1}{6(x-1)^2(x+1)}$$

7. $\dfrac{4}{3xy^2} + \dfrac{9}{2x^5y} - \dfrac{1}{x^3y^3} = \dfrac{4}{3xy^2} \cdot \dfrac{2x^4y}{2x^4y} + \dfrac{9}{2x^5y} \cdot \dfrac{3y^2}{3y^2} - \dfrac{1}{x^3y^3} \cdot \dfrac{6x^2}{6x^2}$

$$= \dfrac{8x^4y}{6x^5y^3} + \dfrac{27y^2}{6x^5y^3} - \dfrac{6x^2}{6x^5y^3} = \dfrac{8x^4y + 27y^2 - 6x^2}{6x^5y^3}$$

Property 5	$a^{-1} = \dfrac{1}{a}$

This property says that a^{-1} is the reciprocal of a. In other words, a^{-1} means "invert a." or "take the reciprocal of a."

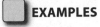 **EXAMPLES**

$$2^{-1} = \frac{1}{2}$$

$$x^{-1} = \frac{1}{x}$$

$$\left(\frac{x}{y}\right)^{-1} = \frac{1}{\frac{x}{y}} = 1 \div \frac{x}{y} = 1 \cdot \frac{y}{x} = \frac{y}{x}$$

$$\left(\frac{2}{3}\right)^{-1} = \frac{3}{2}$$

$$\left(\frac{1}{4}\right)^{-1} = 4$$

$$\left(-\frac{3}{10}\right)^{-1} = -\frac{10}{3}$$

Property 6	$a^{-n} = \dfrac{1}{a^{n}}$

Property 6 allows us to write an expression that has a negative exponent as an expression without an negative exponent.

Property 6 is a combination of Properties 3 and 5: $\dfrac{1}{a^{n}} = \left(\dfrac{1}{a}\right)^{n} = \left(a^{-1}\right)^{n} = a^{-n}$.

EXAMPLES

$$5^{-6} = \frac{1}{5^6}$$

$$x^{-10} = \frac{1}{x^{10}}$$

$$\left(\frac{x}{y}\right)^{-3} = \left(\left(\frac{x}{y}\right)^{-1}\right)^3 = \left(\frac{y}{x}\right)^3$$

$$\left(\frac{2}{5}\right)^{-4} = \left(\frac{5}{2}\right)^4$$

We often need to use a combination of exponent properties to simplify expressions. In the following examples our goal is to rewrite the expression without using a negative exponent.

EXAMPLES
Use Properties 1–6 to rewrite the expression without a negative exponent.

$$(x^6)^{-2} = x^{-12} = \frac{1}{x^{12}}$$

$$x^{-7} \cdot x^6 = x^{-7+6} = x^{-1} = \frac{1}{x}$$

$$\frac{y^3}{y^{-2}} = y^{3-(-2)} = y^{3+2} = y^5$$

$$\frac{x^{-4}}{x^2} = x^{-4-2} = x^{-6} = \frac{1}{x^6}$$

$$(10x - 3)^{-2} = \frac{1}{(10x - 3)^2}$$

$$\left(\frac{x+1}{2x-3}\right)^{-1} = \frac{2x-3}{x+1}$$

Still Struggling

Be careful, in expressions such as $(2x)^{-1}$ the exponent " -1" applies to $2x$, but in $2x^{-1}$ the exponent "-1" applies only to x:

$$(2x)^{-1} = \frac{1}{2x} \text{ and } 2x^{-1} = 2 \cdot x^{-1} = 2 \cdot \frac{1}{x} = \frac{2}{x}$$

 PRACTICE

Use Properties 1 to 6 to rewrite the expression without a negative exponent.

1. $6^{-1} =$

2. $(x^2 y)^{-1} =$

3. $\left(\dfrac{5}{8}\right)^{-1} =$

4. $(x^3)^{-1} =$

5. $\dfrac{x^2}{x^{-1}} =$

6. $x^4 x^{-3} =$

7. $x^8 x^{-11} =$

8. $(x^{-4})^2 =$

9. $\dfrac{y^{-7}}{y^{-2}} =$

10. $\dfrac{x^{-5}}{x^3} =$

11. $(12x - 5)^{-2} =$

12. $(6x)^{-1} =$

13. $\dfrac{(3x - 2)^4}{(3x - 2)^{-1}} =$

14. $(2x^3 + 4)^{-6}(2x^3 + 4)^4 =$

15. $((x - 8)^3)^{-1} =$

16. $\left(\dfrac{x + 7}{2x - 3}\right)^{-1} =$

 SOLUTIONS

1. $6^{-1} = \dfrac{1}{6}$

2. $(x^2 y)^{-1} = \dfrac{1}{x^2 y}$

3. $\left(\dfrac{5}{8}\right)^{-1} = \dfrac{8}{5}$

4. $(x^3)^{-1} = x^{-3} = \dfrac{1}{x^3}$

5. $\dfrac{x^2}{x^{-1}} = x^{2-(-1)} = x^{2+1} = x^3$

6. $x^4 x^{-3} = x^{4+(-3)} = x^{4-3} = x^1 = x$

7. $x^8 x^{-11} = x^{8+(-11)} = x^{-3} = \dfrac{1}{x^3}$

8. $(x^{-4})^2 = x^{(-4)(2)} = x^{-8} = \dfrac{1}{x^8}$

9. $\dfrac{y^{-7}}{y^{-2}} = y^{-7-(-2)} = y^{-7+2} = y^{-5} = \dfrac{1}{y^5}$

10. $\dfrac{x^{-5}}{x^3} = x^{-5-3} = x^{-8} = \dfrac{1}{x^8}$

11. $(12x - 5)^{-2} = \dfrac{1}{(12x - 5)^2}$

12. $(6x)^{-1} = \dfrac{1}{6x}$

13. $\dfrac{(3x - 2)^4}{(3x - 2)^{-1}} = (3x - 2)^{4-(-1)} = (3x - 2)^{4+1} = (3x - 2)^5$

14. $(2x^3 + 4)^{-6}(2x^3 + 4)^4 = (2x^3 + 4)^{-6+4} = (2x^3 + 4)^{-2} = \dfrac{1}{(2x^3 + 4)^2}$

15. $((x-8)^3)^{-1} = \dfrac{1}{(x-8)^3}$

16. $\left(\dfrac{x+7}{2x-3}\right)^{-1} = \dfrac{2x-3}{x+7}$

Properties 7 and 8 allow us to rewrite products and quotients that are raised to powers.

Property 7	$(ab)^n = a^n b^n$

By Property 7 we can take a product followed by the power or take the powers followed by the product.

 EXAMPLES

Use Property 7 to rewrite the expression.

$$(4x)^3 = (4x)(4x)(4x) = (4 \cdot 4 \cdot 4)(x \cdot x \cdot x) = 4^3 x^3 = 64x^3$$

$$[4(x+1)]^2 = 4^2(x+1)^2 = 16(x+1)^2$$

$$(x^2 y)^4 = (x^2)^4 y^4 = x^8 y^4$$

$$(2x)^{-3} = \dfrac{1}{(2x)^3} = \dfrac{1}{2^3 x^3} = \dfrac{1}{8x^3}$$

$$(2x^{-1})^{-3} = 2^{-3}(x^{-1})^{-3} = \dfrac{1}{2^3} x^{(-1)(-3)} = \dfrac{1}{8} x^3$$

$$[(5x+8)^2(x+6)]^4 = [(5x+8)^2]^4(x+6)^4$$
$$= (5x+8)^{(2)(4)}(x+6)^4 = (5x+8)^8(x+6)^4$$

$$(4x^3 y)^2 = 4^2(x^3)^2 y^2 = 16x^{(3)(2)} y^2 = 16x^6 y^2$$

$$4(3x)^3 = 4(3^3 x^3) = 4(27x^3) = 108x^3$$

 Still Struggling

It is *not* true that $(a+b)^n = a^n + b^n$. This mistake is very common.

Property 8	$$\left(\dfrac{a}{b}\right)^n = \dfrac{a^n}{b^n}$$

Property 8 says that we can take the quotient followed by the power or each power followed by the quotient.

◼ EXAMPLES

Use Property 8 to rewrite the expression.

$$\left(\frac{2}{5}\right)^3 = \frac{2}{5} \cdot \frac{2}{5} \cdot \frac{2}{5} = \frac{2^3}{5^3} = \frac{8}{125}$$

$$\left(\frac{x}{y}\right)^4 = \frac{x^4}{y^4}$$

$$\left(\frac{x^2}{y^5}\right)^4 = \frac{(x^2)^4}{(y^5)^4} = \frac{x^8}{y^{20}}$$

We can combine Properties 6 and 8 to help us rewrite a quotient raised to a negative power.

Property 9	$$\left(\dfrac{a}{b}\right)^{-n} = \dfrac{b^n}{a^n}$$

Here is why it is true.

$$\left(\frac{a}{b}\right)^{-n} = \left(\frac{a}{b}\right)^{(-1)(n)} = \left[\left(\frac{a}{b}\right)^{-1}\right]^n = \left(\frac{b}{a}\right)^n = \frac{b^n}{a^n}$$

We will use Property 9 for the rest of the examples and practice problems that involve fractions and negative exponents.

◼ EXAMPLES

Simplify and eliminate any negative exponent.

$$\left(\frac{2}{3}\right)^{-2} = \frac{3^2}{2^2} = \frac{9}{4}$$

$$\left(\frac{1}{2}\right)^{-3} = \frac{2^3}{1^3} = \frac{8}{1} = 8$$

$$\left(\frac{6x+5}{x-1}\right)^{-3} = \frac{(x-1)^3}{(6x+5)^3}$$

$$\left(\frac{1}{(x+2)^3}\right)^{-5} = \frac{[(x+2)^3]^5}{1^5} = (x+2)^{15}$$

$$\left(\frac{(x+7)^3}{x^{-2}}\right)^{-4} = \frac{(x^{-2})^4}{[(x+7)^3]^4} = \frac{x^{-8}}{(x+7)^{12}} = x^{-8} \cdot \frac{1}{(x+7)^{12}} = \frac{1}{x^8} \cdot \frac{1}{(x+7)^{12}} = \frac{1}{x^8(x+7)^{12}}$$

$$\left(\frac{y^{-3}}{x^{-4}}\right)^{-6} = \frac{(x^{-4})^6}{(y^{-3})^6} = \frac{x^{-24}}{y^{-18}} = \frac{\frac{1}{x^{24}}}{\frac{1}{y^{18}}} = \frac{1}{x^{24}} \div \frac{1}{y^{18}} = \frac{1}{x^{24}} \cdot \frac{y^{18}}{1} = \frac{y^{18}}{x^{24}}$$

The last expression can be simplified more quickly using Property 8 followed by Property 3.

$$\left(\frac{y^{-3}}{x^{-4}}\right)^{-6} = \frac{(y^{-3})^{-6}}{(x^{-4})^{-6}} = \frac{y^{(-3)(-6)}}{x^{(-4)(-6)}} = \frac{y^{18}}{x^{24}}$$

PRACTICE

Simplify and eliminate any negative exponent.

1. $(xy^3)^2 =$

2. $(3x)^{-3} =$

3. $(2x)^4 =$

4. $(3(x-4))^2 =$

5. $6(2x)^3 =$

6. $6y^2(3y^4)^2 =$

7. $(5x^2y^4z^6)^2 =$

8. $\left(\dfrac{4}{y^2}\right)^3 =$

9. $\left(\dfrac{2}{x-9}\right)^{-3} =$

10. $\left(\dfrac{(x+8)^2}{x^{-4}}\right)^{-3} =$

11. $\left(\dfrac{(x+3)^{-2}}{y^{-4}}\right)^{-3} =$

✔ SOLUTIONS

1. $(xy^3)^2 = x^2(y^3)^2 = x^2y^6$

2. $(3x)^{-3} = \dfrac{1}{(3x)^3} = \dfrac{1}{3^3x^3} = \dfrac{1}{27x^3}$

3. $(2x)^4 = 2^4x^4 = 16x^4$

4. $(3(x-4))^2 = 3^2\,(x-4)^2 = 9(x-4)^2$

5. $6(2x)^3 = 6(2^3x^3) = 6(8x^3) = 48x^3$

6. $6y^2(3y^4)^2 = 6y^2(3^2(y^4)^2) = 6y^2(9y^8) = 54y^{10}$

7. $(5x^2y^4z^6)^2 = 5^2(x^2)^2(y^4)^2(z^6)^2 = 25x^4y^8z^{12}$

8. $\left(\dfrac{4}{y^2}\right)^3 = \dfrac{4^3}{(y^2)^3} = \dfrac{64}{y^6}$

9. $\left(\dfrac{2}{x-9}\right)^{-3} = \dfrac{(x-9)^3}{2^3} = \dfrac{(x-9)^3}{8}$

10. $\left(\dfrac{(x+8)^2}{x^4}\right)^{-3} = \dfrac{(x^4)^3}{[(x+8)^2]^3} = \dfrac{x^{12}}{(x+8)^6}$

11. $\left(\dfrac{(x+3)^{-2}}{y^{-4}}\right)^{-3} = \dfrac{[(x+3)^{-2}]^{-3}}{(y^{-4})^{-3}} = \dfrac{(x+3)^{(-2)(-3)}}{y^{(-4)(-3)}} = \dfrac{(x+3)^6}{y^{12}}$

Multiplying/Dividing with Exponents

When multiplying (or dividing) quantities that have exponents, we use exponent properties to simplify each factor (or numerator and denominator) and then we multiply (or divide).

EXAMPLE

Simplify the expression.

$$3x^3(4xy^5)^2$$

We begin by simplifying the quantity inside the parentheses, $(4xy^5)^2$.

$$[4^2x^2(y^5)^2] = (16x^2y^{10}),\text{ so}$$

$$3x^3(4xy^5)^2 = 3x^3[4^2x^2(y^5)^2] = 3x^3(16x^2y^{10}) = 3 \cdot 16x^3x^2y^{10} = 48x^5y^{10}$$

$$(2x)^3(3x^3y)^2 = (2^3x^3)(3^2(x^3)^2y^2) = (8x^3)(9x^6y^2) = 8 \cdot 9x^3x^6y^2 = 72x^9y^2$$

$$\frac{(5x^3y^2)^3}{(10x)^2} = \frac{5^3(x^3)^3(y^2)^3}{10^2x^2} = \frac{125x^9y^6}{100x^2} = \frac{125}{100}x^{9-2}y^6 = \frac{5}{4}x^7y^6 = \frac{5x^7y^6}{4}$$

$$(6xy^4)^2(4xy^8)^{-3} = (6^2x^2(y^4)^2)(4^{-3}x^{-3}(y^8)^{-3}) = (36x^2y^8)\left(\frac{1}{64}x^{-3}y^{-24}\right)$$

$$= \frac{36}{64}x^2x^{-3}y^8y^{-24} = \frac{9}{16}x^{-1}y^{-16} = \frac{9}{16} \cdot \frac{1}{x} \cdot \frac{1}{y^{16}} = \frac{9}{16xy^{16}}$$

PRACTICE

Simplify the expression.

1. $\left(\dfrac{x^3}{y^2}\right)^5\left(\dfrac{x}{y}\right)^{-2} =$

2. $\dfrac{(2x^3y^5)^4}{(6x^5y^3)^2} =$

3. $(2x^3)^2(3x^{-1})^3 =$

4. $(3xy^4)^{-2}(12x^2y)^2 =$

5. $(4x^{-1}y^{-2})^2(2x^4y^5)^3 =$

6. $\dfrac{(5x^4y^3)^3}{(15xy^5)^2} =$

7. $\dfrac{(9x^{-2}y^3)^2}{(6xy^2)^3} =$

8. $[9(x+3)^2]^2[2(x+3)]^3 =$

9. $(2xy^2z^4)^4(3x^{-1}z^2)^3(xy^5z^{-4}) =$

10. $\dfrac{2(xy^4)^3(yz^2)^4}{(3xyz)^4} =$

✔ SOLUTIONS

1. $\left(\dfrac{x^3}{y^2}\right)^5\left(\dfrac{x}{y}\right)^{-2} = \dfrac{(x^3)^5}{(y^2)^5}\cdot\dfrac{y^2}{x^2} = \dfrac{x^{15}}{y^{10}}\cdot\dfrac{y^2}{x^2} = x^{15-2}y^{2-10} = x^{13}y^{-8} = \dfrac{x^{13}}{y^8}$

2. $\dfrac{(2x^3y^5)^4}{(6x^5y^3)^2} = \dfrac{2^4(x^3)^4(y^5)^4}{6^2(x^5)^2(y^3)^2} = \dfrac{16x^{12}y^{20}}{36x^{10}y^6} = \dfrac{16}{36}x^{12-10}y^{20-6}$

$\qquad = \dfrac{4}{9}x^2y^{14} = \dfrac{4x^2y^{14}}{9}$

3. $(2x^3)^2(3x^{-1})^3 = [2^2(x^3)^2][3^3(x^{-1})^3] = (4x^6)(27x^{-3})$

$\qquad = 108x^{6+(-3)} = 108x^3$

4. $(3xy^4)^{-2}(12x^2y)^2 = (3^{-2}x^{-2}(y^4)^{-2})(12^2(x^2)^2y^2)$

$\qquad = \left(\dfrac{1}{9}x^{-2}y^{-8}\right)(144x^4y^2) = \dfrac{144}{9}x^{-2+4}y^{-8+2}$

$\qquad = 16x^2y^{-6} = 16x^2\dfrac{1}{y^6} = \dfrac{16x^2}{y^6}$

5. $(4x^{-1}y^{-2})^2(2x^4y^5)^3 = [4^2(x^{-1})^2(y^{-2})^2][2^3(x^4)^3(y^5)^3]$

$\qquad = (16x^{-2}y^{-4})(8x^{12}y^{15}) = 128x^{-2+12}y^{-4+15}$

$\qquad = 128x^{10}y^{11}$

6. $\dfrac{(5x^4y^3)^3}{(15xy^5)^2} = \dfrac{5^3(x^4)^3(y^3)^3}{15^2x^2(y^5)^2} = \dfrac{125x^{12}y^9}{225x^2y^{10}} = \dfrac{125}{225}x^{12-2}y^{9-10} = \dfrac{5}{9}x^{10}y^{-1}$

$\qquad = \dfrac{5x^{10}}{9}\cdot\dfrac{1}{y} = \dfrac{5x^{10}}{9y}$

7. $\dfrac{(9x^{-2}y^3)^2}{(6xy^2)^3} = \dfrac{9^2(x^{-2})^2(y^3)^2}{6^3x^3(y^2)^3} = \dfrac{81x^{-4}y^6}{216x^3y^6} = \dfrac{81}{216}x^{-4-3}y^{6-6}$

$\qquad = \dfrac{3}{8}x^{-7}y^0 = \dfrac{3}{8}\cdot\dfrac{1}{x^7}\cdot 1 = \dfrac{3}{8x^7}$

8. $[9(x+3)^2]^2[2(x+3)]^3 = [9^2((x+3)^2)^2][2^3(x+3)^3]$

$\qquad = [81(x+3)^4][8(x+3)^3] = 648(x+3)^{4+3}$

$\qquad = 648(x+3)^7$

9. $(2xy^2z^4)^4(3x^{-1}z^2)^3(xy^5z^{-4}) = [2^4x^4(y^2)^4(z^4)^4][3^3(x^{-1})^3(z^2)^3](xy^5z^4)$

$\qquad = (16x^4y^8z^{16})(27x^{-3}z^6)(x^1y^5z^4)$

$\qquad = 432x^{4+(-3)+1}y^{8+5}z^{16+6+4} = 432x^2y^{13}z^{26}$

10. $\dfrac{2(xy^4)^3(yz^2)^4}{(3xyz)^4} = \dfrac{2x^3(y^4)^3\,y^4(z^2)^4}{3^4\,x^4 y^4 z^4} = \dfrac{2x^3 y^{12} y^4 z^8}{81 x^4 y^4 z^4} = \dfrac{2x^3 y^{16} z^8}{81 x^4 y^4 z^4}$

$= \dfrac{2}{81} x^{3-4} y^{16-4} z^{8-4} = \dfrac{2}{81} x^{-1} y^{12} z^4 = \dfrac{2}{81} \cdot \dfrac{1}{x} y^{12} z^4 = \dfrac{2 y^{12} z^4}{81 x}$

There are times in algebra, and especially in calculus, when we must write a fraction as a product. Using the property $\dfrac{1}{a} = a^{-1}$, we can rewrite a fraction as a product of the numerator and the denominator raised to the -1 power. Here is the idea: $\dfrac{numerator}{denominator} = (numerator)(denominator)^{-1}$.

EXAMPLES

Write the fraction as a product.

$$\frac{3}{x} = 3x^{-1}$$

$$\frac{4}{x+3} = 4(x+3)^{-1}$$

$$\frac{x^n}{y^m} = x^n (y^m)^{-1} = x^n y^{-m}$$

$$\frac{5x-8}{(2x+3)^3} = (5x-8)(2x+3)^{-3}$$

PRACTICE

Write the fraction as a product.

1. $\dfrac{4x^2}{y^5} =$

2. $\dfrac{2x(x-3)}{(x+1)^2} =$

3. $\dfrac{x}{y} =$

4. $\dfrac{2x}{(3y)^2} =$

5. $\dfrac{2x-3}{2x+5} =$

 **SOLUTIONS**

1. $\dfrac{4x^2}{y^5} = 4x^2 y^{-5}$

2. $\dfrac{2x(x-3)}{(x+1)^2} = 2x(x-3)(x+1)^{-2}$

3. $\dfrac{x}{y} = xy^{-1}$

4. $\dfrac{2x}{(3y)^2} = 2x(3y)^{-2}$

5. $\dfrac{2x-3}{2x+5} = (2x-3)(2x+5)^{-1}$

Roots

The square root of a number is the positive number whose square is the root. For example, 3 is the square root of 9 because $3^2 = 9$. It might seem that negative numbers could be square roots. It is true that $(-3)^2 = 9$, but $\sqrt{9}$ is the symbol for the nonnegative number whose square is 9. Sometimes we say that 3 is the *principal* square root of 9.

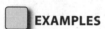 **EXAMPLES**

$$\sqrt{16} = 4 \text{ because } 4^2 = 16$$
$$\sqrt{81} = 9 \text{ because } 9^2 = 81$$

In general, $\sqrt[n]{a} = b$ if $b^n = a$. If n is even, we assume b is the nonnegative root. In this book, we assume even roots will be taken only of nonnegative numbers. That is, if we are taking the square root, fourth root, sixth root, etc., we must assume that the quantity under the radical symbol, $\sqrt[n]{\ }$, is not negative. For instance, in the expression $\sqrt{x}$ we assume that x is not negative. Note that there is no problem with odd roots being negative numbers. For example $\sqrt[3]{-64} = -4$ because $(-4)^3 = (-4)(-4)(-4) = -64$.

Root properties are similar to exponent properties.

Property 10	$\sqrt[n]{ab} = \sqrt[n]{a}\,\sqrt[k]{b}$

Property 10 allows us to take the product followed by the root or we can take the individual roots followed by the product.

EXAMPLES

Rewrite using Property 10.

$$\sqrt{64} = \sqrt{4 \cdot 16} = \sqrt{4} \cdot \sqrt{16} = 2 \cdot 4 = 8$$

$$\sqrt[5]{3}\sqrt[5]{4x} = \sqrt[5]{12x}$$

$$\sqrt[4]{6x}\sqrt[4]{4y} = \sqrt[4]{24xy}$$

 ## Still Struggling

Property 10 only applies to multiplication. There is no similar property for addition (nor subtraction). A common mistake is to "simplify" the sum of two squares. For example, $\sqrt{x^2 + 9} = x + 3$ is incorrect. The following example should give you an idea of why these two expressions are not equal. If there *were* the property $\sqrt[n]{a+b} = \sqrt[n]{a} + \sqrt[n]{b}$, then we would have $\sqrt{58} = \sqrt{49+9} = \sqrt{49} + \sqrt{9} = 7 + 3 = 10$. This could only be true if $10^2 = 58$, which of course, it isn't.

Property 11	$\sqrt[n]{\dfrac{a}{b}} = \dfrac{\sqrt[n]{a}}{\sqrt[n]{b}}$

Property 11 allows us to take the quotient followed by the root or the individual roots followed by the quotient.

 EXAMPLE

$$\sqrt{\frac{4}{9}} = \frac{\sqrt{4}}{\sqrt{9}} = \frac{2}{3}$$

Property 12 $\left(\sqrt[n]{a}\right)^m = \sqrt[n]{a^m}$ (Remember that if n is even, then a must not be negative.)

By Property 12, we can take the root followed by the power or the power followed by the root, as in $\sqrt{5^3} = \left(\sqrt{5}\right)^3$.

Property 13 $\left(\sqrt[n]{a}\right)^n = \sqrt[n]{a^n} = a$

We can think of Property 13 as a root-power cancellation property. We will use this property later to simplify expressions involving a root and, in later chapters, to solve equations.

EXAMPLES

Use Properties 10 to 12 to rewrite the expression.

$$\sqrt[3]{27} = \sqrt[3]{3^3} = 3$$

$$\left(\sqrt{5}\right)^2 = 5$$

$$\sqrt[3]{8x^3} = \sqrt[3]{(2x)^3} = 2x$$

PRACTICE

Use Properties 10 to 12 to rewrite the expression.

1. $\sqrt{25x^2} =$

2. $\sqrt[3]{8y^3} =$

3. $\sqrt{(4-x)^2} =$

4. $\sqrt[3]{[5(x-1)]^3} =$

SOLUTIONS

1. $\sqrt{25x^2} = \sqrt{(5x)^2} = 5x$

2. $\sqrt[3]{8y^3} = \sqrt[3]{(2y)^3} = 2y$

3. $\sqrt{(4-x)^2} = 4-x$

4. $\sqrt[3]{[5(x-1)]^3} = 5(x-1)$

Simplifying Roots

Exponent and root properties can be used to simplify roots in the same way canceling is used to simplify fractions. For instance, we normally wouldn't leave $\sqrt{25}$ without simplifying it as 5, any more than we would leave $\frac{12}{4}$ without simplifying it as 3. In $\sqrt[n]{a^m}$, if m is at least as large as n, then $\sqrt[n]{a^m}$ can be simplified using Property 10 $\left(\sqrt[n]{ab} = \sqrt[n]{a}\sqrt[n]{b}\right)$ and Property 13 $\left(\sqrt[n]{a^n} = a\right)$. For example, if we are simplifying a square root, we write the quantity under the radical symbol as the product of a square and something else.

EXAMPLES

Simplify the expression.

$$\sqrt{27}$$

We can write 27 as the product of a square and something else: $27 = 3^2 \cdot 3$.

$$\sqrt{27} = \sqrt{3^2 3^1} = \sqrt{3^2}\sqrt{3} = 3\sqrt{3}$$

$$\sqrt{32x^3} = \sqrt{2^2 2^2 2^1 x^2 x^1} = \sqrt{2^2}\sqrt{2^2}\sqrt{x^2}\sqrt{2x} = 2 \cdot 2x\sqrt{2x} = 4x\sqrt{2x}$$

$$\sqrt{\frac{8}{9}} = \frac{\sqrt{8}}{\sqrt{9}} = \frac{\sqrt{2^2 2^1}}{\sqrt{9}} = \frac{\sqrt{2^2}\sqrt{2}}{3} = \frac{2\sqrt{2}}{3}$$

$$\sqrt[3]{625x^5 y^4} = \sqrt[3]{5^3 5^1 x^3 x^2 y^3 y^1} = \sqrt[3]{5^3}\sqrt[3]{x^3}\sqrt[3]{y^3}\sqrt[3]{5x^2 y} = 5xy\sqrt[3]{5x^2 y}$$

$$\sqrt[4]{(x-6)^9} = \sqrt[4]{(x-6)^4 (x-6)^4 (x-6)^1} = \sqrt[4]{(x-6)^4}\sqrt[4]{(x-6)^4}\sqrt[4]{x-6}$$
$$= (x-6)(x-6)\sqrt[4]{x-6} = (x-6)^2\sqrt[4]{x-6}$$

 PRACTICE

Simplify the expression.

1. $\sqrt[3]{x^7} =$

2. $\sqrt{x^{10}} =$

3. $\sqrt[3]{16x^7 y^5} =$

4. $\sqrt[5]{(4x-1)^8} =$

5. $\sqrt{25(x+4)^2} =$

6. $\sqrt[4]{x^9 y^6} =$

7. $\sqrt[50]{\dfrac{x^{100}}{y^{200}}} =$

 SOLUTIONS

1. $\sqrt[3]{x^7} = \sqrt[3]{x^3 x^3 x^1} = \sqrt[3]{x^3}\sqrt[3]{x^3}\sqrt[3]{x^1} = xx\sqrt[3]{x} = x^2\sqrt[3]{x}$

2. $\sqrt{x^{10}} = \sqrt{x^5 x^5} = \sqrt{(x^5)^2} = x^5$

3. $\sqrt[3]{16x^7y^5} = \sqrt[3]{2^3 2x^3 x^3 xy^3 y^2} = \sqrt[3]{2^3}\sqrt[3]{x^3}\sqrt[3]{x^3}\sqrt[3]{y^3}\sqrt[3]{2xy^2}$

$\qquad = 2xxy\sqrt[3]{2xy^2} = 2x^2y\sqrt[3]{2xy^2}$

4. $\sqrt[5]{(4x-1)^8} = \sqrt[5]{(4x-1)^5(4x-1)^3} = \sqrt[5]{(4x-1)^5}\sqrt[5]{(4x-1)^3}$

$\qquad = (4x-1)\sqrt[5]{(4x-1)^3}$

5. $\sqrt{25(x+4)^2} = \sqrt{5^2(x+4)^2} = \sqrt{5^2}\sqrt{(x+4)^2} = 5(x+4)$

6. $\sqrt[4]{x^9y^6} = \sqrt[4]{x^4x^4x^1y^4y^2} = \sqrt[4]{x^4}\sqrt[4]{x^4}\sqrt[4]{y^4}\sqrt[4]{xy^2} = xxy\sqrt[4]{x^1y^2} = x^2y\sqrt[4]{xy^2}$

7. $\sqrt[50]{\dfrac{x^{100}}{y^{200}}} = \dfrac{\sqrt[50]{x^{100}}}{\sqrt[50]{y^{100}}} = \dfrac{\sqrt[50]{x^{50}x^{50}}}{\sqrt[50]{y^{50}y^{50}y^{50}y^{50}}} = \dfrac{\sqrt[50]{x^{50}}\sqrt[50]{x^{50}}}{\sqrt[50]{y^{50}}\sqrt[50]{y^{50}}\sqrt[50]{y^{50}}\sqrt[50]{y^{50}}}$

$\qquad = \dfrac{xx}{yyyy} = \dfrac{x^2}{y^4}$

We can use the same radical properties to simplify roots of numbers that are not perfect squares, cubes, etc. If the number under the root (also called the radical symbol), $\sqrt[n]{}$, has a *factor* that is a perfect power of n, then the radical can be simplified. For example, $\sqrt{18}$ be simplified because it has a perfect square as a factor. We separate the perfect power (in this case, 9) from the other factors and use root properties. In the following examples, we use the same properties, $\sqrt[n]{ab} = \sqrt[n]{a}\sqrt[n]{b}$ and $\sqrt[n]{a^n} = a$, to simplify quantities such as $\sqrt{18}$.

EXAMPLES

Simplify the expression.

$\sqrt{18}$

18 has a perfect square, 9, as a factor. We write 18 as the product $9 \cdot 2$ and then use the property $\sqrt[n]{ab} = \sqrt[n]{a}\sqrt[n]{b}$ to separate 9 from 2 and then the property $\sqrt[n]{a^n} = a$ to eliminate the radical from $\sqrt{9}$. Thus, $\sqrt{18} = \sqrt{9 \cdot 2} = \sqrt{9}\sqrt{2} = 3\sqrt{2}$.

$\sqrt{48} = \sqrt{4^2 3} = \sqrt{4^2}\sqrt{3} = 4\sqrt{3}$

$\sqrt[3]{162} = \sqrt[3]{3^3 \cdot 3 \cdot 2} = \sqrt[3]{3^3}\sqrt[3]{3 \cdot 2} = 3\sqrt[3]{6}$

$\sqrt[5]{64x^6y^3} = \sqrt[5]{2^5 \cdot 2x^5xy^3} = \sqrt[5]{2^5}\sqrt[5]{x^5}\sqrt[5]{2xy^3} = 2x\sqrt[5]{2xy^3}$

$\sqrt[3]{(2x-7)^5} = \sqrt[3]{(2x-7)^3(2x-7)^2} = \sqrt[3]{(2x-7)^3}\sqrt[3]{(2x-7)^2} = (2x-7)\sqrt[3]{(2x-7)^2}$

$\sqrt{\dfrac{48x^3}{25}}$

In addition to the two root properties that we have been using, we will use the property $\sqrt[n]{\dfrac{a}{b}} = \dfrac{\sqrt[n]{a}}{\sqrt[n]{b}}$ to simplify the fraction.

$$\sqrt{\dfrac{48x^3}{25}} = \dfrac{\sqrt{48x^3}}{\sqrt{25}} = \dfrac{\sqrt{4^2 3x^2 x}}{\sqrt{5^2}} = \dfrac{\sqrt{4^2}\sqrt{x^2}\sqrt{3x}}{5} = \dfrac{4x\sqrt{3x}}{5}$$

PRACTICE

Simplify the expression.

1. $\sqrt[3]{54x^5} =$

2. $\sqrt{50x^3 y} =$

3. $\sqrt{\dfrac{8}{9}} =$

4. $\sqrt[4]{32x^7 y^5} =$

5. $\sqrt[3]{\dfrac{40(3x-1)^4}{x^6}} =$

SOLUTIONS

1. $\sqrt[3]{54x^5} = \sqrt[3]{3^3 2x^3 x^2} = \sqrt[3]{3^3}\sqrt[3]{x^3}\sqrt[3]{2x^2} = 3x\sqrt[3]{2x^2}$

2. $\sqrt{50x^3 y} = \sqrt{5^2 2x^2 xy} = \sqrt{5^2}\sqrt{x^2}\sqrt{2xy} = 5x\sqrt{2xy}$

3. $\sqrt{\dfrac{8}{9}} = \dfrac{\sqrt{8}}{\sqrt{9}} = \dfrac{\sqrt{2^2 2}}{3} = \dfrac{\sqrt{2^2}\sqrt{2}}{3} = \dfrac{2\sqrt{2}}{3}$

4. $\sqrt[4]{32x^7 y^5} = \sqrt[4]{2^4 2x^4 x^3 y^4 y} = \sqrt[4]{2^4}\sqrt[4]{x^4}\sqrt[4]{y^4}\sqrt[4]{2x^3 y} = 2xy\sqrt[4]{2x^3 y}$

5. $\sqrt[3]{\dfrac{40(3x-1)^4}{x^6}} = \dfrac{\sqrt[3]{40(3x-1)^4}}{\sqrt[3]{x^6}} = \dfrac{\sqrt[3]{2^3 5(3x-1)^3 (3x-1)}}{\sqrt[3]{x^3 x^3}}$

$$= \dfrac{\sqrt[3]{2^3}\sqrt[3]{(3x-1)^3}\sqrt[3]{5(3x-1)}}{\sqrt[3]{x^3}\sqrt[3]{x^3}} = \dfrac{2(3x-1)\sqrt[3]{5(3x-1)}}{xx}$$

$$= \dfrac{2(3x-1)\sqrt[3]{5(3x-1)}}{x^2}$$

Roots of fractions or fractions with a root in the denominator are not simplified. To eliminate a root in a denominator, we use two facts, $\sqrt[n]{a^n} = a$ and $\dfrac{b}{b} = 1$. This process is called *rationalizing the denominator*. We begin with square roots. If the denominator is a square root, we multiply the fraction by the denominator over itself. This forces the new denominator to be in the form $\sqrt{a^2}$.

 EXAMPLES

Rationalize the denominator.

$$\frac{1}{\sqrt{2}} = \frac{1}{\sqrt{2}} \cdot \frac{\sqrt{2}}{\sqrt{2}} = \frac{\sqrt{2}}{\sqrt{2^2}} = \frac{\sqrt{2}}{2}$$

$$\frac{4}{\sqrt{x}} = \frac{4}{\sqrt{x}} \cdot \frac{\sqrt{x}}{\sqrt{x}} = \frac{4\sqrt{x}}{\sqrt{x^2}} = \frac{4\sqrt{x}}{x}$$

$$\sqrt{\frac{2}{3}} = \frac{\sqrt{2}}{\sqrt{3}} \cdot \frac{\sqrt{3}}{\sqrt{3}} = \frac{\sqrt{6}}{\sqrt{3^2}} = \frac{\sqrt{6}}{3}$$

$$\frac{6}{\sqrt{2x}} = \frac{6}{\sqrt{2x}} \cdot \frac{\sqrt{2x}}{\sqrt{2x}} = \frac{6\sqrt{2x}}{\sqrt{(2x)^2}} = \frac{6\sqrt{2x}}{2x}$$

PRACTICE

Rationalize the denominator.

1. $\dfrac{3}{\sqrt{5}} =$

2. $\dfrac{7}{\sqrt{y}} =$

3. $\sqrt{\dfrac{6}{7}} =$

4. $\dfrac{8x}{\sqrt{3}} =$

5. $\sqrt{\dfrac{7xy}{11}} =$

6. $\dfrac{2}{\sqrt{10y}} =$

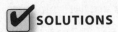

SOLUTIONS

1. $\dfrac{3}{\sqrt{5}} = \dfrac{3}{\sqrt{5}} \cdot \dfrac{\sqrt{5}}{\sqrt{5}} = \dfrac{3\sqrt{5}}{\sqrt{5^2}} = \dfrac{3\sqrt{5}}{5}$

2. $\dfrac{7}{\sqrt{y}} = \dfrac{7}{\sqrt{y}} \cdot \dfrac{\sqrt{y}}{\sqrt{y}} = \dfrac{7\sqrt{y}}{\sqrt{y^2}} = \dfrac{7\sqrt{y}}{y}$

3. $\sqrt{\dfrac{6}{7}} = \dfrac{\sqrt{6}}{\sqrt{7}} \cdot \dfrac{\sqrt{7}}{\sqrt{7}} = \dfrac{\sqrt{42}}{\sqrt{7^2}} = \dfrac{\sqrt{42}}{7}$

4. $\dfrac{8x}{\sqrt{3}} = \dfrac{8x}{\sqrt{3}} \cdot \dfrac{\sqrt{3}}{\sqrt{3}} = \dfrac{8x\sqrt{3}}{\sqrt{3^2}} = \dfrac{8x\sqrt{3}}{3}$

5. $\sqrt{\dfrac{7xy}{11}} = \dfrac{\sqrt{7xy}}{\sqrt{11}} \cdot \dfrac{\sqrt{11}}{\sqrt{11}} = \dfrac{\sqrt{77xy}}{\sqrt{11^2}} = \dfrac{\sqrt{77xy}}{11}$

6. $\dfrac{2}{\sqrt{10y}} = \dfrac{2}{\sqrt{10y}} \cdot \dfrac{\sqrt{10y}}{\sqrt{10y}} = \dfrac{2\sqrt{10y}}{\sqrt{(10y)^2}} = \dfrac{2\sqrt{10y}}{10y}$

In the case of a cube (or higher) root, multiplying the fraction by the denominator over itself usually does not work. To eliminate the nth root in the denominator, we need to write the denominator as the nth root of some quantity to the nth power. For example, to simplify $\dfrac{1}{\sqrt[3]{5}}$ we need 5^3 under the radical. To get 5^3 under the radical, we multiply the numerator and denominator by $\sqrt[3]{5^2}$ over itself. Thus,

$$\dfrac{1}{\sqrt[3]{5}} \cdot \dfrac{\sqrt[3]{5^2}}{\sqrt[3]{5^2}} = \dfrac{\sqrt[3]{5^2}}{\sqrt[3]{5^3}} = \dfrac{\sqrt[3]{25}}{5}.$$

When the denominator is written as a power (often the power is 1) subtract this power from the root. The factor will include this number as an exponent.

EXAMPLES

Rationalize the denominator.

$\dfrac{8}{\sqrt[4]{x^3}}$

The root minus the power is $4 - 3 = 1$, so we need another x^1 under the root.

$$\frac{8}{\sqrt[4]{x^3}} \cdot \frac{\sqrt[4]{x^1}}{\sqrt[4]{x^1}} = \frac{8\sqrt[4]{x}}{\sqrt[4]{x^4}} = \frac{8\sqrt[4]{x}}{x}$$

Rationalize the denominator.

$$\frac{4x}{\sqrt[5]{x^2}}$$

The root minus the power is $5 - 2 = 3$, so we need another x^3 under the root.

$$\frac{4x}{\sqrt[5]{x^2}} \cdot \frac{\sqrt[5]{x^3}}{\sqrt[5]{x^3}} = \frac{4x\sqrt[5]{x^3}}{\sqrt[5]{x^5}} = \frac{4x\sqrt[5]{x^3}}{x} = 4\sqrt[5]{x^3}$$

$$\sqrt[7]{\frac{2}{x^3}} = \frac{\sqrt[7]{2}}{\sqrt[7]{x^3}} \cdot \frac{\sqrt[7]{x^4}}{\sqrt[7]{x^4}} = \frac{\sqrt[7]{2x^4}}{\sqrt[7]{x^7}} = \frac{\sqrt[7]{2x^4}}{x}$$

$$\frac{1}{\sqrt[3]{2x}} = \frac{1}{\sqrt[3]{2x}} \cdot \frac{\sqrt[3]{(2x)^2}}{\sqrt[3]{(2x)^2}} = \frac{\sqrt[3]{2^2 x^2}}{\sqrt[3]{(2x)^3}} = \frac{\sqrt[3]{4x^2}}{2x}$$

$$\frac{y}{\sqrt[5]{9}} = \frac{y}{\sqrt[5]{3^2}} \cdot \frac{\sqrt[5]{3^3}}{\sqrt[5]{3^3}} = \frac{y\sqrt[5]{27}}{\sqrt[5]{3^5}} = \frac{y\sqrt[5]{27}}{3}$$

$$\frac{2}{\sqrt[4]{xy^2}} = \frac{2}{\sqrt[4]{xy^2}} \cdot \frac{\sqrt[4]{x^3y^2}}{\sqrt[4]{x^3y^2}} = \frac{2\sqrt[4]{x^3y^2}}{\sqrt[4]{x^4y^4}} = \frac{2\sqrt[4]{x^3y^2}}{\sqrt[4]{(xy)^4}} = \frac{2\sqrt[4]{x^3y^2}}{xy}$$

PRACTICE

Rationalize the denominator.

1. $\dfrac{6}{\sqrt[3]{x^2}}$

2. $\sqrt[5]{\dfrac{3}{2}}$

3. $\dfrac{1}{\sqrt[5]{(x-4)^2}}$

4. $\dfrac{x}{\sqrt[4]{x}}$

5. $\dfrac{9}{\sqrt[5]{8}}$

6. $\dfrac{x}{\sqrt[4]{9}}$

7. $\dfrac{1}{\sqrt[5]{x^2 y^4}}$

8. $\sqrt[8]{\dfrac{12}{x^5 y^6}}$

9. $\dfrac{1}{\sqrt[4]{8xy^2}}$

10. $\sqrt[5]{\dfrac{4}{27x^3 y}}$

 **SOLUTIONS**

1. $\dfrac{6}{\sqrt[3]{x^2}} = \dfrac{6}{\sqrt[3]{x^2}} \cdot \dfrac{\sqrt[3]{x^1}}{\sqrt[3]{x^1}} = \dfrac{6\sqrt[3]{x}}{\sqrt[3]{x^3}} = \dfrac{6\sqrt[3]{x}}{x}$

2. $\sqrt[5]{\dfrac{3}{2}} = \dfrac{\sqrt[5]{3}}{\sqrt[5]{2^1}} \cdot \dfrac{\sqrt[5]{2^4}}{\sqrt[5]{2^4}} = \dfrac{\sqrt[5]{3 \cdot 2^4}}{\sqrt[5]{2^5}} = \dfrac{\sqrt[5]{48}}{2}$

3. $\dfrac{1}{\sqrt[5]{(x-4)^2}} = \dfrac{1}{\sqrt[5]{(x-4)^2}} \cdot \dfrac{\sqrt[5]{(x-4)^3}}{\sqrt[5]{(x-4)^3}} = \dfrac{\sqrt[5]{(x-4)^3}}{\sqrt[5]{(x-4)^5}} = \dfrac{\sqrt[5]{(x-4)^3}}{x-4}$

4. $\dfrac{x}{\sqrt[4]{x}} = \dfrac{x}{\sqrt[4]{x^1}} \cdot \dfrac{\sqrt[4]{x^3}}{\sqrt[4]{x^3}} = \dfrac{x\sqrt[4]{x^3}}{x} = \sqrt[4]{x^3}$

5. $\dfrac{9}{\sqrt[5]{8}} = \dfrac{9}{\sqrt[5]{2^3}} = \dfrac{9}{\sqrt[5]{2^3}} \cdot \dfrac{\sqrt[5]{2^2}}{\sqrt[5]{2^2}} = \dfrac{9\sqrt[5]{4}}{\sqrt[5]{2^5}} = \dfrac{9\sqrt[5]{4}}{2}$

6. $\dfrac{x}{\sqrt[4]{9}} = \dfrac{x}{\sqrt[4]{3^2}} = \dfrac{x}{\sqrt[4]{3^2}} \cdot \dfrac{\sqrt[4]{3^2}}{\sqrt[4]{3^2}} = \dfrac{x\sqrt[4]{9}}{\sqrt[4]{3^4}} = \dfrac{x\sqrt[4]{9}}{3}$

7. $\dfrac{1}{\sqrt[5]{x^2 y^4}} = \dfrac{1}{\sqrt[5]{x^2 y^4}} \cdot \dfrac{\sqrt[5]{x^3 y^1}}{\sqrt[5]{x^3 y^1}} = \dfrac{\sqrt[5]{x^3 y}}{\sqrt[5]{x^5 y^5}} = \dfrac{\sqrt[5]{x^3 y}}{\sqrt[5]{(xy)^5}} = \dfrac{\sqrt[5]{x^3 y}}{xy}$

8. $\sqrt[8]{\dfrac{12}{x^5 y^6}} = \dfrac{\sqrt[8]{12}}{\sqrt[8]{x^5 y^6}} = \dfrac{\sqrt[8]{12}}{\sqrt[8]{x^5 y^6}} \cdot \dfrac{\sqrt[8]{x^3 y^2}}{\sqrt[8]{x^3 y^2}} = \dfrac{\sqrt[8]{12 x^3 y^2}}{\sqrt[8]{x^8 y^8}} = \dfrac{\sqrt[8]{12 x^3 y^2}}{\sqrt[8]{(xy)^8}} = \dfrac{\sqrt[8]{12 x^3 y^2}}{xy}$

9. $\dfrac{1}{\sqrt[4]{8xy^2}} = \dfrac{1}{\sqrt[4]{2^3 x^1 y^2}} \cdot \dfrac{\sqrt[4]{2x^3 y^2}}{\sqrt[4]{2x^3 y^2}} = \dfrac{\sqrt[4]{2x^3 y^2}}{\sqrt[4]{2^4 x^4 y^4}} = \dfrac{\sqrt[4]{2x^3 y^2}}{\sqrt[4]{(2xy)^4}} = \dfrac{\sqrt[4]{2x^3 y^2}}{2xy}$

10. $\sqrt[5]{\dfrac{4}{27x^3 y}} = \dfrac{\sqrt[5]{4}}{\sqrt[5]{3^3 x^3 y^1}} \cdot \dfrac{\sqrt[5]{3^2 x^2 y^4}}{\sqrt[5]{3^2 x^2 y^4}} = \dfrac{\sqrt[5]{4 \cdot 9x^2 y^4}}{\sqrt[5]{3^5 x^5 y^5}} = \dfrac{\sqrt[5]{36x^2 y^4}}{\sqrt[5]{(3xy)^5}} = \dfrac{\sqrt[5]{36x^2 y^4}}{3xy}$

Roots Expressed as Exponents

Roots can be written as exponents with the following two properties. Rewriting radical expressions as expressions written to powers is a useful skill in algebra and especially in calculus.

> **Property 14** $\sqrt[n]{a} = a^{1/n}$

The exponent is a fraction whose numerator is 1 and whose denominator is the root.

EXAMPLE

Rewrite the expression using a fraction exponent.

$$\sqrt{x} = x^{1/2}$$

$$\sqrt[3]{2x+1} = (2x+1)^{1/3}$$

$$\frac{1}{\sqrt{x}} = \frac{1}{x^{1/2}} = x^{-1/2}$$

> **Property 15** $\left(\sqrt[n]{a}\right)^m = \sqrt[n]{a^m} = a^{m/n}$ (If n is even, a must be nonnegative.)

The exponent is a fraction whose numerator is the power and whose denominator is the root.

EXAMPLES

Rewrite the expression using a fraction exponent.

$$\sqrt[5]{x^3} = x^{3/5}$$

$$\sqrt[5]{x^6} = x^{6/5}$$

$$\sqrt{(2x^2 - 1)^7} = (2x^2 - 1)^{7/2}$$

$$\sqrt[3]{(12x)^2} = (12x)^{2/3}$$

$$\frac{4}{\sqrt{x}} = \frac{4}{x^{1/2}} = 4x^{-1.2}$$

$$\frac{15}{\sqrt{x^3}} = \frac{15}{x^{3/2}} = 15x^{-3/2}$$

PRACTICE

Rewrite the expression using a fraction exponent.

1. $\sqrt{14x} =$

2. $\dfrac{3}{\sqrt{2x}} =$

3. $\sqrt[6]{(x+4)^5} =$

4. $\dfrac{3x-5}{\sqrt{x-5}} =$

5. $\dfrac{1}{\sqrt[3]{(10x)^4}} =$

6. $2x^2 \sqrt{(x-y)^3} =$

7. $\dfrac{3x+8}{\sqrt[7]{(12x+5)^3}} =$

8. $\sqrt{\dfrac{x-3}{y^5}} =$

9. $\sqrt[4]{\dfrac{16x^3}{3x+1}} =$

10. $\sqrt[5]{\dfrac{(x-1)^4}{(x+1)^3}} =$

SOLUTIONS

1. $\sqrt{14x} = (14x)^{1/2}$

2. $\dfrac{3}{\sqrt{2x}} = \dfrac{3}{(2x)^{1/2}} = 3(2x)^{-1/2}$

3. $\sqrt[6]{(x+4)^5} = (x+4)^{5/6}$

4. $\dfrac{3x-5}{\sqrt{x-5}} = \dfrac{3x-5}{(x-5)^{1/2}} = (3x-5)(x-5)^{-1/2}$

5. $\dfrac{1}{\sqrt[3]{(10x)^4}} = \dfrac{1}{(10x)^{4/3}} = (10x)^{-4/3}$

6. $2x^2\sqrt{(x-y)^3} = 2x^2(x-y)^{3/2}$

7. $\dfrac{3x+8}{\sqrt[7]{(12x+5)^3}} = \dfrac{3x+8}{(12x+5)^{3/7}} = (3x+8)(12x+5)^{-3/7}$

8. $\sqrt{\dfrac{x-3}{y^5}} = \dfrac{\sqrt{x-3}}{\sqrt{y^5}} = \dfrac{(x-3)^{1/2}}{y^{5/2}} = (x-3)^{1/2}\,y^{-5/2}$

9. $\sqrt[4]{\dfrac{16x^3}{3x+1}} = \dfrac{\sqrt[4]{16x^3}}{\sqrt[4]{3x+1}} = \dfrac{(16x^3)^{1/4}}{(3x+1)^{1/4}} = (16x^3)^{1/4}(3x+1)^{-1/4}$

10. $\sqrt[5]{\dfrac{(x-1)^4}{(x+1)^3}} = \dfrac{\sqrt[5]{(x-1)^4}}{\sqrt[5]{(x+1)^3}} = \dfrac{(x-1)^{4/5}}{(x+1)^{3/5}} = (x-1)^{4/5}(x+1)^{-3/5}$

Simplifying Multiple Roots

The exponent-root properties are useful for simplifying multiple roots. With the properties $\sqrt[n]{a^m} = a^{m/n}$ and $(a^m)^n = a^{mn}$ we can gradually rewrite the multiple root as a single root. We rewrite each root as a power, one root at a time, and then we multiply all of the exponents. This gives us an expression containing a single exponent. Once we have the expression written with a single exponent, we rewrite it as a root.

EXAMPLES

Write the expression as a single root.

$\sqrt[4]{\sqrt[5]{x}}$

We begin with $\sqrt[5]{x}$.

$$\sqrt[4]{\sqrt[5]{x}} = \sqrt[4]{x^{1/5}} = (x^{1/5})^{1/4} = x^{(1/5)(1/4)} = x^{1/20} = \sqrt[20]{x}$$

$$\sqrt[6]{\sqrt[3]{y^5}} = \sqrt[6]{y^{5/3}} = (y^{5/3})^{1/6} = y^{(5/3)(1/6)} = y^{5/18}$$

PRACTICE

Write the expression as a single root.

1. $\sqrt{\sqrt{10}} =$

2. $\sqrt{\sqrt[4]{x^3}} =$

3. $\sqrt[5]{\sqrt[7]{2x^4}} =$

4. $\sqrt[2]{\sqrt[15]{(x-8)^4}} =$

5. $\sqrt{\sqrt{\sqrt[3]{y}}} =$

SOLUTIONS

1. $\sqrt{\sqrt{10}} = \sqrt{10^{1/2}} = (10^{1/2})^{1/2} = 10^{1/4} = \sqrt[4]{10}$

2. $\sqrt{\sqrt[4]{x^3}} = \sqrt{x^{3/4}} = (x^{3/4})^{1/2} = x^{3/8} = \sqrt[8]{x^3}$

3. $\sqrt[5]{\sqrt[7]{2x^4}} = \sqrt[5]{(2x^4)^{1/7}} = ((2x^4)^{1/7})^{1/5} = (2x^4)^{1/35} = \sqrt[35]{2x^4}$

4. $\sqrt[2]{\sqrt[15]{(x-8)^4}} = \sqrt[2]{(x-8)^{4/15}} = ((x-8)^{4/15})^{1/2} = (x-8)^{4/30}$
   $= (x-8)^{2/15} = \sqrt[15]{(x-8)^2}$

5. $\sqrt{\sqrt{\sqrt[3]{y}}} = \sqrt{\sqrt{y^{1/3}}} = \sqrt{(y^{1/3})^{1/2}} = \sqrt{y^{1/6}} = (y^{1/6})^{1/2} = y^{1/12} = \sqrt[12]{y}$

Summary

In this chapter, we learned how to use the exponent and root properties below to simplify and rewrite expressions.

TABLE 5-1		
Exponent Properties	**Root Properties**	**Roots Expressed as Exponents**
$a^m a^n = a^{m+n}$	$\sqrt[n]{ab} = \sqrt[n]{a}\sqrt[n]{b}$	$\sqrt[n]{a} = a^{1/n}$
$\dfrac{a^m}{a^n} = a^{m-n}$	$\sqrt[n]{\dfrac{a}{b}} = \dfrac{\sqrt[n]{a}}{\sqrt[n]{b}}$	$\sqrt[n]{a^m} = a^{m/n}$
$(a^m)^n = a^{mn}$	$\left(\sqrt[n]{a}\right)^m = \sqrt[n]{a^m}$ If n is even, $a \geq 0$	$\left(\sqrt[n]{a}\right)^m = a^{m/n}$ If n is even, $a \geq 0$
$a^0 = 1$	$\left(\sqrt[n]{a}\right)^n = \sqrt[n]{a^n} = a$ If n is even, $a \geq 0$	
$a^{-n} = \dfrac{1}{a^n}$		
$(ab)^n = a^n b^n$		
$\left(\dfrac{a}{b}\right)^m = \dfrac{a^m}{b^m}$		
$\left(\dfrac{a}{b}\right)^{-n} = \dfrac{b^n}{a^n}$		

We also learned how to

- *Find the LCD of fractions whose denominators contained exponents.* The LCD includes expressions raised to the highest power in each denominator.
- *Simplify square roots.* If the quantity under that radical is the product of a square and something else, the square root can be simplified with the properties $\sqrt[n]{ab} = \sqrt[n]{a}\sqrt[n]{b}$ and $\sqrt[n]{a^n} = a$.
- *Rationalize a denominator.* If the denominator includes a square root as a factor, we multiply the numerator and denominator by this factor. This removes the square root from the denominator because of the property $\left(\sqrt{a}\right)^2 = a$.

We will use exponent and root properties later when solving some equations.

QUIZ

1. Simplify $\dfrac{(5x)^3}{5x}$.

   A. x^2

   B. $5x^2$

   C. $\dfrac{25}{x^2}$

   D. $25x^2$

2. $\dfrac{8}{xy}+\dfrac{3}{x^2}=$

   A. $\dfrac{8x+3y}{x^2y}$

   B. $\dfrac{11}{xy+x^2}$

   C. $\dfrac{8+3y}{x^2y}$

   D. $\dfrac{11y}{x^2+y}$

3. Simplify $\left(\dfrac{4x^3}{x^5}\right)^{-2}$.

   A. $\dfrac{16}{x^4}$

   B. $\dfrac{y^3}{x^3}$

   C. $\dfrac{x^4}{16}$

   D. $\dfrac{2}{x}$

4. Simplify $\left(\dfrac{x}{y}\right)^{-3}$.

   A. x^3y^3

   B. $\dfrac{y^3}{x^3}$

   C. $-\dfrac{x^3}{y^3}$

   D. $-\dfrac{y^3}{x^3}$

5. Simplify $(3x^{-2}y^4)^2$.

    A. $\dfrac{y^8}{9x^4}$

    B. $\dfrac{9y^8}{x^4}$

    C. $\dfrac{9y^4}{x^2}$

    D. $\dfrac{3y^8}{x^4}$

6. Write $\dfrac{7x}{3y}$ as a product.

    A. $(7x)(3y)^{-1}$

    B. $21xy^{-1}$

    C. $-21xy$

    D. $\dfrac{7x}{3y}$ cannot be written as a product.

7. Simplify $\dfrac{6x^{-2}y^3}{15x^4y^{-5}}$.

    A. $\dfrac{2x^2y^2}{5}$

    B. $\dfrac{2y^8}{5x^6}$

    C. $\dfrac{2y^2}{5x^2}$

    D. $\dfrac{2y^2}{5x^6}$

8. Rewrite $10x^{2/3}$ without using a fraction exponent.

    A. $\sqrt{10x^3}$

    B. $10\sqrt{x^3}$

    C. $\sqrt[3]{10x^2}$

    D. $10\sqrt[3]{x^2}$

9. $\dfrac{4}{(x+3)^2(x+2y)} + \dfrac{1}{(x+3)(x+2y)} =$

A. $\dfrac{4+x+2y}{(x+3)^2(x+2y)}$

B. $\dfrac{4+x+2y}{(x^2+9)(x+2y)}$

C. $\dfrac{7+x}{(x+3)^2(x+2y)}$

D. $\dfrac{7+x}{(x^2+9)(x+2y)}$

10. $\dfrac{1}{xy^3} - \dfrac{3}{x^2} + \dfrac{4x}{y^2} =$

A. $\dfrac{x-3y^3+4x^3y}{x^2y^3}$

B. $\dfrac{x-3y+4xy}{x^2y^3}$

C. $\dfrac{1-3y^3+4y}{x^2y^3}$

D. $\dfrac{x^2-3xy^3+4y^3}{x^2y^3}$

11. Simplify $\sqrt{18y^3}$.

A. $3y\sqrt{6y}$

B. $3\sqrt{2y}$

C. $3y^2\sqrt{2y}$

D. $3y\sqrt{2y}$

12. Rewrite $\sqrt{14x}$ using a fraction exponent.

A. $14x^{2/1}$

B. $(14x)^{2/1}$

C. $(14x)^{1/2}$

D. $14x^{1/2}$

13. **Rationalize the denominator:** $\dfrac{15}{\sqrt{3}}$.

    A. $\dfrac{\sqrt{15}}{3}$

    B. $\dfrac{\sqrt{45}}{3}$

    C. $\dfrac{5\sqrt{3}}{3}$

    D. $5\sqrt{3}$

14. **Write** $\dfrac{6x-7}{4x+5}$ **as a product.**

    A. $(6x-7)(4x+5)^{-1}$

    B. $(6x-7)[-(4x+5)]$

    C. $(6x-7)(-4x-5)$

    D. $\dfrac{6x-7}{4x+5}$ cannot be written as a product.

15. $\sqrt{\dfrac{7x^2}{3}} =$

    A. $\dfrac{7x^2\sqrt{3}}{3}$

    B. $\dfrac{\sqrt{63x^2}}{3}$

    C. $\dfrac{\sqrt{7x^2}}{3}$

    D. $\dfrac{\sqrt{21x^2}}{3}$

16. $\sqrt[3]{48x^4y^6} =$

    A. $2xy^2\sqrt[3]{6xy}$

    B. $2xy^2\sqrt[3]{6x}$

    C. $2xy\sqrt[3]{3xy}$

    D. $2xy\sqrt[3]{3x}$

17. $\dfrac{4}{\sqrt[3]{y}} =$

    A. $\dfrac{4\sqrt[3]{y}}{y^2}$

    B. $\dfrac{4\sqrt[3]{y^5}}{y}$

    C. $\dfrac{4\sqrt[3]{y}}{y}$

    D. $\dfrac{4\sqrt[3]{y^2}}{y}$

18. $((5x)^2)^{-1} =$

    A. $\dfrac{5}{x^2}$

    B. $\dfrac{x^2}{5}$

    C. $\dfrac{1}{25x^2}$

    D. $\dfrac{x^2}{25}$

19. $\sqrt[3]{\sqrt[2]{10y}} =$

    A. $\sqrt[6]{10y}$

    B. $10y^{1/6}$

    C. $10\sqrt[6]{y}$

    D. $\sqrt{10}\sqrt[6]{y}$

chapter **6**

Factoring and the Distributive Property

Distributing multiplication over addition (and subtraction) and factoring (the opposite of distributing) are extremely important properties in algebra. We use the distributive property of multiplication over addition both to expand expressions and to factor them (to write an expression as a product of two or more numbers and/or algebraic expressions). After developing our factoring skills, we will work with fractions having algebraic expressions in their numerators and/or denominators. In later chapters, we will use our ability to factor expressions when solving equations.

CHAPTER OBJECTIVES

In this chapter, you will

- Use the distributive property to expand expressions
- Combine like terms
- Use various techniques to factor expressions
- Use the FOIL method to expand expressions
- Use factoring techniques to simplify fractions
- Use factoring techniques to add fractions

The distributive property of multiplication over addition, $a(b + c) = ab + ac$, says that we can first take the sum $(b + c)$ and then the product (a times the sum of b and c) or the individual products (ab and ac) and then the sum (the sum of ab and ac). For instance, $12(6 + 4)$ could be computed as $12(6 + 4) = 12(6) + 12(4) = 72 + 48 = 120$ or as $12(6 + 4) = 12(10) = 120$. The distributive property of multiplication over subtraction, $a(b - c) = ab - ac$, says the same about a product and difference. Rather than working with numbers, we will use the distributive property to work with algebraic expressions. We begin with simple variables.

EXAMPLES

Use the distributive property to rewrite the expression.
In distributing variables, we often use the exponent property $a^m a^n = a^{m+n}$.

$7(x - y) = 7x - 7y$

$4(3x + 1) = 12x + 4$

$x^2(3x - 5y) = 3x^3 - 5x^2y$

$8xy(x^3 + 4y) = 8x^4y + 32xy^2$

$6x^2y^3(5x - 2y^2) = 30x^3y^3 - 12x^2y^5$

$\sqrt{x}(x^2 + 12) = x^2\sqrt{x} + 12\sqrt{x}$

$y^{-2}(y^4 + 6) = y^{-2}y^4 + 6y^{-2} = y^2 + 6y^{-2}$

$3xy(2x^2 + 5y - 7) = 6x^3y + 15xy^2 - 21xy$

PRACTICE

Use the distributive property to rewrite the expression.

1. $3(14 - 2) =$

2. $\dfrac{1}{2}(6 + 8) =$

3. $4(6 - 2x) =$

4. $9x(4y + x) =$

5. $3xy^4(9x + 2y) =$

6. $3\sqrt[3]{x}(6y - 2x) =$

7. $\sqrt{x}(1 + \sqrt{x}) =$

8. $10y^{-3}(xy^4 - 8) =$

9. $4x^2(2y - 5x + 6) =$

✔ **SOLUTIONS**

1. $3(14 - 2) = 3(14) - 3(2) = 42 - 6 = 36$

2. $\dfrac{1}{2}(6 + 8) = \dfrac{1}{2}(6) + \dfrac{1}{2}(8) = 3 + 4 = 7$

3. $4(6 - 2x) = 4(6) - 4(2x) = 24 - 8x$

4. $9x(4y + x) = 36xy + 9x^2$

5. $3xy^4(9x + 2y) = 27x^2y^4 + 6xy^5$

6. $3\sqrt[3]{x}(6y - 2x) = 18\sqrt[3]{x}\,y - 6x\sqrt[3]{x}$

7. $\sqrt{x}\left(1 + \sqrt{x}\right) = \sqrt{x} + \left(\sqrt{x}\right)\left(\sqrt{x}\right) = \sqrt{x} + \left(\sqrt{x}\right)^2 = \sqrt{x} + x$

8. $10y^{-3}(xy^4 - 8) = 10xy^4y^{-3} - 80y^{-3} = 10xy - 80y^{-3}$

9. $4x^2(2y - 5x + 6) = 8x^2y - 20x^3 + 24x^2$

Distributing Negative Numbers

Distributing a minus sign or a negative number over addition or subtraction changes the sign of every term inside the parentheses: $-(a + b) = -a - b$ and $-(a - b) = -a + b$. Using the distributive property, we can think of $-(a + b)$ as $(-1)(a + b)$ and $-(a - b)$ as $(-1)(a - b)$. Therefore,

$$-(a + b) = (-1)(a + b) = (-1)a + (-1)b = -a + (-b) = -a - b$$

and

$$-(a - b) = (-1)(a - b) = (-1)a - (-1)b = -a - (-1)b = -a - (-b) = -a + b.$$

Still Struggling

Two common mistakes are to write $-(a + b) = -a + b$ and $-(a - b) = -a - b$. The minus sign and negative sign in front of the parentheses changes the signs of *every* term inside the parentheses, not just the first term.

EXAMPLES

Use the distributive property to rewrite the expression.

$-(3 + x) = -3 - x$

$-(2 + x - 3y) = -2 - x + 3y$

$-(y - x^2) = -y + x^2$

$-(x^2 - x - 2) = -x^2 + x + 2$

$-(-2 + y) = 2 - y$

$-(-4x - 7y - 2) = 4x + 7y + 2$

$-(-9 - y) = 9 + y$

PRACTICE

Use the distributive property to rewrite the expression.

1. $-(4 + x) =$

2. $-(-x - y) =$

3. $-(2x^2 - 5) =$

4. $-(-18 + xy^2) =$

5. $-(2x - 16y + 5) =$

6. $-(x^2 - 5x - 6) =$

SOLUTIONS

1. $-(4 + x) = -4 - x$

2. $-(-x - y) = x + y$

3. $-(2x^2 - 5) = -2x^2 + 5$

4. $-(-18 + xy^2) = 18 - xy^2$

5. $-(2x - 16y + 5) = -2x + 16y - 5$

6. $-(x^2 - 5x - 6) = -x^2 + 5x + 6$

Distributing negative quantities has the same effect on signs as distributing a minus sign: every sign in the parentheses changes.

EXAMPLES

Use the distributive property to rewrite the expression.

$$-8(4 + 5x) = -32 - 40x$$

$$-xy(1 - x) = -xy + x^2y$$

$$-3x^2(-2y + 9x) = 6x^2y - 27x^3$$

$$-100(-4 - x) = 400 + 100x$$

PRACTICE

Use the distributive property to rewrite the expression.

1. $-2(16 + y) =$

2. $-50(3 - x) =$

3. $-12xy(-2x + y) =$

4. $-7x^2(-x - 4y) =$

5. $-6y(-3x - y + 4) =$

SOLUTIONS

1. $-2(16 + y) = -32 - 2y$

2. $-50(3 - x) = -150 + 50x$

3. $-12xy(-2x + y) = 24x^2y - 12xy^2$

4. $-7x^2(-x - 4y) = 7x^3 + 28x^2y$

5. $-6y(-3x - y + 4) = 18xy + 6y^2 - 24y$

Combining Like Terms

Two or more terms are alike if they have the same variables and the exponents (or roots) on those variables are the same: $3x^2y$ and $5x^2y$ are like terms but $6xy$ and $4xy^2$ are not. Constants are terms with no variables. The number in front of the variable(s) is the *coefficient*—in the expression $4x^2y^3$, 4 is the coefficient. If no number appears in front of the variable, then the coefficient is 1. We add and subtract like terms by adding or subtracting their coefficients.

EXAMPLES

Combine like terms.

$$3x^2y + 5x^2y$$

Both terms are alike because they have the same variables with the same exponents, so we simply add the coefficients 3 and 5, to obtain, $3x^2y + 5x^2y = (3 + 5)x^2y = 8x^2y$.

$$14\sqrt{x} - 10\sqrt{x} = (14 - 10)\sqrt{x} = 4\sqrt{x}$$

$$8xyz + 9xyz - 6xyz = (8 + 9 - 6)xyz = 11xyz$$

$$3x + x = 3x + 1x = (3 + 1)x = 4x$$

$$7x\sqrt{y} - x\sqrt{y} = 7x\sqrt{y} - 1x\sqrt{y} = (7-1)x\sqrt{y} = 6x\sqrt{y}$$

$$\frac{2}{3}x^2 - 4xy + \frac{3}{4}x^2 + \frac{5}{2}xy = \left(\frac{2}{3} + \frac{3}{4}\right)x^2 + \left(-4 + \frac{5}{2}\right)xy = \left(\frac{8}{12} + \frac{9}{12}\right)x^2 + \left(\frac{-8}{2} + \frac{5}{2}\right)xy$$

$$= \frac{17}{12}x^2 - \frac{3}{2}xy$$

$$3x^2 + 4xy - 8xy^2 - (2x^2 - 3xy - 4xy^2 + 6) = 3x^2 + 4xy - 8xy^2 - 2x^2 + 3xy + 4xy^2 - 6$$

$$= 3x^2 - 2x^2 - 8xy^2 + 4xy^2 + 4xy + 3xy - 6$$

$$= (3-2)x^2 + (-8+4)xy^2 + (4+3)xy - 6$$

$$= x^2 - 4xy^2 + 7xy - 6$$

PRACTICE

Combine like terms.

1. $3xy + 7xy =$

2. $4x^2 - 6x^2 =$

3. $-\dfrac{3}{5}xy^2 + 2xy^2 =$

4. $8\sqrt{x} - \sqrt{x} =$

5. $2xy^2 - 4x^2y - 7xy^2 + 17x^2y =$

6. $14x + 8 - (2x - 4) =$

7. $16x^{-4} + 3x^{-2} - 4x + 9x^{-4} - x^{-2} + 5x - 6 =$

8. $5x\sqrt{y} + 7\sqrt{xy} + 1 - \left(3x\sqrt{y} - 7\sqrt{xy} + 4\right) =$

9. $x^2y + xy^2 + 6x + 4 - (4x^2y + 3xy^2 - 2x + 5) =$

SOLUTIONS

1. $3xy + 7xy = (3+7)xy = 10xy$

2. $4x^2 - 6x^2 = (4-6)x^2 = -2x^2$

3. $\dfrac{-3}{5}xy^2 + 2xy^2 = \left(\dfrac{-3}{5} + 2\right)xy^2 = \left(\dfrac{-3}{5} + \dfrac{10}{5}\right)xy^2 = \dfrac{7}{5}xy^2$

4. $8\sqrt{x} - \sqrt{x} = (8-1)\sqrt{x} = 7\sqrt{x}$

5. $2xy^2 - 4x^2y - 7xy^2 + 17x^2y = 2xy^2 - 7xy^2 - 4x^2y + 17x^2y$

$$= (2-7)xy^2 + (-4+17)x^2y$$

$$= -5xy^2 + 13x^2y$$

6. $14x + 8 - (2x - 4) = 14x + 8 - 2x + 4 = 14x - 2x + 8 + 4$

$$= (14 - 2)x + 12 = 12x + 12$$

7. $16x^{-4} + 3x^{-2} - 4x + 9x^{-4} - x^{-2} + 5x - 6$

$$= 16x^{-4} + 9x^{-4} + 3x^{-2} - x^{-2} - 4x + 5x - 6$$

$$= (16 + 9)x^{-4} + (3 - 1)x^{-2} + (-4 + 5)x - 6$$

$$= 25x^{-4} + 2x^{-2} + x - 6$$

8. $5x\sqrt{y} + 7\sqrt{xy} + 1 - \left(3x\sqrt{y} - 7\sqrt{xy} + 4\right)$

$$= 5x\sqrt{y} + 7\sqrt{xy} + 1 - 3x\sqrt{y} + 7\sqrt{xy} - 4$$

$$= 5x\sqrt{y} - 3x\sqrt{y} + 7\sqrt{xy} + 7\sqrt{xy} + 1 - 4$$

$$= (5 - 3)x\sqrt{y} + (7 + 7)\sqrt{xy} - 3$$

$$= 2x\sqrt{y} + 14\sqrt{xy} - 3$$

9. $x^2y + xy^2 + 6x + 4 - (4x^2y + 3xy^2 - 2x + 5)$

$$= x^2y + xy^2 + 6x + 4 - 4x^2y - 3xy^2 + 2x - 5$$

$$= x^2y - 4x^2y + xy^2 - 3xy^2 + 6x + 2x + 4 - 5$$

$$= (1 - 4)x^2y + (1 - 3)xy^2 + (6 + 2)x - 1$$

$$= -3x^2y - 2xy^2 + 8x - 1$$

Adding/Subtracting Fractions

With the distributive property and our ability to combine like terms, we can simplify the sum and difference of fractions. Remember that the denominators must be the same before we add or subtract fractions. Once we have the fractions written with the LCD, we add or subtract their numerators. This involves combining terms.

We generally leave the denominators factored.

EXAMPLES

Find the sum or difference.

$$\frac{2}{x - 4} + \frac{x}{x + 1} = \frac{2}{x - 4} \cdot \frac{x + 1}{x + 1} + \frac{x}{x + 1} \cdot \frac{x - 4}{x - 4}$$

$$= \frac{2(x + 1) + x(x - 4)}{(x + 1)(x - 4)} = \frac{2x + 2 + x^2 - 4x}{(x + 1)(x - 4)} = \frac{x^2 - 2x + 2}{(x + 1)(x - 4)}$$

$$4 - \frac{2x+1}{x+3} = \frac{4}{1} - \frac{2x+1}{x+3} = \frac{4}{1} \cdot \frac{x+3}{x+3} - \frac{2x+1}{x+3}$$

$$= \frac{4(x+3)-(2x+1)}{x+3} = \frac{4x+12-2x-1}{x+3} = \frac{2x+11}{x+3}$$

$$\frac{x}{x-5} - \frac{x}{x+2} = \frac{x}{x-5} \cdot \frac{x+2}{x+2} - \frac{x}{x+2} \cdot \frac{x-5}{x-5}$$

$$= \frac{x(x+2)-x(x-5)}{(x+2)(x-5)} = \frac{x^2+2x-x^2+5x}{(x+2)(x-5)} = \frac{7x}{(x+2)(x-5)}$$

PRACTICE

Find the sum or difference.

1. $\dfrac{7}{2x+3} + \dfrac{4}{x-2} =$

2. $\dfrac{1}{x-1} + \dfrac{x}{x+2} =$

3. $\dfrac{3x-4}{x+5} - 2 =$

4. $\dfrac{x}{2x+y} + \dfrac{y}{3x-4y} =$

5. $\dfrac{x}{6x+3} + \dfrac{x}{6x-3} =$

SOLUTIONS

1. $\dfrac{7}{2x+3} + \dfrac{4}{x-2} = \dfrac{7}{2x+3} \cdot \dfrac{x-2}{x-2} + \dfrac{4}{x-2} \cdot \dfrac{2x+3}{2x+3}$

$$= \frac{7(x-2)+4(2x+3)}{(x-2)(2x+3)} = \frac{7x-14+8x+12}{(x-2)(2x+3)}$$

$$= \frac{15x-2}{(x-2)(2x+3)}$$

2. $\dfrac{1}{x-1} + \dfrac{x}{x+2} = \dfrac{1}{x-1} \cdot \dfrac{x+2}{x+2} + \dfrac{x}{x+2} \cdot \dfrac{x-1}{x-1} = \dfrac{1(x+2)+x(x-1)}{(x+2)(x-1)}$

$$= \frac{x+2+x^2-x}{(x+2)(x-1)} = \frac{x^2+2}{(x+2)(x-1)}$$

3. $\dfrac{3x-4}{x+5}-2=\dfrac{3x-4}{x+5}-\dfrac{2}{1}=\dfrac{3x-4}{x+5}-\dfrac{2}{1}\cdot\dfrac{x+5}{x+5}=\dfrac{3x-4-2(x+5)}{x+5}$

$$=\dfrac{3x-4-2x-10}{x+5}=\dfrac{x-14}{x+5}$$

4. $\dfrac{x}{2x+y}+\dfrac{y}{3x-4y}=\dfrac{x}{2x+y}\cdot\dfrac{3x-4y}{3x-4y}+\dfrac{y}{3x-4y}\cdot\dfrac{2x+y}{2x+y}$

$$=\dfrac{x(3x-4y)+y(2x+y)}{(3x-4y)(2x+y)}=\dfrac{3x^2-4xy+2xy+y^2}{(3x-4y)(2x+y)}$$

$$=\dfrac{3x^2-2xy+y^2}{(3x-4y)(2x+y)}$$

5. $\dfrac{x}{6x+3}+\dfrac{x}{6x-3}=\dfrac{x}{6x+3}\cdot\dfrac{6x-3}{6x-3}+\dfrac{x}{6x-3}\cdot\dfrac{6x+3}{6x+3}$

$$=\dfrac{x(6x-3)+x(6x+3)}{(6x-3)(6x+3)}=\dfrac{6x^2-3x+6x^2+3x}{(6x-3)(6x+3)}$$

$$=\dfrac{12x^2}{(6x-3)(6x+3)}$$

Factoring

The distributive property, $a(b + c) = ab + ac$, can be used to factor a quantity from two or more terms. In the formula $ab + ac = a(b + c)$, a is factored from (or divided into) ab and ac on the right side of the equation. The first step in factoring is to decide what quantity to factor from each term. The second step is to write each term as a product of the factor and something else (this step will become unnecessary once you are experienced). The last step is to apply the distributive property in reverse. An expression is "completely factored" if that the terms inside the parentheses have no common factor (other than 1). For instance, $8(2x + 3)$ is completely factored because $2x$ and 3 have no common factors (other than 1). The expression $4(15x + 10)$ is not completely factored because 5 divides both $15x$ and 10.

EXAMPLES

Completely factor the expression.

$$4 + 6x$$

Each term is divisible by 2, so factor 2 from 4 and 6x: $4 + 6x = 2 \cdot 2 + 2 \cdot 3x = 2(2 + 3x)$.

$$2x + 5x^2 = x \cdot 2 + x \cdot 5x = x(2 + 5x)$$

$$8x + 8 = 8 \cdot x + 8 \cdot 1 = 8(x + 1)$$

$$4xy + 6x^2 + 2xy^2 = 2x \cdot 2y + 2x \cdot 3x + 2x \cdot y^2 = 2x(2y + 3x + y^2)$$

$$3x^2 + 6x = 3x \cdot x + 3x \cdot 2 = 3x(x + 2)$$

Complicated expressions can be factored in several steps. Take for example $48x^5y^3z^6 + 60x^4yz^3 + 36x^6y^2z$. We can begin by dividing $12xyz$ from each term, to obtain:

$$48x^5y^3z^6 + 60x^4yz^3 + 36x^6y^2z = 12xyz \cdot 4x^4y^2z^5 + 12xyz \cdot 5x^3z^2 + 12xyz \cdot 3x^5y$$

$$= 12xyz(4x^4y^2z^5 + 5x^3z^2 + 3x^5y)$$

Each term in the parentheses is divisible by x^2, so it is not completely factored. We now factor x^2 from each term, and then we will simplify it, thus obtaining:

$$4x^4y^2z^5 + 5x^3z^2 + 3x^5y = x^2 \cdot 4x^2y^2z^5 + x^2 \cdot 5xz^2 + x^2 \cdot 3x^3y = x^2(4x^2y^2z^5 + 5xz^2 + 3x^3y)$$

We now factor the original problem, and obtain:

$$48x^5y^3z^6 + 60x^4yz^3 + 36x^6y^2z = 12xyz \cdot x^2(4x^2y^2z^5 + 5xz^2 + 3x^3y)$$

$$= 12x^3yz(4x^2y^2z^5 + 5xz^2 + 3x^3y)$$

 PRACTICE

Completely factor the expression.

1. $4x - 10y =$
2. $3x + 6y - 12 =$
3. $5x^2 + 15 =$
4. $4x^2 + 4x =$
5. $4x^3 - 6x^2 + 12x =$
6. $-24xy^2 + 6x^2 + 18x =$
7. $30x^4 - 6x^2 =$
8. $15x^3y^2z^7 - 30xy^2z^4 + 6x^4y^2z^6 =$

SOLUTIONS

1. $4x - 10y = 2 \cdot 2x - 2 \cdot 5y = 2(2x - 5y)$

2. $3x + 6y - 12 = 3 \cdot x + 3 \cdot 2y - 3 \cdot 4 = 3(x + 2y - 4)$

3. $5x^2 + 15 = 5 \cdot x^2 + 5 \cdot 3 = 5(x^2 + 3)$

4. $4x^2 + 4x = 4x \cdot x + 4x \cdot 1 = 4x(x + 1)$

5. $4x^3 - 6x^2 + 12x = 2x \cdot 2x^2 - 2x \cdot 3x + 2x \cdot 6 = 2x(2x^2 - 3x + 6)$

6. $-24xy^2 + 6x^2 + 18x = 6x \cdot (-4y^2) + 6x \cdot x + 6x \cdot 3$
$$= 6x(-4y^2 + x + 3)$$

7. $30x^4 - 6x^2 = 6x^2 \cdot 5x^2 - 6x^2 \cdot 1 = 6x^2(5x^2 - 1)$

8. $15x^3y^2z^7 - 30xy^2z^4 + 6x^4y^2z^6 = 3xy^2z^4 \cdot 5x^2z^3 - 3xy^2z^4 \cdot 10 + 3xy^2z^4 \cdot 2x^3z^2$
$$= 3xy^2z^4(5x^2z^3 - 10 + 2x^3z^2)$$

Factoring with Negative Numbers

Factoring a negative number from two or more terms has the same effect on signs within parentheses as distributing a negative number does—every sign changes. Negative quantities are factored in the next examples and practice problems.

EXAMPLES

Factor a negative quantity.

$$-2 - 3x = -(2 + 3x)$$

$$x + y$$

As both terms are positive (that is, have positive 1 as a coefficient), we will have to use the fact that $1 = (-1)(-1)$ to write each term with a negative coefficient. Thus, $x + y = (-1)(-1)x + (-1)(-1)y = (-1)[(-1)x + (-1)y] = (-1)(-x - y) = -(-x - y)$.

In general, $x = -(-x)$, for any x.

$$-4 + x = -(4 - x)$$
$$2x^2 + 4x = -2x(-x - 2)$$
$$12xy - 25x = -x(-12y + 25)$$
$$x - y - z + 5 = -(-x + y + z - 5)$$

PRACTICE

Factor a negative quantity from the expression.

1. $28xy^2 - 14x =$
2. $4x + 16xy =$
3. $-18y^2 + 6xy =$
4. $25 + 15y =$
5. $-8x^2y^2 - 4xy^2$
6. $-18x^2y^2 - 24xy^3 =$
7. $20xyz^2 - 5yz =$

SOLUTIONS

1. $28xy^2 - 14x = -7x(-4y^2 + 2)$
2. $4x + 16xy = -4x(-1 - 4y)$
3. $-18y^2 + 6xy = -6y(3y - x)$
4. $25 + 15y = -5(-5 - 3y)$
5. $-8x^2y^2 - 4xy^2 = -4xy^2(2x + 1)$
6. $-18x^2y^2 - 24xy^3 = -6xy^2(3x + 4y)$
7. $20xyz^2 - 5yz = -5yz(-4xz + 1)$

 Still Struggling

The distributive property and the associative property of multiplication can be confusing because both involve parentheses and multiplication. The associative property involves the product of three quantities, $(ab)c = a(bc)$. This property says that when multiplying three quantities we can multiply the first two then the third, or multiply the second two then the first. For example, it might be tempting to write $5(x + 1)(y - 3) = (5x + 5)(5y - 15)$. But $(5x + 5)(5y - 15) = [5(x + 1)][5(y - 3)] = 25(x + 1)(y - 3)$. The "5" can be grouped either with "$x + 1$" or with "$y - 3$" but not both. The correct computation is either $[5(x + 1)](y - 3) = (5x + 5)(y - 3)$ or $(x + 1)[5(y - 3)] = (x + 1)(5y - 15)$.

Factors themselves can have more than one term. For instance $3(x + 4) - x(x + 4)$ has $x + 4$ as a factor in each term, so $x + 4$ can be factored from $3(x + 4)$ and from $x(x + 4)$: $3(x + 4) - x(x + 4) = (3 - x)(x + 4)$.

EXAMPLES

Completely factor the expression.

$$2x(3x + y) + 5y(3x + y)$$

We can factor $3x + y$ from each term, leaving $2x$ and $5y$ inside the parentheses.

$$2x(3x + y) + 5y(3x + y) = (2x + 5y)(3x + y)$$

$$10y(x - y) + x - y = 10y(x - y) + 1(x - y) = (10y + 1)(x - y)$$

$$8(2x - 1) + 2x(2x - 1) - 3y(2x - 1) = (8 + 2x - 3y)(2x - 1)$$

PRACTICE

Completely factor the expression.

1. $2(x - y) + 3y(x - y) =$

2. $4(2 + 7x) - x(2 + 7x) =$

3. $3(3 + x) + x(3 + x) =$

4. $6x(4 - 3x) - 2y(4 - 3x) - 5(4 - 3x) =$

5. $2x + 1 + 9x(2x + 1) =$

6. $3(x - 2y)^4 + 2x(x - 2y)^4 =$

SOLUTIONS

1. $2(x - y) + 3y(x - y) = (2 + 3y)(x - y)$

2. $4(2 + 7x) - x(2 + 7x) = (4 - x)(2 + 7x)$

3. $3(3 + x) + x(3 + x) = (3 + x)(3 + x) = (3 + x)^2$

4. $6x(4 - 3x) - 2y(4 - 3x) - 5(4 - 3x) = (6x - 2y - 5)(4 - 3x)$

5. $2x + 1 + 9x(2x + 1) = 1(2x + 1) + 9x(2x + 1) = (1 + 9x)(2x + 1)$

6. $3(x - 2y)^4 + 2x(x - 2y)^4 = (3 + 2x)(x - 2y)^4$

More Factoring

An algebraic expression raised to different powers might appear in different terms. The common factor is the expression raised to the lowest power.

EXAMPLES

Completely factor the expression.

$$6(x + 1)^2 - 5(x + 1)$$

Each term has $x + 1$ as a factor. The smallest power on this factor is 1, so the common factor is $(x + 1)^1$.

$$6(x + 1)^2 - 5(x + 1) = [6(x + 1)](x + 1) - 5(x + 1)$$
$$= [6(x + 1) - 5](x + 1) = (6x + 6 - 5)(x + 1) = (6x + 1)(x + 1)$$

$$10(2x - 3)^3 + 3(2x - 3)^2 = [10(2x - 3)](2x - 3)^2 + 3(2x - 3)^2 = [10(2x - 3) + 3](2x - 3)^2$$
$$= (20x - 30 + 3)(2x - 3)^2 = (20x - 27)(2x - 3)^2$$

$$9(14x + 5)^4 + 6x(14x + 5) - (14x + 5) = [9(14x + 5)^3](14x + 5) + 6x(14x + 5) - 1(14x + 5)$$
$$= [9(14x + 5)^3 + 6x - 1](14x + 5)$$

PRACTICE

Completely factor the expression.

1. $8(x + 2)^3 + 5(x + 2)^2 =$

2. $-4(x + 16)^4 + 9(x + 16)^2 + x + 16 =$

3. $(x + 2y)^3 - 4(x + 2y) =$

4. $2(x^2 - 6)^9 + (x^2 - 6)^4 + 4(x^2 - 6)^3 + (x^2 - 6)^2 =$

5. $(15xy - 1)(2x - 1)^3 - 8(2x - 1)^2 =$

SOLUTIONS

1. $8(x + 2)^3 + 5(x + 2)^2 = [8(x + 2)](x + 2)^2 + 5(x + 2)^2$
$$= [8(x + 2) + 5](x + 2)^2$$
$$= (8x + 16 + 5)(x + 2)^2 = (8x + 21)(x + 2)^2$$

2. $-4(x + 16)^4 + 9(x + 16)^2 + x + 16$
$$= [-4(x + 16)^3](x + 16) + 9(x + 16)(x + 16) + 1(x + 16)$$
$$= [-4(x + 16)^3 + 9(x + 16) + 1](x + 16)$$
$$= [-4(x + 16)^3 + 9x + 144 + 1](x + 16)$$
$$= [-4(x + 16)^3 + 9x + 145)(x + 16)$$

3. $(x + 2y)^3 - 4(x + 2y) = (x + 2y)^2(x + 2y) - 4(x + 2y)$
$$= [(x + 2y)^2 - 4](x + 2y)$$

4. $2(x^2 - 6)^9 + (x^2 - 6)^4 + 4(x^2 - 6)^3 + (x^2 - 6)^2$

$\qquad = 2(x^2 - 6)^7(x^2 - 6)^2 + (x^2 - 6)^2(x^2 - 6)^2 + 4(x^2 - 6)(x^2 - 6)^2 + 1(x^2 - 6)^2$

$\qquad = [2(x^2 - 6)^7 + (x^2 - 6)^2 + 4(x^2 - 6) + 1](x^2 - 6)^2$

$\qquad = [2(x^2 - 6)^7 + (x^2 - 6)^2 + 4x^2 - 24 + 1](x^2 - 6)^2$

$\qquad = [2(x^2 - 6)^7 + (x^2 - 6)^2 + 4x^2 - 23](x^2 - 6)^2$

5. $(15xy - 1)(2x - 1)^3 - 8(2x - 1)^2 = (15xy - 1)(2x - 1)(2x - 1)^2 - 8(2x - 1)^2$

$\qquad\qquad\qquad\qquad\qquad\qquad = [(15xy - 1)(2x - 1) - 8](2x - 1)^2$

Factoring by Grouping

Some expressions having four terms can be factored with a technique called *factoring by grouping*. Factoring by grouping works to factor an expression containing four terms if we can factor the first two terms and the last two terms so that these pairs have a common factor. We begin by factoring the first two terms; next we see if we can factor the last two terms in a similar way.

 EXAMPLES

Use factoring by grouping to factor the expression.

$$3x^2 - 3 + x^3 - x$$

Let us begin by factoring 3 from the first two terms.

$$3x^2 - 3 + x^3 - x = 3(x^2 - 1) + x^3 - x$$

We now see if we can factor the last two terms so that $x^2 - 1$ is a common factor. We will factor x from each of the last two terms.

$$3x^2 - 3 + x^3 - x = 3(x^2 - 1) + x^3 - x$$
$$= 3(x^2 - 1) + x(x^2 - 1)$$

Now we can factor $x^2 - 1$ from each term.

$$3x^2 - 3 + x^3 - x = 3(x^2 - 1) + x^3 - x$$
$$= 3(x^2 - 1) + x(x^2 - 1)$$
$$= (x^2 - 1)(3 + x)$$

$$3xy - 2y + 3x^2 - 2x = y(3x - 2) + x(3x - 2) = (y + x)(3x - 2)$$
$$5x^2 - 25 - x^2y + 5y = 5(x^2 - 5) - y(x^2 - 5) = (5 - y)(x^2 - 5)$$
$$4x^4 + x^3 - 4x - 1 = x^3(4x + 1) - (4x + 1) = (x^3 - 1)(4x + 1)$$

PRACTICE

Use factoring by grouping to factor the expression.

1. $6xy^2 + 4xy + 9xy + 6x =$

2. $x^3 + x^2 - x - 1 =$

3. $15xy + 5x + 6y + 2 =$

4. $2x^4 - 6x - x^3y + 3y =$

5. $9x^3 + 18x^2 - x - 2 =$

 SOLUTIONS

1. $6xy^2 + 4xy + 9xy + 6x = 2xy(3y + 2) + 3x(3y + 2)$

$$= (2xy + 3x)(3y + 2) = x(2y + 3)(3y + 2)$$

2. $x^3 + x^2 - x - 1 = x^2(x + 1) - 1(x + 1) = (x^2 - 1)(x + 1)$

3. $15xy + 5x + 6y + 2 = 5x(3y + 1) + 2(3y + 1)$

$$= (5x + 2)(3y + 1)$$

4. $2x^4 - 6x - x^3y + 3y = 2x(x^3 - 3) - y(x^3 - 3) = (2x - y)(x^3 - 3)$

5. $9x^3 + 18x^2 - x - 2 = 9x^2(x + 2) - 1(x + 2) = (9x^2 - 1)(x + 2)$

Factoring to Simplify Fractions

Factoring algebraic expressions serves many purposes in mathematics, and one of the more important purposes is in working with fractions. For now, we will simplify fractions by factoring the numerator and denominator and dividing out common factors. Later, we will use what we learned to perform fraction arithmetic.

 EXAMPLES

Factor the numerator and denominator, and then write the fraction in lowest terms.

$$\frac{6x^2 + 2xy}{4x^2y - 10xy}$$

Each term in the numerator and denominator has a factor of 2x.

$$\frac{6x^2 + 2xy}{4x^2y - 10xy} = \frac{2x(3x + y)}{2x(2xy - 5y)} = \frac{3x + y}{2xy - 5y}$$

$$\frac{xy - x}{16x^2 - xy} = \frac{x(y - 1)}{x(16x - y)} = \frac{y - 1}{16x - y}$$

PRACTICE

Factor the numerator and denominator, and then write the fraction in lowest terms.

1. $\dfrac{18x - 24y}{6} =$

2. $\dfrac{8xy - 9x^2}{2x} =$

3. $\dfrac{14x^2y^2 + 21xy}{3x^2} =$

4. $\dfrac{28x - 14y}{7x} =$

5. $\dfrac{16x^3y^2 + 4xy}{12xy^2 - 8x^2y} =$

6. $\dfrac{15xyz^2 + 5x^2z}{30x^2y + 25x} =$

7. $\dfrac{24xyz^4 + 6x^2yz^3 - 18xz^2}{54xy^3z^3 + 48x^3y^2z^5} =$

SOLUTIONS

1. $\dfrac{18x - 24y}{6} = \dfrac{6(3x - 4y)}{6} = 3x - 4y$

2. $\dfrac{8xy - 9x^2}{2x} = \dfrac{x(8y - 9x)}{2x} = \dfrac{8y - 9x}{2}$

3. $\dfrac{14x^2y^2 + 21xy}{3x^2} = \dfrac{x(14xy^2 + 21y)}{3x^2} = \dfrac{14xy^2 + 21y}{3x}$

4. $\dfrac{28x - 14y}{7x} = \dfrac{7(4x - 2y)}{7x} = \dfrac{4x - 2y}{x}$

5. $\dfrac{16x^3y^2 + 4xy}{12xy^2 - 8x^2y} = \dfrac{4xy(4x^2y + 1)}{4xy(3y - 2x)} = \dfrac{4x^2y + 1}{3y - 2x}$

6. $\dfrac{15xyz^2 + 5x^2z}{30x^2y + 25x} = \dfrac{5x(3yz^2 + xz)}{5x(6xy + 5)} = \dfrac{3yz^2 + xz}{6xy + 5}$

7. $\dfrac{24xyz^4 + 6x^2yz^3 - 18xz^2}{54xy^3z^3 + 48x^3y^2z^5} = \dfrac{6xz^2(4yz^2 + xyz - 3)}{6xz^2(9y^3z + 8x^2y^2z^3)} = \dfrac{4yz^2 + xyz - 3}{9y^3z + 8x^2y^2z^3}$

Simplifying a fraction or adding two fractions sometimes only requires that we factor -1 from one or more denominators. For instance, in $\dfrac{y-x}{x-y}$ the numerator and denominator are only off by a factor of -1. To simplify this fraction, we factor -1 from the numerator or denominator:

$$\frac{y-x}{x-y}=\frac{-(-y+x)}{x-y}=\frac{-(x-y)}{x-y}=\frac{-1}{1}=-1 \text{ or } \frac{y-x}{x-y}=\frac{y-x}{-(-x+y)}=\frac{y-x}{-(y-x)}=\frac{1}{-1}=-1.$$

In the sum $\dfrac{3}{y-x}+\dfrac{x}{x-y}$ the denominators are off by a factor of -1. Factor -1 from one of the denominators and use the fact that $\dfrac{a}{-b}=\dfrac{-a}{b}$ to write both terms with the same denominator.

$$\frac{3}{y-x}+\frac{x}{x-y}=\frac{3}{-(-y+x)}+\frac{x}{x-y}=\frac{3}{-(x-y)}+\frac{x}{x-y}=\frac{-3}{x-y}+\frac{x}{x-y}=\frac{-3+x}{x-y}$$

EXAMPLES

Factor -1 from the denominator and move the negative sign to the numerator.

$$\frac{1}{1-x}=\frac{1}{-(-1+x)}=\frac{1}{-(x-1)}=\frac{-1}{x-1}$$

$$\frac{3}{-x-6}=\frac{3}{-(x+6)}=\frac{-3}{x+6}$$

$$\frac{-3x}{14+9x}=\frac{-3x}{-(-14-9x)}=\frac{-(-3x)}{-14-9x}=\frac{3x}{-14-9x}$$

$$\frac{16x-5}{7x-3}=\frac{16x-5}{-(-7x+3)}=\frac{16x-5}{-(3-7x)}=\frac{-(16x-5)}{3-7x}=\frac{-16x+5}{3-7x}$$

PRACTICE

Factor -1 from the denominator and move the negative sign to the numerator.

1. $\dfrac{1}{y-x}=$

2. $\dfrac{16}{4-x}=$

3. $\dfrac{-10x^2}{7-3x} =$

4. $\dfrac{9+8y}{-6-x} =$

5. $\dfrac{8xy-5}{5-8xy} =$

6. $\dfrac{5xy-4+3x}{9x-16} =$

✔ SOLUTIONS

1. $\dfrac{1}{y-x} = \dfrac{1}{-(-y+x)} = \dfrac{1}{-(x-y)} = \dfrac{-1}{x-y}$

2. $\dfrac{16}{4-x} = \dfrac{16}{-(-4+x)} = \dfrac{16}{-(x-4)} = \dfrac{-16}{x-4}$

3. $\dfrac{-10x^2}{7-3x} = \dfrac{-10x^2}{-(-7+3x)} = \dfrac{-10x^2}{-(3x-7)} = \dfrac{-(-10x^2)}{3x-7} = \dfrac{10x^2}{3x-7}$

4. $\dfrac{9+8y}{-6-x} = \dfrac{9+8y}{-(6+x)} = \dfrac{-(9+8y)}{6+x} = \dfrac{-9-8y}{6+x}$

5. $\dfrac{8xy-5}{5-8xy} = \dfrac{8xy-5}{-(-5+8xy)} = \dfrac{8xy-5}{-(8xy-5)} = \dfrac{-(8xy-5)}{8xy-5} = \dfrac{-1}{1} = -1$

6. $\dfrac{5xy-4+3x}{9x-16} = \dfrac{5xy-4+3x}{-(-9x+16)} = \dfrac{5xy-4+3x}{-(16-9x)} = \dfrac{-(5xy-4+3x)}{16-9x}$

$\qquad = \dfrac{-5xy+4-3x}{16-9x}$

The FOIL Method

The FOIL method helps us to use the distributive property to expand expressions such as $(x + 4)(2x - 1)$. The letters in "FOIL" describe the sums and products in these expansions.

$$F(\text{first} \times \text{first}) + O(\text{outer} \times \text{outer}) + I(\text{inner} \times \text{inner}) + L(\text{last} \times \text{last})$$

$$\overset{F}{}\quad\overset{F}{}$$
$$F \quad \text{First} \times \text{First}\,(x + 4)(2x - 1) : x(2x) = 2x^2$$

$$\overset{O}{}\quad\overset{O}{}$$
$$+O \quad \text{Outer} \times \text{Outer}\,(x + 4)(2x - 1) : x(-1) = -x$$

$$\overset{I}{}\ \overset{I}{}$$
$$+I \quad \text{Inner} \times \text{Inner}(x + 4)(2x - 1) : 4(2x) = 8x$$

$$\overset{L}{}\quad\overset{L}{}$$
$$+L \quad \text{Last} \times \text{Last}(x + 4)(2x - 1) : 4(-1) = -4$$

$$(x + 4)(2x - 1) = 2x^2 - x + 8x - 4 = 2x^2 + 7x - 4$$

EXAMPLES

Use the FOIL method to expand the product.

$$(x + 16)(x - 4) = \overbrace{x \cdot x}^{F} + \overbrace{x(-4)}^{O} + \overbrace{16 \cdot x}^{I} + \overbrace{16(-4)}^{L} = x^2 - 4x + 16x - 64$$
$$= x^2 + 12x - 64$$

$$(2x + 3)(7x - 6) = 2x(7x) + 2x(-6) + 3(7x) + 3(-6) = 14x^2 - 12x + 21x - 18$$
$$= 14x^2 + 9x - 18$$

$$(2x + 1)^2 = (2x + 1)(2x + 1) = 2x(2x) + 2x(1) + 1(2x) + 1(1)$$
$$= 4x^2 + 2x + 2x + 1 = 4x^2 + 4x + 1$$

$$(x - 7)(x + 7) = x \cdot x + 7x + (-7)x + (-7)(7) = x^2 + 7x - 7x - 49 = x^2 - 49$$

PRACTICE

Use the FOIL method to expand the product.

1. $(5x - 1)(2x + 3) =$

2. $(4x + 2)(x - 6) =$

3. $(2x + 1)(9x + 4) =$

4. $(12x - 1)(2x - 5) =$

5. $(x^2 + 2)(x - 1) =$

6. $(x^2 - y)(x + 2y) =$

7. $\left(\sqrt{x} - 3\right)\left(\sqrt{x} + 4\right) =$

8. $(x - 5)(x + 5) =$

9. $(x - 6)(x + 6) =$

10. $\left(\sqrt{x} + 2\right)\left(\sqrt{x} - 2\right) =$

11. $(x + 8)^2 =$

12. $(x - y)^2 =$

13. $(2x + 3y)^2 =$

14. $\left(\sqrt{x} + \sqrt{y}\right)^2 =$

15. $\left(\sqrt{x} + \sqrt{y}\right)\left(\sqrt{x} - \sqrt{y}\right) =$

SOLUTIONS

1. $(5x - 1)(2x + 3) = 5x(2x) + 5x(3) + (-1)(2x) + (-1)(3)$

$$= 10x^2 + 15x - 2x - 3 = 10x^2 + 13x - 3$$

2. $(4x + 2)(x - 6) = 4x(x) + 4x(-6) + 2x + 2(-6)$

$$= 4x^2 - 24x + 2x - 12 = 4x^2 - 22x - 12$$

3. $(2x + 1)(9x + 4) = 2x(9x) + 2x(4) + 1(9x) + 1(4)$

$$= 18x^2 + 8x + 9x + 4 = 18x^2 + 17x + 4$$

4. $(12x - 1)(2x - 5) = 12x(2x) + 12x(-5) + (-1)(2x) + (-1)(-5)$

$$= 24x^2 - 60x - 2x + 5 = 24x^2 - 62x + 5$$

5. $(x^2 + 2)(x - 1) = x^2(x) + x^2(-1) + 2x + 2(-1) = x^3 - x^2 + 2x - 2$

6. $(x^2 - y)(x + 2y) = x^2(x) + x^2(2y) + (-y)x + (-y)(2y)$

$$= x^3 + 2x^2y - xy - 2y^2$$

7. $\left(\sqrt{x} - 3\right)\left(\sqrt{x} + 4\right) = \sqrt{x} \cdot \sqrt{x} + 4\sqrt{x} + (-3)\sqrt{x} + (-3)(4)$

$$= \left(\sqrt{x}\right)^2 + 1\sqrt{x} - 12 = x + \sqrt{x} - 12$$

8. $(x - 5)(x + 5) = x(x) + 5x + (-5)x + (-5)(5)$

$$= x^2 + 5x - 5x - 25 = x^2 - 25$$

9. $(x - 6)(x + 6) = x(x) + 6x + (-6)x + (-6)(6)$

$$= x^2 + 6x - 6x - 36 = x^2 - 36$$

10. $\left(\sqrt{x} + 2\right)\left(\sqrt{x} - 2\right) = \left(\sqrt{x}\right)\left(\sqrt{x}\right) + (-2)\sqrt{x} + 2\sqrt{x} + 2(-2)$

$$= \left(\sqrt{x}\right)^2 - 2\sqrt{x} + 2\sqrt{x} - 4 = x - 4$$

11. $(x + 8)^2 = (x + 8)(x + 8) = x(x) + 8x + 8x + 8(8)$

$= x^2 + 16x + 64$

12. $(x - y)^2 = (x - y)(x - y) = x(x) + x(-y) + x(-y) + (-y)(-y)$

$= x^2 - xy - xy + y^2 = x^2 - 2xy + y^2$

13. $(2x + 3y)^2 = (2x + 3y)(2x + 3y)$

$= 2x(2x) + 2x(3y) + 3y(2x) + (3y)(3y)$

$= 4x^2 + 6xy + 6xy + 9y^2 = 4x^2 + 12xy + 9y^2$

14. $\left(\sqrt{x} + \sqrt{y}\right)^2 = \left(\sqrt{x} + \sqrt{y}\right)\left(\sqrt{x} + \sqrt{y}\right)$

$= \sqrt{x}\left(\sqrt{x}\right) + \sqrt{x}\left(\sqrt{y}\right) + \sqrt{x}\left(\sqrt{y}\right) + \sqrt{y}\left(\sqrt{y}\right)$

$= \left(\sqrt{x}\right)^2 + 2\sqrt{x}\sqrt{y} + \left(\sqrt{y}\right)^2$

$= x + 2\sqrt{xy} + y$

15. $\left(\sqrt{x} + \sqrt{y}\right)\left(\sqrt{x} - \sqrt{y}\right) = \sqrt{x}\left(\sqrt{x}\right) + \sqrt{x}\left(-\sqrt{y}\right) + \sqrt{x}\sqrt{y} + \sqrt{y}\left(\sqrt{y}\right)$

$= \left(\sqrt{x}\right)^2 + \sqrt{x}\sqrt{y} - \sqrt{x}\sqrt{y} + \left(\sqrt{y}\right)^2 = x - y$

Factoring Quadratic Polynomials

We now work in the opposite direction with the distributive property—factoring. First we will factor *quadratic polynomials*, expressions of the form $ax^2 + bx + c$ (where a is not 0). For example, $x^2 + 5x + 6$ is factored as $(x + 2)$ $(x + 3)$. Quadratic polynomials whose first factors are x^2 are the easiest to factor. Their factorization always begins as $(x \pm \underline{\quad})(x \pm \underline{\quad})$. This forces the first term to be x^2 when the FOIL method is used. All we need to do is to fill in the two blanks and to decide when to use plus and minus signs. All quadratic polynomials factor, though some do not factor "nicely." We will only concern ourselves with "nicely" factorable polynomials in this chapter.

If the second sign is minus, then the signs in the factors will be different (one plus and one minus). If the second sign is plus then both of the signs will be the same. If the first sign in the quadratic polynomial is a plus sign, both signs in the factors will be plus. If the first sign is a minus sign (and the second is a plus sign), both signs in the factors will be minus.

EXAMPLES

Determine whether to begin the factoring as $(x + \_\_)(x + \_\_)$, $(x - \_\_)(x - \_\_)$, or $(x - \_\_)(x + \_\_)$.

$$x^2 - 4x - 5 = (x - \_\_)(x + \_\_) \text{ or } (x + \_\_)(x - \_\_)$$
$$x^2 + x - 12 = (x + \_\_)(x - \_\_) \text{ or } (x - \_\_)(x + \_\_)$$
$$x^2 - 6x + 8 = (x - \_\_)(x - \_\_)$$
$$x^2 + 4x + 3 = (x + \_\_)(x + \_\_)$$

PRACTICE

Determine whether to begin the factoring as $(x + \_\_)(x + \_\_)$, $(x - \_\_)(x - \_\_)$, or $(x - \_\_)(x + \_\_)$.

1. $x^2 - 5x - 6 =$

2. $x^2 + 2x + 1 =$

3. $x^2 + 3x - 10 =$

4. $x^2 - 6x + 8 =$

5. $x^2 - 11x - 12 =$

6. $x^2 - 9x + 14 =$

7. $x^2 + 7x + 10 =$

8. $x^2 + 4x - 21 =$

SOLUTIONS

1. $x^2 - 5x - 6 = (x - \_\_)(x + \_\_)$

2. $x^2 + 2x + 1 = (x + \_\_)(x + \_\_)$

3. $x^2 + 3x - 10 = (x - \_\_)(x + \_\_)$

4. $x^2 - 6x + 8 = (x - \_\_)(x - \_\_)$

5. $x^2 - 11x - 12 = (x - \_\_)(x + \_\_)$

6. $x^2 - 9x + 14 = (x - \_\_)(x - \_\_)$

7. $x^2 + 7x + 10 = (x + \_\_)(x + \_\_)$

8. $x^2 + 4x - 21 = (x - \_\_)(x + \_\_)$

Once the signs are determined all that remains is to fill in the two blanks. We look at all of the pairs of factors of the constant term. These pairs will be the candidates for the blanks. For example, if the constant term is 12, we need to consider 1 and 12, 2 and 6, and 3 and 4. If both signs in the factors are the same, these are the only ones we need to try. If the signs are different, we need to reverse the order: 1 and 12, as well as 12 and 1; 2 and 6, as well as 6 and 2; 3 and 4, as well as 4 and 3. We try the FOIL method on these pairs. (Not every quadratic polynomial can be factored in this way.)

EXAMPLE

Write the possible factors, and then check to see which of them are the correct factors.

$x^2 + x - 12$ Factors to check: $(x + 1)(x - 12)$, $(x - 1)(x + 12)$, $(x + 2)(x - 6)$,

$(x - 2)(x + 6)$, $(x - 4)(x + 3)$ and $(x + 4)(x - 3)$

$(x + 1)(x - 12) = x^2 - 11x - 12$

$(x - 1)(x + 12) = x^2 + 11x - 12$

$(x + 2)(x - 6) = x^2 - 4x - 12$

$(x - 2)(x + 6) = x^2 + 4x - 12$

$(x - 4)(x + 3) = x^2 - x - 12$

$(x + 4)(x - 3) = x^2 + x - 12$ **(This works.)**

EXAMPLES

$x^2 - 2x - 15$ Factors to check: $(x + 15)(x - 1)$, $(x - 15)(x + 1)$, $(x + 5)(x - 3)$
and $(x - 5)(x + 3)$ **(This works)**

$x^2 - 11x + 18$ Factors to check: $(x - 1)(x - 18)$, $(x - 3)(x - 6)$ and $(x - 2)(x - 9)$
(This works)

$x^2 + 8x + 7$ Factors to check: $(x + 1)(x + 7)$ **(This works)**

PRACTICE

Factor the expression.

1. $x^2 - 5x - 6 =$

2. $x^2 + 2x + 1 =$

3. $x^2 + 3x - 10 =$

4. $x^2 - 6x + 8 =$

5. $x^2 - 11x - 12 =$

6. $x^2 - 9x + 14 =$

7. $x^2 + 7x + 10 =$

8. $x^2 + 4x - 21 =$

9. $x^2 + 13x + 36 =$

10. $x^2 + 5x - 24 =$

 SOLUTIONS

1. $x^2 - 5x - 6 = (x - 6)(x + 1)$

2. $x^2 + 2x + 1 = (x + 1)(x + 1) = (x + 1)^2$

3. $x^2 + 3x - 10 = (x + 5)(x - 2)$

4. $x^2 - 6x + 8 = (x - 4)(x - 2)$

5. $x^2 - 11x - 12 = (x - 12)(x + 1)$

6. $x^2 - 9x + 14 = (x - 7)(x - 2)$

7. $x^2 + 7x + 10 = (x + 5)(x + 2)$

8. $x^2 + 4x - 21 = (x + 7)(x - 3)$

9. $x^2 + 13x + 36 = (x + 4)(x + 9)$

10. $x^2 + 5x - 24 = (x + 8)(x - 3)$

We can use a factoring shortcut when the first term is x^2. If the second sign is plus, choose the factors whose *sum* is the coefficient of the second term. For example, in the expression $x^2 - 7x + 6$ we need the factors of 6 to sum to 7: $x^2 - 7x + 6 = (x - 1)(x - 6)$. The factors of 6 we need for $x^2 + 5x + 6$ need to sum to 5: $x^2 + 5x + 6 = (x + 2)(x + 3)$.

If the second sign is minus, the *difference* of the factors needs to be the coefficient of the middle term. If the first sign is plus, the bigger factor has the plus sign. If the first sign is minus, the bigger factor has the minus sign.

 EXAMPLES

Factor the quadratic polynomial.

$x^2 + 3x - 10$: The factors of 10 whose difference is 3 are 2 and 5. The first sign is plus, so the plus sign goes with 5, the bigger factor: $x^2 + 3x - 10 = (x + 5)(x - 2)$.

$x^2 - 5x - 14$: The factors of 14 whose difference is 5 are 2 and 7. The first sign is minus, so the minus sign goes with 7, the bigger factor: $x^2 - 5x - 14 = (x - 7)(x + 2)$.

$x^2 + 11x + 24$: $3 \cdot 8 = 24$ and $3 + 8 = 11$ $x^2 + 11x + 24 = (x + 3)(x + 8)$

$x^2 - 9x + 18$: $3 \cdot 6 = 18$ and $3 + 6 = 9$ $x^2 - 9x + 18 = (x - 3)(x - 6)$

$x^2 + 9x - 36$: $3 \cdot 12 = 36$ and $12 - 3 = 9$ $x^2 + 9x - 36 = (x + 12)(x - 3)$

$x^2 - 2x - 8$: $2 \cdot 4 = 8$ and $4 - 2 = 2$ $x^2 - 2x - 8 = (x + 2)(x - 4)$

 PRACTICE

Factor the quadratic polynomial.

1. $x^2 - 6x + 9 =$

2. $x^2 - x - 12 =$

3. $x^2 + 9x - 22 =$

4. $x^2 + x - 20 =$

5. $x^2 + 13x + 36 =$

6. $x^2 - 19x + 34 =$

7. $x^2 - 18x + 17 =$

8. $x^2 + 24x - 25 =$

9. $x^2 - 14x + 48 =$

10. $x^2 + 16x + 64 =$

11. $x^2 - 49 =$

(*Hint:* $x^2 - 49 = x^2 + 0x - 49$)

 SOLUTIONS

1. $x^2 - 6x + 9 = (x - 3)(x - 3) = (x - 3)^2$

2. $x^2 - x - 12 = (x - 4)(x + 3)$

3. $x^2 + 9x - 22 = (x + 11)(x - 2)$

4. $x^2 + x - 20 = (x + 5)(x - 4)$

5. $x^2 + 13x + 36 = (x + 4)(x + 9)$

6. $x^2 - 19x + 34 = (x - 2)(x - 17)$

7. $x^2 - 18x + 17 = (x - 1)(x - 17)$

8. $x^2 + 24x - 25 = (x + 25)(x - 1)$

9. $x^2 - 14x + 48 = (x - 6)(x - 8)$

10. $x^2 + 16x + 64 = (x + 8)(x + 8) = (x + 8)^2$

11. $x^2 - 49 = (x - 7)(x + 7)$

The shortcut for factoring a quadratic polynomial where first term is x^2 can help you identify quadratic polynomials that do not factor "nicely." The next three examples are quadratic polynomials that do not factor "nicely."

$$x^2 + x + 1$$

$$x^2 + 14x + 19$$

$$x^2 - 5x + 10$$

Factoring the Difference of Two Squares

A quadratic polynomial of the form $x^2 - c^2$ is called the *difference of two squares*. We can use the shortcut described above on $x^2 - c^2 = x^2 + 0x - c^2$. The factors of c^2 must have a difference of 0. This can only happen if they are the same, so the factors of c^2 we want are c and c, as shown in $x^2 - c^2 = (x - c)(x + c)$.

 EXAMPLES

Use the formula to factor the difference of two squares.

$$x^2 - 9 = (x - 3)(x + 3)$$

$$x^2 - 100 = (x - 10)(x + 10)$$

$$x^2 - 49 = (x - 7)(x + 7)$$

$$16 - x^2 = (4 - x)(4 + x)$$

When the sign between x^2 and c^2 is plus, the quadratic cannot be factored using real numbers. For example, the quadratic polynomial $x^2 + 25$ cannot be factored using real numbers.

 PRACTICE

Use the formula to factor the difference of two squares.

1. $x^2 - 4 =$

2. $x^2 - 81 =$

3. $x^2 - 25 =$

4. $x^2 - 64 =$

5. $x^2 - 1 =$

6. $x^2 - 15 =$

7. $25 - x^2 =$

SOLUTIONS

1. $x^2 - 4 = (x - 2)(x + 2)$
2. $x^2 - 81 = (x - 9)(x + 9)$
3. $x^2 - 25 = (x - 5)(x + 5)$
4. $x^2 - 64 = (x - 8)(x + 8)$
5. $x^2 - 1 = (x - 1)(x + 1)$
6. $x^2 - 15 = \left(x - \sqrt{15}\right)\left(x + \sqrt{15}\right)$
7. $25 - x^2 = (5 - x)(5 + x)$

The difference of two squares can come in the form $x^{2n} - c^{2n}$, where n is any nonzero number. The factorization is $x^{2n} - c^{2n} = (x^n - c^n)(x^n + c^n)$. (When n is odd, we can factor $x^n - c^n$ as well, but this factorization will not be covered here.) In the following problems, we make use of the fact that $1^n = 1$ for any exponent n.

EXAMPLES

Use the formula to factor the difference of two squares.

$$x^6 - 1$$

The powers are $2 \cdot 3 = 6$. Thus,

$$x^6 - 1 = x^6 - 1^6 = x^{2(3)} - 1^{2(3)} = (x^3 - 1^3)(x^3 + 1^3) = (x^3 - 1)(x^3 - 1).$$

$$x^{10} - 1 = x^{10} - 1^{10} = (x^5 - 1)(x^5 + 1)$$

$$x^6 - \frac{1}{64} = (x^3)^2 - \left(\frac{1}{8}\right)^2 = \left(x^3 - \frac{1}{8}\right)\left(x^3 + \frac{1}{8}\right)$$

Sometimes, we need to use the formula twice.

$$16 - x^4 = 2^4 - x^4 = (2^2 - x^2)(2^2 + x^2) = (4 - x^2)(4 + x^2)$$

Notice that the factor $4 - x^2$ is *also* the difference of two squares, so it, too, can be factored, as in $(4 - x^2)(4 + x^2) = (2 - x)(2 + x)(4 + x^2)$.

$$16x^4 - 1 = (2x)^4 - 1^4 = (4x^2 - 1)(4x^2 + 1) = (2x - 1)(2x + 1)(4x^2 + 1)$$

$$x^8 - 1 = x^8 - 1^8 = (x^4 - 1)(x^4 + 1) = (x^2 - 1)(x^2 + 1)(x^4 + 1)$$

$$= (x - 1)(x + 1)(x^2 + 1)(x^4 + 1)$$

PRACTICE

Use the formula to factor the difference of two squares.

1. $x^4 - 1 =$

2. $x^8 - 16 =$

3. $x^8 - \dfrac{1}{16} =$

4. $256x^4 - 1 =$

5. $x^4 - 81 =$

6. $81x^4 - 1 =$

7. $\dfrac{1}{64}x^6 - 1 =$

8. $16x^4 - 81 =$

9. $\dfrac{16}{81}x^4 - 16 =$

10. $x^{12} - 1 =$

 SOLUTIONS

1. $x^4 - 1 = (x^2 - 1)(x^2 + 1) = (x - 1)(x + 1)(x^2 + 1)$

2. $x^8 - 16 = (x^4 - 4)(x^4 + 4) = (x^2 - 2)(x^2 + 2)(x^4 + 4)$

3. $x^8 - \dfrac{1}{16} = \left(x^4 - \dfrac{1}{4}\right)\left(x^4 + \dfrac{1}{4}\right) = \left(x^2 - \dfrac{1}{2}\right)\left(x^2 + \dfrac{1}{2}\right)\left(x^4 + \dfrac{1}{4}\right)$

4. $256x^4 - 1 = (16x^2 - 1)(16x^2 + 1) = (4x - 1)(4x + 1)(16x^2 + 1)$

5. $x^4 - 81 = (x^2 - 9)(x^2 + 9) = (x - 3)(x + 3)(x^2 + 9)$

6. $81x^4 - 1 = (9x^2 - 1)(9x^2 + 1) = (3x - 1)(3x + 1)(9x^2 + 1)$

7. $\dfrac{1}{64}x^6 - 1 = \left(\dfrac{1}{8}x^3 - 1\right)\left(\dfrac{1}{8}x^3 + 1\right)$

8. $16x^4 - 81 = (4x^2 - 9)(4x^2 + 9) = (2x - 3)(2x + 3)(4x^2 + 9)$

9. $\dfrac{16}{81}x^4 - 16 = \left(\dfrac{4}{9}x^2 - 4\right)\left(\dfrac{4}{9}x^2 + 4\right)$

$= \left(\dfrac{2}{3}x - 2\right)\left(\dfrac{2}{3}x + 2\right)\left(\dfrac{4}{9}x^2 + 4\right)$

10. $x^{12} - 1 = (x^6 - 1)(x^6 + 1) = (x^3 - 1)(x^3 + 1)(x^6 + 1)$

More on Factoring Quadratic Polynomials

When the first term is not x^2, we look to see if we can factor the coefficient of x^2 from each term. If we can, then we are left with a quadratic whose first term is x^2. For example each term in $2x^2 + 16x - 18$ is divisible by 2: $2x^2 + 16x - 18 = 2(x^2 + 8x - 9) = 2(x + 9)(x - 1)$.

 PRACTICE

Factor the quadratic polynomial by first dividing each term by the coefficient of x^2.

1. $4x^2 + 28x + 48 =$

2. $3x^2 - 9x - 54 =$

3. $9x^2 - 9x - 18 =$

4. $15x^2 - 60 =$

5. $6x^2 + 24x + 24 =$

 SOLUTIONS

1. $4x^2 + 28x + 48 = 4(x^2 + 7x + 12) = 4(x + 4)(x + 3)$

2. $3x^2 - 9x - 54 = 3(x^2 - 3x - 18) = 3(x - 6)(x + 3)$

3. $9x^2 - 9x - 18 = 9(x^2 - x - 2) = 9(x - 2)(x + 1)$

4. $15x^2 - 60 = 15(x^2 - 4) = 15(x - 2)(x + 2)$

5. $6x^2 + 24x + 24 = 6(x^2 + 4x + 4) = 6(x + 2)(x + 2) = 6(x + 2)^2$

The coefficient of the x^2 term will not always factor away. In order to factor quadratic polynomials such as $4x^2 + 8x + 3$, we need to try all combinations of factors of 4 and of 3: $(4x + \underline{\ \ })(x + \underline{\ \ })$ and $(2x + \underline{\ \ })(2x + \underline{\ \ })$. The blanks will be filled in with the factors of 3. These are all the possibilities for which the F and L in FOIL will work. We then need to see which one (if any) make O + I in FOIL work.

Of all the possibilities $(4x + 1)(x + 3)$, $(4x + 3)(x + 1)$, and $(2x + 1)(2x + 3)$, the last one is the one that works.

EXAMPLE

Factor the quadratic polynomial.

$$4x^2 - 4x - 15$$

The possibilities that give us $4x^2$ _____ -15 are

(a) $(4x + 15)(x - 1)$

(b) $(4x - 15)(x + 1)$

(c) $(4x - 1)(x + 15)$

(d) $(4x + 1)(x - 15)$

(e) $(4x + 5)(x - 3)$

(f) $(4x - 5)(x + 3)$

(g) $(4x + 3)(x - 5)$

(h) $(4x - 3)(x + 5)$

(i) $(2x + 15)(2x - 1)$

(j) $(2x - 15)(2x + 1)$

(k) $(2x + 5)(2x - 3)$

(l) $(2x - 5)(2x + 3)$

We have chosen these combinations to force the first and last terms of the quadratic to be $4x^2$ and -15, respectively. We only need to check the combination that will give a middle term of $-4x$ (if there is one).

(a) $-4x + 15x = 11x$

(b) $4x - 15x = -11x$

(c) $60x - x = 59x$

(d) $-60x + x = -59x$

(e) $-12x + 5x = -7x$

(f) $12x - 5x = 7x$

(g) $-20x + 3x = -17x$

(h) $20x - 3x = 17x$

(i) $-2x + 30x = 28x$

(j) $2x - 30x = -28x$

(k) $-6x + 10x = 4x$

(l) $6x - 10x = -4x$

Because the sum of the middle terms is $-4x$, Combination (l) is the correct factorization: $4x^2 - 4x - 15 = (2x - 5)(2x + 3)$.

You can see that when the constant term and the coefficient of x^2 have many factors, this list of factorizations to check can grow rather long. Fortunately there is a way around this problem as we shall see in chapter 10.

 PRACTICE

Factor the quadratic polynomial.

1. $6x^2 + 25x - 9 =$

2. $18x^2 + 21x + 5 =$

3. $8x^2 - 35x + 12 =$

4. $25x^2 + 25x - 14 =$

5. $4x^2 - 9 =$

6. $4x^2 + 20x + 25 =$

7. $12x^2 + 32x - 35 =$

 SOLUTIONS

1. $6x^2 + 25x - 9 = (2x + 9)(3x - 1)$

2. $18x^2 + 21x + 5 = (3x + 1)(6x + 5)$

3. $8x^2 - 35x + 12 = (8x - 3)(x - 4)$

4. $25x^2 + 25x - 14 = (5x - 2)(5x + 7)$

5. $4x^2 - 9 = (2x - 3)(2x + 3)$

6. $4x^2 + 20x + 25 = (2x + 5)(2x + 5) = (2x + 5)^2$

7. $12x^2 + 32x - 35 = (6x - 5)(2x + 7)$

Quadratic-Type Expressions

An expression with three terms where the power of the first term is twice that of the second and the third term is a constant is called a *quadratic-type* expression. These expressions factor in the same way as quadratic polynomials. The power on x in the factorization will be the power on x in the middle term. To see the effect of changing the exponents, let us look at the factors of $x^2 - 2x - 3 = (x - 3)(x + 1)$. Each of the expressions below, factors the same way.

$$x^4 - 2x^2 - 3 = (x^2 - 3)(x^2 + 1)$$
$$x^6 - 2x^3 - 3 = (x^3 - 3)(x^3 + 1)$$

$$x^{10} - 2x^5 - 3 = (x^5 - 3)(x^5 + 1)$$

$$x^{-4} - 2x^{-2} - 3 = (x^{-2} - 3)(x^{-2} + 1)$$

$$x^{2/3} - 2x^{1/3} - 3 = (x^{1/3} - 3)(x^{1/3} + 1)$$

$$x^1 - 2x^{1/2} - 3 = (x^{1/2} - 3)(x^{1/2} + 1)$$

For any nonzero power, $x^{2 \cdot \text{Power}} - 2x^{\text{Power}} - 3$ factors as $(x^{\text{Power}} - 3)(x^{\text{Power}} + 1)$. In general, we factor the expression $ax^{2 \cdot \text{Power}} + bx^{\text{Power}} + c$ in the same way that we factor $ax^2 + bx + c$, except that we work with x^{Power} instead of x.

EXAMPLES

Factor the expression.

$$4x^6 + 20x^3 + 21 = (2x^3 + 3)(2x^3 + 7)$$

$$x^{2/3} - 5x^{1/3} + 6 = (x^{1/3} - 2)(x^{1/3} - 3)$$

$$x^4 + x^2 - 2 = (x^2 + 2)(x^2 - 1) = (x^2 + 2)(x - 1)(x + 1)$$

$$x - 2\sqrt{x} - 8 = x^1 - 2x^{1/2} - 8 = (x^{1/2} - 4)(x^{1/2} + 2) = (\sqrt{x} - 4)(\sqrt{x} + 2)$$

$$\sqrt{x} - 2\sqrt[4]{x} - 15 = x^{1/2} - 2x^{1/4} - 15 = (x^{1/4} - 5)(x^{1/4} + 3) = (\sqrt[4]{x} - 5)(\sqrt[4]{x} + 3)$$

PRACTICE

Factor the expression.

1. $x^4 - 3x^2 + 2 =$

2. $x^{10} - 3x^5 + 2 =$

3. $x^{2/5} - 3x^{1/5} + 2 =$

4. $x^{-6} - 3x^{-3} + 2 =$

5. $x^{1/2} - 3x^{1/4} + 2 =$

6. $x^4 + 10x^2 + 9 =$

7. $x^6 - 4x^3 - 21 =$

8. $4x^6 + 4x^3 - 35 =$

9. $10x^{10} + 23x^5 + 6 =$

10. $9x^4 - 6x^2 + 1 =$

11. $x^{2/7} - 3x^{1/7} - 18 =$

12. $6x^{2/3} - 7x^{1/3} - 3 =$

13. $x^{1/3} + 11x^{1/6} + 10 =$

14. $15x^{1/2} - 8x^{1/4} + 1 =$

15. $14x - \sqrt{x} - 3 =$

16. $25x^6 + 20x^3 + 4 =$

17. $x + 6\sqrt{x} + 9 =$

SOLUTIONS

1. $x^4 - 3x^2 + 2 = (x^2 - 2)(x^2 - 1)$

2. $x^{10} - 3x^5 + 2 = (x^5 - 2)(x^5 - 1)$

3. $x^{2/5} - 3x^{1/5} + 2 = (x^{1/5} - 2)(x^{1/5} - 1)$

4. $x^{-6} - 3x^{-3} + 2 = (x^{-3} - 2)(x^{-3} - 1)$

5. $x^{1/2} - 3x^{1/4} + 2 = (x^{1/4} - 2)(x^{1/4} - 1)$

6. $x^4 + 10x^2 + 9 = (x^2 + 9)(x^2 + 1)$

7. $x^6 - 4x^3 - 21 = (x^3 - 7)(x^3 + 3)$

8. $4x^6 + 4x^3 - 35 = (2x^3 + 7)(2x^3 - 5)$

9. $10x^{10} + 23x^5 + 6 = (10x^5 + 3)(x^5 + 2)$

10. $9x^4 - 6x^2 + 1 = (3x^2 - 1)(3x^2 - 1) = (3x^2 - 1)^2$

11. $x^{2/7} - 3x^{1/7} - 18 = (x^{1/7} - 6)(x^{1/7} + 3)$

12. $6x^{2/3} - 7x^{1/3} - 3 = (2x^{1/3} - 3)(3x^{1/3} + 1)$

13. $x^{1/3} + 11x^{1/6} + 10 = (x^{1/6} + 10)(x^{1/6} + 1)$

14. $15x^{1/2} - 8x^{1/4} + 1 = (3x^{1/4} - 1)(5x^{1/4} - 1)$

15. $14x - \sqrt{x} - 3 = 14x^1 - x^{1/2} - 3 = (2x^{1/2} - 1)(7x^{1/2} + 3)$
$$= (2\sqrt{x} - 1)(7\sqrt{x} + 3)$$

16. $25x^6 + 20x^3 + 4 = (5x^3 + 2)(5x^3 + 2) = (5x^3 + 2)^2$

17. $x + 6\sqrt{x} + 9 = x^1 + 6x^{1/2} + 9 = (x^{1/2} + 3)(x^{1/2} + 3)$
$$= (x^{1/2} + 3)^2 = (\sqrt{x} + 3)^2$$

Factoring to Simplify a Larger Family of Fractions

Now that we have more factoring skills, we can use them to simplify a larger family of fractions. To write a fraction in its lowest terms, we factor the numerator and denominator and divide out any like factors.

EXAMPLES

Factor the numerator and denominator and then write the fraction in lowest terms.

$$\frac{x^2-1}{x+1} = \frac{(x+1)(x-1)}{x+1} = \frac{x-1}{1} = x-1$$

$$\frac{y-x}{x^2-y^2} = \frac{y-x}{(x-y)(x+y)} = \frac{-(x-y)}{(x-y)(x+y)} = \frac{-1}{x+y}$$

$$\frac{x^2-5x+6}{x^2-2x-3} = \frac{(x-3)(x-2)}{(x-3)(x+1)} = \frac{x-2}{x+1}$$

$$\frac{3x^3-3x^2-6x}{6x^3-12x^2} = \frac{3x(x^2-x-2)}{6x^2(x-2)} = \frac{3x(x+1)(x-2)}{6x^2(x-2)} = \frac{x+1}{2x}$$

$$\frac{x^2+10x+25}{x^2+6x+5} = \frac{(x+5)(x+5)}{(x+5)(x+1)} = \frac{x+5}{x+1}$$

PRACTICE

Factor the numerator and denominator and then write the fraction in lowest terms. You may leave the numerator and denominator factored.

1. $\dfrac{16x^3-24x^2}{8x^4-12x^3} =$

2. $\dfrac{x^2+2x-8}{x^2+7x+12} =$

3. $\dfrac{x^2-7x+6}{x^2+4x-5} =$

4. $\dfrac{2x^2-5x-12}{2x^2-x-6} =$

5. $\dfrac{3x^2-7x+2}{6x^2+x-1} =$

6. $\dfrac{2x^3+6x^2+4x}{3x^3+3x^2-36x} =$

7. $\dfrac{4x^3y+28x^2y+40xy}{6x^3y^2-6x^2y^2-36xy^2} =$

8. $\dfrac{x-4}{16-x^2} =$

9. $\dfrac{2x^2 - 5x + 2}{1 - 2x} =$

10. $\dfrac{x^2 - y^2}{x^4 - y^4} =$

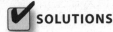 **SOLUTIONS**

1. $\dfrac{16x^3 - 24x^2}{8x^4 - 12x^3} = \dfrac{8x^2(2x - 3)}{4x^3(2x - 3)} = \dfrac{2}{x}$

2. $\dfrac{x^2 + 2x - 8}{x^2 + 7x + 12} = \dfrac{(x + 4)(x - 2)}{(x + 4)(x + 3)} = \dfrac{x - 2}{x + 3}$

3. $\dfrac{x^2 - 7x + 6}{x^2 + 4x - 5} = \dfrac{(x - 1)(x - 6)}{(x - 1)(x + 5)} = \dfrac{x - 6}{x + 5}$

4. $\dfrac{2x^2 - 5x - 12}{2x^2 - x - 6} = \dfrac{(2x + 3)(x - 4)}{(2x + 3)(x - 2)} = \dfrac{x - 4}{x - 2}$

5. $\dfrac{3x^2 - 7x + 2}{6x^2 + x - 1} = \dfrac{(3x - 1)(x - 2)}{(3x - 1)(2x + 1)} = \dfrac{x - 2}{2x + 1}$

6. $\dfrac{2x^3 + 6x^2 + 4x}{3x^3 + 3x^2 - 36x} = \dfrac{2x(x^2 + 3x + 2)}{3x(x^2 + x - 12)}$

$\qquad = \dfrac{2x(x + 1)(x + 2)}{3x(x + 4)(x - 3)} = \dfrac{2(x + 1)(x + 2)}{3(x + 4)(x - 3)}$

7. $\dfrac{4x^3y + 28x^2y + 40xy}{6x^3y^2 - 6x^2y^2 - 36xy^2} = \dfrac{4xy(x^2 + 7x + 10)}{6xy^2(x^2 - x - 6)}$

$\qquad = \dfrac{4xy(x + 2)(x + 5)}{6xy^2(x + 2)(x - 3)} = \dfrac{2(x + 5)}{3y(x - 3)}$

8. $\dfrac{x - 4}{16 - x^2} = \dfrac{x - 4}{(4 - x)(4 + x)} = \dfrac{-(4 - x)}{(4 - x)(4 + x)} = \dfrac{-1}{4 + x}$

9. $\dfrac{2x^2 - 5x + 2}{1 - 2x} = \dfrac{(2x - 1)(x - 2)}{1 - 2x} = \dfrac{(2x - 1)(x - 2)}{-(2x - 1)} = \dfrac{x - 2}{-1}$

$\qquad = -(x - 2)$

10. $\dfrac{x^2 - y^2}{x^4 - y^4} = \dfrac{x^2 - y^2}{(x^2 - y^2)(x^2 + y^2)} = \dfrac{1}{x^2 + y^2}$

Adding/Subtracting Fractions

We now use what we learned about factoring algebraic expressions to add/subtract fractions. Remember that we can only add/subtract fractions that have the same denominator. If the denominators are different, we can factor them so that we can identify the LCD (least common denominator). Once we have the LCD, we rewrite the fractions so that they have the same denominator and then perform the arithmetic on the numerators. For now, we will practice finding the LCD.

 EXAMPLES

Factor each denominator completely and find the LCD.

$$\frac{4}{x^2-3x-4}+\frac{2x}{x^2-1}=\frac{4}{(x-4)(x+1)}+\frac{2x}{(x-1)(x+1)}$$

From the first fraction, we see that the LCD needs $x-4$ and $x+1$ as factors. From the second fraction, we see that the LCD includes $x-1$ and $x+1$, but $x+1$ has been accounted for by the first fraction. The LCD is $(x-4)(x-1)(x+1)$.

$$\frac{7x+5}{2x^2-6x-36}-\frac{10x-1}{x^2+x-6}=\frac{7x+5}{2(x+3)(x-6)}-\frac{10x-1}{(x+3)(x-2)}$$

$$LCD=2(x+3)(x-6)(x-2)$$

$$\frac{x-2}{x^2+6x+5}+\frac{1}{3x^2+18x+15}=\frac{x-2}{(x+5)(x+1)}+\frac{1}{3(x+5)(x+1)}$$

$$LCD=3(x+5)(x+1)$$

$$\frac{1}{x-1}+\frac{3}{1-x}=\frac{1}{x-1}+\frac{3}{-(x-1)}=\frac{1}{x-1}+\frac{-3}{x-1}$$

$$LCD=x-1$$

$$4-\frac{2x+9}{x-5}=\frac{4}{1}-\frac{2x+9}{x-5}$$

$$LCD=x-5$$

$$\frac{3x}{x^2+8x+16}+\frac{2}{x^2+6x+8}=\frac{3x}{(x+4)(x+4)}+\frac{2}{(x+4)(x+2)}$$

$$LCD=(x+4)(x+4)(x+2)=(x+4)^2(x+2)$$

PRACTICE

Factor each denominator completely and find the LCD.

1. $\dfrac{x-10}{x^2+8x+7}+\dfrac{5}{2x^2-2}$

2. $\dfrac{3x+22}{x^2-5x-24}-\dfrac{x+14}{x^2+x-6}$

3. $\dfrac{7}{x-3}+\dfrac{1}{3-x}$

4. $\dfrac{x+1}{6x^2+21x-12}+\dfrac{4}{9x^2+27x+18}$

5. $\dfrac{3}{2x^2+4x-48}+\dfrac{7}{6x-24}-\dfrac{1}{4x^2+20x-24}$

6. $\dfrac{2x-4}{x^2-7x+12}-\dfrac{x}{x^2-6x+9}$

7. $\dfrac{6x-7}{x^2-5}+3$

SOLUTIONS

1. $\dfrac{x-10}{x^2+8x+7}+\dfrac{5}{2x^2-2}=\dfrac{x-10}{(x+7)(x+1)}+\dfrac{5}{2(x-1)(x+1)}$

   $LCD=2(x+7)(x+1)(x-1)$

2. $\dfrac{3x+22}{x^2-5x-24}-\dfrac{x+14}{x^2+x-6}=\dfrac{3x+22}{(x-8)(x+3)}-\dfrac{x+14}{(x+3)(x-2)}$

   $LCD=(x-8)(x+3)(x-2)$

3. $\dfrac{7}{x-3}+\dfrac{1}{3-x}=\dfrac{7}{x-3}+\dfrac{1}{-(x-3)}=\dfrac{7}{x-3}+\dfrac{-1}{x-3}$

   $LCD=x-3$

4. $\dfrac{x+1}{6x^2+21x-12}+\dfrac{4}{9x^2+27x+18}=\dfrac{x+1}{3(2x-1)(x+4)}+\dfrac{4}{9(x+1)(x+2)}$

   $LCD=9(2x-1)(x+4)(x+1)(x+2)$

5. $\dfrac{3}{2x^2+4x-48}+\dfrac{7}{6x-24}-\dfrac{1}{4x^2+20x-24}$

$=\dfrac{2}{2(x-4)(x+6)}+\dfrac{7}{6(x-4)}-\dfrac{1}{4(x+6)(x-1)}$

LCD $=12(x-4)(x+6)(x-1)$

6. $\dfrac{2x-4}{x^2-7x+12}-\dfrac{x}{x^2-6x+9}=\dfrac{2x-4}{(x-3)(x-4)}-\dfrac{x}{(x-3)(x-3)}$

LCD $=(x-3)(x-3)(x-4)=(x-3)^2(x-4)$

7. $\dfrac{6x-7}{x^2-5}+3=\dfrac{6x-7}{x^2-5}+\dfrac{3}{1}$

LCD $=x^2-5$

Once we find the LCD, we rewrite each fraction in terms of the LCD, that is, we multiply each fraction by the "missing" factors over themselves. Once we do this, each fraction has the same denominator, so we can add or subtract the numerators.

EXAMPLES

Find the sum or difference.

$$\frac{1}{x^2+2x-3}+\frac{x}{x^2-9}=\frac{1}{(x+3)(x-1)}+\frac{x}{(x-3)(x+3)}$$

$$\text{LCD}=(x+3)(x-1)(x-3)$$

The factor $x-3$ is "missing" in the first denominator so multiply the first fraction by $\dfrac{x-3}{x-3}$. An $x-1$ is "missing" from the second denominator so multiply the second fraction by $\dfrac{x-1}{x-1}$.

$$\frac{1}{(x+3)(x-1)}\cdot\frac{x-3}{x-3}+\frac{x}{(x-3)(x+3)}\cdot\frac{x-1}{x-1}$$

$$=\frac{x-3}{(x+3)(x-1)(x-3)}+\frac{x(x-1)}{(x+3)(x-1)(x-3)}$$

$$\frac{x-3+x(x-1)}{(x+3)(x-1)(x-3)}=\frac{x-3+x^2-x}{(x+3)(x-1)(x-3)}=\frac{x^2-3}{(x+3)(x-1)(x-3)}$$

$$\frac{6x}{x^2+2x+1}-\frac{2}{x^2+4x+3}=\frac{6x}{(x+1)(x+1)}-\frac{2}{(x+1)(x+3)}$$

$$=\frac{6x}{(x+1)^2}\cdot\frac{x+3}{x+3}-\frac{2}{(x+1)(x+3)}\cdot\frac{x+1}{x+1}=\frac{6x(x+3)-2(x+1)}{(x+1)^2(x+3)}$$

$$=\frac{6x^2+18x-2x-2}{(x+1)^2(x+3)}=\frac{6x^2+16x-2}{(x+1)^2(x+3)}$$

$$6+\frac{1}{2x+5}=\frac{6}{1}+\frac{1}{2x+5}=\frac{6}{1}\cdot\frac{2x+5}{2x+5}+\frac{1}{2x+5}$$

$$=\frac{6(2x+5)+1}{2x+5}=\frac{12x+30+1}{2x+5}=\frac{12x+31}{2x+5}$$

$$\frac{3}{x^2+4x-5}+\frac{x-2}{x+5}=\frac{3}{(x+5)(x-1)}+\frac{x-2}{x+5}=\frac{3}{(x+5)(x-1)}+\frac{x-2}{x+5}\cdot\frac{x-1}{x-1}$$

$$=\frac{3+(x-2)(x-1)}{(x+5)(x-1)}=\frac{3+x^2-3x+2}{(x+5)(x-1)}=\frac{x^2-3x+5}{(x+5)(x-1)}$$

PRACTICE

Rewrite each fraction so that they have the same denominator, and then find the sum or difference.

1. $\dfrac{1}{x^2+5x+6}+\dfrac{1}{x^2+2x-3}=$

2. $\dfrac{5}{2x^2-5x-12}+\dfrac{2x}{3x-12}=$

3. $\dfrac{1}{x^2+x-20}-\dfrac{2}{x^2+x-12}=$

4. $\dfrac{1}{6x^2+24x-30}+\dfrac{5}{2x^2-2x-60}=$

5. $\dfrac{2}{x^2-4x}+\dfrac{1}{2x^2-32}-\dfrac{3}{2x^2+10x+8}=$

6. $1-\dfrac{2x-3}{x+4}=$

7. $\dfrac{1}{x-3}+\dfrac{1}{x^2-2x-3}=$

✓ SOLUTIONS

1. $\dfrac{1}{x^2+5x+6}+\dfrac{1}{x^2+2x-3}=\dfrac{1}{(x+2)(x+3)}+\dfrac{1}{(x-1)(x+3)}$

$$=\dfrac{1}{(x+2)(x+3)}\cdot\dfrac{x-1}{x-1}+\dfrac{1}{(x-1)(x+3)}\cdot\dfrac{x+2}{x+2}$$

$$=\dfrac{1(x-1)+1(x+2)}{(x+2)(x+3)(x-1)}$$

$$=\dfrac{2x+1}{(x+2)(x+3)(x-1)}$$

2. $\dfrac{5}{2x^2-5x-12}+\dfrac{2x}{3x-12}=\dfrac{5}{(2x+3)(x-4)}+\dfrac{2x}{3(x-4)}$

$$=\dfrac{5}{(2x+3)(x-4)}\cdot\dfrac{3}{3}+\dfrac{2x}{3(x-4)}\cdot\dfrac{2x+3}{2x+3}$$

$$=\dfrac{5\cdot3+2x(2x+3)}{3(2x+3)(x-4)}=\dfrac{15+4x^2+6x}{3(2x+3)(x-4)}$$

$$=\dfrac{4x^2+6x+15}{3(2x+3)(x-4)}$$

3. $\dfrac{1}{x^2+x-20}-\dfrac{2}{x^2+x-12}=\dfrac{1}{(x+5)(x-4)}-\dfrac{2}{(x+4)(x-3)}$

$$=\dfrac{1}{(x+5)(x-4)}\cdot\dfrac{(x+4)(x-3)}{(x+4)(x-3)}$$

$$-\dfrac{2}{(x+4)(x-3)}\cdot\dfrac{(x+5)(x-4)}{(x+5)(x-4)}$$

$$=\dfrac{1(x+4)(x-3)-2(x+5)(x-4}{(x+5)(x-4)(x+4)(x-3)}$$

$$=\dfrac{x^2+x-12-2(x^2+x-20)}{(x+5)(x-4)(x+4)(x-3)}$$

$$=\dfrac{x^2+x-12-2x^2-2x+40}{(x+5)(x-4)(x+4)(x-3)}$$

$$=\dfrac{-x^2-x+28}{(x+5)(x-4)(x+4)(x-3)}$$

4. $\dfrac{1}{6x^2+24x-30}+\dfrac{5}{2x^2-2x-60}=\dfrac{1}{6(x-1)(x+5)}+\dfrac{5}{2(x+5)(x-6)}$

$$=\frac{1}{6(x-1)(x+5)}\cdot\frac{x-6}{x-6}+\frac{5}{2(x+5)(x-6)}\cdot\frac{3(x-1)}{3(x-1)}$$

$$=\frac{1(x-6)+5(3)(x-1)}{6(x-1)(x+5)(x-6)}$$

$$=\frac{x-6+15(x-1)}{6(x-1)(x+5)(x-6)}$$

$$=\frac{x-6+15x-15}{6(x-1)(x+5)(x-6)}$$

$$=\frac{16x-21}{6(x-1)(x+5)(x-6)}$$

5. $\dfrac{2}{x^2-4x}+\dfrac{1}{2x^2-32}-\dfrac{3}{2x^2+10x+8}$

$$=\frac{2}{x(x-4)}+\frac{1}{2(x-4)(x+4)}-\frac{3}{2(x+4)(x+1)}$$

$$=\frac{2}{x(x-4)}\cdot\frac{2(x+4)(x+1)}{2(x+4)(x+1)}+\frac{1}{2(x-4)(x+4)}\cdot\frac{x(x+1)}{x(x+1)}$$

$$-\frac{3}{2(x+4)(x+1)}\cdot\frac{x(x-4)}{x(x-4)}$$

$$=\frac{2(2)(x+4)(x+1)+x(x+1)-3x(x-4)}{2x(x-4)(x+4)(x+1)}$$

$$=\frac{4(x^2+5x+4)+x^2+x-3x^2+12x}{2x(x-4)(x+4)(x+1)}$$

$$=\frac{4x^2+20x+16+x^2+x-3x^2+12x}{2x(x-4)(x+4)(x+1)}$$

$$=\frac{2x^2+33x+16}{2x(x-4)(x+4)(x+1)}$$

6. $1 - \dfrac{2x-3}{x+4} = \dfrac{1}{1} - \dfrac{2x-3}{x+4} = \dfrac{1}{1} \cdot \dfrac{x+4}{x+4} - \dfrac{2x-3}{x+4} = \dfrac{x+4-(2x-3)}{x+4}$

$$= \dfrac{x+4-2x+3}{x+4} = \dfrac{-x+7}{x+4}$$

7. $\dfrac{1}{x-3} + \dfrac{1}{x^2-2x-3} = \dfrac{1}{x-3} + \dfrac{1}{(x-3)(x+1)}$

$$= \dfrac{1}{x-3} \cdot \dfrac{x+1}{x+1} + \dfrac{1}{(x-3)(x+1)}$$

$$= \dfrac{x+1+1}{(x-3)(x+1)} = \dfrac{x+2}{(x-3)(x+1)}$$

Summary

In this chapter, we learned how to:

- *Use the distributive property of multiplication over addition to rewrite expressions.* The distributive property is $a(b \pm c) = ab \pm ac$. We generally use this property where one or more of *a*, *b*, and *c* are variables.

- *Combine like terms.* Terms are alike if they have the same variables raised to the same powers. Combine like terms by adding their coefficients.

- *Distribute negative numbers and other negative symbols.* Distributing a negative number, negative sign, or minus sign changes the sign of every term inside the parentheses.

- *Perform basic factoring.* Begin by writing each term as a product of numbers and variables, dividing out the common factor(s) from each term. The common factor goes outside of the parentheses, the terms go inside the parentheses.

- *Factor by grouping.* Some expressions having four terms can be factored with a technique called factoring by grouping. Factor the first two terms and look at the last pair of terms. If the last pair of terms can be factored so that it has a common factor with the first pair of terms, factor out that common factor. You are then left with two terms having a common factor. Finish by factoring this common factor.

- *Simplifying fractions.* As before, simplify a fraction by factoring its numerator and denominator and dividing out any common factor.

- *Add/subtract fractions.* As before, begin by factoring each denominator and identifying the LCD. After rewriting each fraction so that it has the LCD as its denominator, add/subtract their numerators.

- *Perform the FOIL method to expand expressions.* To perform the FOIL method on $(a+b)(c+d)$, multiply the First terms, a and c, the Outer terms, a and d, the Inner terms, b and c, and the Last terms, b and d, and then add these products together as in $(a + b)(c + d) = ac + ad + bc + bd$.

- *Factor quadratic polynomials.* Factor a quadratic polynomial, $ax^2 + bx + c$, in such a way that the FOIL method gives you the original expression. If the first term is x^2, begin with $(x \pm \_\_)(x \pm \_\_)$. Fill in the blanks with the factors of c whose sum is b. If c is positive, then its factors are either both positive or both negative. If c is negative, then its factors have different signs.

- *Factor the difference of two squares.* The difference of two squares can be factored as $a^2 - b^2 = (a - b)(a + b)$.

Factoring and distributing are important, basic skills in algebra and calculus. We will use them throughout the rest of the book when solving equations.

QUIZ

1. $24x^3y^2 + 18xy =$
   A. $6x(x^2y + 3)$
   B. $6xy(4x^2 + 3y)$
   C. $6xy(4x^2y + 3)$
   D. $6x^2y(4y + 3)$

2. $50xy^2 - 15xy + 20x^2y =$
   A. $-5xy(10y - 3 + 4x)$
   B. $5x(10y^2 - 3y + 4x)$
   C. $-5xy(-10y - 3 + 4x)$
   D. $5xy(10y - 3 + 4x)$

3. $x^2 - 2x - 24 =$
   A. $(x + 6)(x - 4)$
   B. $(x - 6)(x + 4)$
   C. $(x - 8)(x + 3)$
   D. $(x + 8)(x - 3)$

4. $3x^2 + 6x - 9 =$
   A. $3(x + 3)(x - 1)$
   B. $3(x - 3)(x + 1)$
   C. $(3x + 3)(x - 3)$
   D. $(3x - 3)(x + 3)$

5. $4y^2 - 16 =$
   A. $4(y - 2)^2$
   B. $2(y - 2)(y + 2)$
   C. $(2y - 4)^2$
   D. $4(y - 2)(y + 2)$

6. $3x^2 + x - 2 =$
   A. $(3x + 2)(x - 1)$
   B. $3(x - 2)(x + 1)$
   C. $(3x - 2)(x + 1)$
   D. $3(x + 2)(x - 1)$

7. $-10x^3 - 6x^2y + 10x =$

   A. $-2x(5x^2 + 3xy - 5)$

   B. $-2x(5x^2 - 3x^2y + 5)$

   C. $-2x(5x^2 - 3xy - 5)$

   D. $-2x(5x^2 + 3x^2y + 5)$

8. $x^2 - \dfrac{1}{16} =$

   A. $\left(x - \dfrac{1}{4}\right)^2$

   B. $\left(x - \dfrac{1}{4}\right)\left(x + \dfrac{1}{4}\right)$

   C. $\dfrac{1}{2}\left(x - \dfrac{1}{2}\right)\left(x + \dfrac{1}{2}\right)$

   D. $\dfrac{1}{2}\left(x - \dfrac{1}{2}\right)^2$

9. $81x^2 - 4 =$

   A. $(9x - 2)(9x + 2)$

   B. $(9x - 2)^2$

   C. $2(3x - 1)^2$

   D. $(3x - 2)(3x + 2)$

10. $\dfrac{3x}{6x^2 - 9y} =$

   A. $\dfrac{1}{2x - 3y}$

   B. $\dfrac{1}{2x - 9y}$

   C. $\dfrac{x}{2x^2 - 3y}$

   D. $\dfrac{-1}{3y - 2x}$

11. $\dfrac{x^4 - 64}{x^2 - 8} =$

    A. $(x-4)(x+4)$

    B. $(x-2)(x+2)$

    C. $x^2 + 8$

    D. $(x+8)^2$

12. $(2x - 5)^2 =$

    A. $4x^2 - 20x + 25$

    B. $4x^2 + 25$

    C. $4x^2 - 25$

    D. $4x^2 + 20x - 25$

13. $\dfrac{x^2 + 3x - 4}{3x^2 + 11x - 4} =$

    A. $\dfrac{x+1}{3x+1}$

    B. $\dfrac{x+1}{3x-1}$

    C. $\dfrac{x-1}{3x-1}$

    D. $\dfrac{x-1}{3x+1}$

14. $\dfrac{5x - 1}{1 - 5x} =$

    A. $5x+1$

    B. $1-5x$

    C. 1

    D. -1

15. $\dfrac{4y - 3x}{6x - 8y} =$

    A. $-\dfrac{1}{4}$

    B. $-\dfrac{1}{2}$

    C. $\dfrac{1}{2}$

    D. $\dfrac{4y-3x}{6x-8y}$ cannot be simplified.

16. $\dfrac{3t^3 - 6t^2 + 3t}{3t} =$

   A. $(t-1)^2$

   B. $t^2 -1$

   C. $\dfrac{(t-1)^2}{t}$

   D. $-5t^2 +3t$

17. $\dfrac{4}{x^2+x-6} + \dfrac{3}{2x^2-3x-2} =$

   A. $\dfrac{13x+11}{(x-3)(x+2)(2x+1)}$

   B. $\dfrac{11x+13}{(x+3)(x-2)(2x+1)}$

   C. $\dfrac{11x+13}{(x-3)(x+2)(2x+1)}$

   D. $\dfrac{13x+11}{(x+3)(x-2)(2x+1)}$

18. $(7x + 1)(3x - 1) =$

   A. $21x^2 -1$

   B. $21x^2 -4x+1$

   C. $21x^2 +4x-1$

   D. $21x^2 -4x-1$

19. $\dfrac{6}{x^2-9x+20} + \dfrac{1}{x^2-25} =$

   A. $\dfrac{7x+24}{(x-5)(x-4)(x+5)}$

   B. $\dfrac{7x+24}{(x-5)(x+4)(x+5)}$

   C. $\dfrac{7x+26}{(x-5)(x-4)(x+5)}$

   D. $\dfrac{7x+26}{(x-5)(x+4)(x+5)}$

20. $-3y(2x^2 + x - 6y) =$

    A. $-6y^2 - 3xy - 18y^2$

    B. $-6y^2 + 3xy - 18y^2$

    C. $-6x^2y - 3xy + 18y^2$

    D. $-6y^2 + 3xy + 18y^2$

21. $x^{2/5} + x^{1/5} - 6 =$

    A. $(x^{1/5} - 6)(x^{1/5} + 1)$

    B. $(x^{1/5} + 3)(x^{1/5} - 2)$

    C. $(x^{1/5} - 1)(x^{1/5} + 6)$

    D. $x^{2/5} + x^{1/5} - 6$ cannot be factored.

22. $x^3 + 5x^2 - x - 5 =$

    A. $(x - 1)(x + 1)(x + 5)$

    B. $(x - 1)^2(x + 5)$

    C. $(x - 1)(x + 5)^2$

    D. $(x + 1)^2(x - 5)$

23. $4x(x - 3)(x + 3) =$

    A. $4x^3 + 36x$

    B. $4x^3 - 24x^2 + 36x$

    C. $4x^3 + 24x^2 + 36x$

    D. $4x^3 - 36x$

24. $\dfrac{4}{x^2 - 1} - \dfrac{2}{x^2 + 2x + 1} =$

    A. $\dfrac{2x + 6}{(x - 1)(x + 1)^2}$

    B. $\dfrac{2}{(x - 1)(x + 1)}$

    C. $\dfrac{2x + 2}{(x - 1)(x + 1)^2}$

    D. $\dfrac{4}{(x - 1)(x + 1)}$

chapter 7

Linear Equations

The algebra we've done up to now involved rewriting expressions using various fraction, exponent, and arithmetic properties. In this chapter, we will use some of the skills we developed to solve *linear equations*. We will use the skills that we develop in this chapter throughout the rest of the book, and you will use them in just about any math course. In fact, you will solve so many of these equations that solving them could become second nature.

CHAPTER OBJECTIVES

In this chapter, you will

- Use the order of operations to compute complex expressions
- Solve linear equations
- Solve equations that lead to linear equations
- Clear an equation of fractions/decimals before solving it
- Solve a formula for a given variable

In a linear equation, the variables are raised to the first power—there are no variables in denominators, no variables to any power (other than one), and no variables under root signs. A linear equation contains an unknown, usually only one but possibly several. What is meant by the phrase, "solve for x" is to isolate x on one side of the equation and to move everything else on the other side. Usually, although not always, the last line is the sentence: "x = (number)," where the number satisfies the original equation. That is, when the number is substituted for x, the equation is true.

In the equation $3x + 7 = 1$; $x = -2$ is the solution because $3(-2) + 7 = 1$ is a true statement. For any other number, the statement would be false. For instance, if we were to say that $x = 4$, the sentence would be $3(4) + 7 = 1$, which is false. Not every equation has a solution. For example, $x + 3 = x + 10$ has no solution. Why not? There is no number that can be added to 3 and be the same quantity as when it is added to 10. If you were to try to solve for x, you would end up with the false statement $3 = 10$.

In order to solve equations and to verify solutions, we must follow the *order of operations*. For example, in the formula:

$$s = \sqrt{\frac{(x-y)^2 + (z-y)^2}{n-1}}$$

what operation is done first? Second? Third? A mnemonic for remembering operation order is "**P**lease **e**xcuse **m**y **d**ear **A**unt **S**ally."

P—parentheses first

E—exponents (and roots) second

M—multiplication third

D—division third (multiplication and division should be done together, working from left to right)

A—addition fourth

S—subtraction fourth (addition and subtraction should be done together, working from left to right)

When working with fractions, think of numerators and denominators as being in parentheses.

EXAMPLES

Compute the following.

$$3^2 - 2(4)$$

This expression has three operations—multiplication, subtraction, and exponentiation. According to PEMDAS, exponentiation is done first, so we square 3 first.

$$3^2 - 2(4) = 9 - 2(4)$$

Multiplication follows exponentiation, so we multiply 2 and 4 next.

$$3^2 - 2(4) = 9 - 2(4) = 9 - 8$$

Subtraction is performed last.

$$3^2 - 2(4) = 9 - 2(4) = 9 - 8 = 1$$

$$2(3+1)^2 = 2(4)^2 = 2(16) = 32$$

$$\frac{5(6-2)}{3(3+1)^2} = \frac{5(4)}{3(4)^2} = \frac{5(4)}{3(16)} = \frac{5}{12}$$

$$\sqrt{\frac{4(10+6)}{10+3(5)}} = \sqrt{\frac{4(16)}{10+15}} = \sqrt{\frac{64}{25}} = \frac{8}{5}$$

In the next example, we will use the following root and exponent properties to simplify some of the expressions.

$$a^{m/n} = \sqrt[n]{a^m} \qquad \sqrt{ab} = \sqrt{a}\sqrt{b} \qquad \sqrt{a^2} = a$$

$$[6(3)^2 - 8 + 2]^{3/2}$$

We begin by simplifying the expression inside the brackets.

$$[6(3)^2 - 8 + 2]^{3/2} = [6(9) - 8 + 2]^{3/2} = (54 - 8 + 2)^{3/2} = 48^{3/2}$$

We now rewrite $48^{3/2}$ as a root of a number raised to a power, and then we will simplify the root using the root properties above.

$$[6(3)^2 - 8 + 2]^{3/2} = [6(9) - 8 + 2]^{3/2} = (54 - 8 + 2)^{3/2} = 48^{3/2}$$
$$= \sqrt{48^3} = \sqrt{(48)^2(48)} = \sqrt{(48)^2} \cdot \sqrt{48} = 48\sqrt{48}$$
$$= 48\sqrt{(4^2)(3)} = 48\sqrt{(16)(3)} = 48(4)\sqrt{3} = 192\sqrt{3}$$

PRACTICE

Compute the following.

1. $\dfrac{4(3^2) - 6(2)}{6 + 2^2}$

2. $\sqrt{\dfrac{8^2 - 3^2 - 5(6)}{2(5) + 6}}$

3. $5^2 - [2(3) + 1]$

4. $\dfrac{8 + 2(3^2 - 4^2)}{15 - 6(3 + 1)}$

5. $\dfrac{3(20 + 4)}{2(9) - 4^2} \cdot \sqrt[3]{\dfrac{5(8) + 41}{12 - 3(16 - 13)}}$

 **SOLUTIONS**

1. $\dfrac{4(3^2) - 6(2)}{6 + 2^2} = \dfrac{4(9) - 6(2)}{6 + 4} = \dfrac{36 - 12}{10} = \dfrac{24}{10} = \dfrac{12}{5}$ or $2\frac{2}{5}$

2. $\sqrt{\dfrac{8^2 - 3^2 - 5(6)}{2(5) + 6}} = \sqrt{\dfrac{64 - 9 - 5(6)}{2(5) + 6}} = \sqrt{\dfrac{64 - 9 - 30}{10 + 6}} = \sqrt{\dfrac{25}{16}} = \dfrac{5}{4}$

3. $5^2 - [2(3) + 1] = 5^2 - (6 + 1) = 5^2 - 7 = 25 - 7 = 18$

4. $\dfrac{8 + 2(3^2 - 4^2)}{15 - 6(3 + 1)} = \dfrac{8 + 2(9 - 16)}{15 - 6(4)} = \dfrac{8 + 2(-7)}{15 - 24} = \dfrac{8 - 14}{-9} = \dfrac{-6}{-9} = \dfrac{2}{3}$

5. $\dfrac{3(20 + 4)}{2(9) - 4^2} \cdot \sqrt[3]{\dfrac{5(8) + 41}{12 - 3(16 - 13)}} = \dfrac{3(20 + 4)}{2(9) - 16} \cdot \sqrt[3]{\dfrac{40 + 41}{12 - 3(3)}}$

$= \dfrac{3(24)}{18 - 16} \sqrt[3]{\dfrac{40 + 41}{12 - 9}} = \dfrac{72}{2} \sqrt[3]{\dfrac{81}{3}}$

$= 36\sqrt[3]{27} = 36(3) = 108$

Solving Linear Equations

To solve equations for the unknown, we use inverse operations to isolate the variable on one side of the equation. These inverse operations "undo" what has been done to the variable. That is, inverse operations are used to move quantities across the equal sign. For instance, in the equation $5x = 10$, x is multiplied by 5, so to move 5 across the equal sign, we need to "unmultiply" the 5. That is, divide both sides of the equation by 5 (equivalently, multiply each side of the equation by $\frac{1}{5}$). In the equation $5 + x = 10$, to move 5 across the equal sign by "unadding" 5. That is, subtract 5 from both sides of the equation (equivalently, add –5 to both sides of the equation).

In short, what is added must be subtracted; what is subtracted must be added; what is multiplied must be divided; and what is divided must be multiplied. There are other operation pairs (an operation and its inverse); some will be discussed later.

In much of this book, when the coefficient of x (the number multiplying x) is an integer, both sides of the equation will be divided by that integer. And when the coefficient is a fraction, both sides of the equation will be multiplied by the reciprocal of that fraction.

 EXAMPLES

Solve the equation.

$$5x = 2$$

$5x = 2$ Divide both sides by 5 or multiply both sides by $\frac{1}{5}$.

$$\frac{5x}{5} = \frac{2}{5}$$

$$x = \frac{2}{5}$$

Solve the equation.

$$\frac{2}{3}x = 6$$

We want to isolate x, so we eliminate the coefficient of x, $\frac{2}{3}$, by dividing each side of the equation by this number. Dividing each side by $\frac{2}{3}$ is the same as multiplying each side by $\frac{3}{2}$.

$$\frac{3}{2} \cdot \frac{2}{3}x = \frac{3}{2} \cdot 6$$

$$x = 9$$

Solve the equation.

$$-3x = 18$$

$$-3x = 18$$

$$\frac{-3}{-3}x = \frac{18}{-3}$$

$$x = -6$$

Solve the equation.

$$-x = 2$$

$$\frac{-x}{-1} = \frac{2}{-1}$$ Remember that $-x$ can be written as $-1x$; that is,
-1 is its own reciprocal.

$$x = -2$$

Solve the equation.

$$-\frac{1}{4}x = 7$$

$$-\frac{1}{4}x = 7$$

$$-4\left(-\frac{1}{4}x\right) = -4(7)$$ Reciprocals have the same sign.

$$x = -28$$

 PRACTICE ⸻⸻⸻⸻⸻⸻⸻⸻⸻⸻⸻⸻

Solve the equation.

1. $4x = 36$

2. $-2x = -26$

3. $\dfrac{3}{4}x = 24$

4. $\dfrac{-1}{3}x = 5$

 SOLUTIONS ⸻⸻⸻⸻⸻⸻⸻⸻⸻⸻⸻⸻

1. $4x = 36$

$$\frac{4}{4}x = \frac{36}{4}$$

$$x = 9$$

2. $-2x = -26$

$$\frac{-2}{-2}x = \frac{-26}{-2}$$

$$x = 13$$

3. $\frac{3}{4}x = 24$

$$\frac{4}{3} \cdot \frac{3}{4}x = \frac{4}{3} \cdot 24$$

$$x = 32$$

4. $\frac{-1}{3}x = 5$

$$(-3)\frac{-1}{3}x = (-3)5$$

$$x = -15$$

A Strategy for Solving Linear Equations

Some equations can be solved in a number of ways. However, the general method in this book will be the same. This method works for most of the linear equations that you will need to solve in an algebra course.

1. Simplify both sides of the equation.
2. Collect all terms with variables in them on one side of the equation and all non-variable terms on the other (this is done by adding/subtracting terms).
3. Factor out the variable.
4. Divide both sides of the equation by the variable's coefficient (this is what has been factored out in step 3).

Of course, you might need only one or two of these steps. In the previous examples and practice problems, only Step 4 was used.

In the following examples, the step number used will be in parentheses. Although it will not normally be done here, it is a good idea to verify the solution. We can verify the solution by substituting it in the original equation to see if it makes the equation a true statement.

EXAMPLES

Solve the equation.

$$2(x - 3) + 7 = 5x - 8$$

$$2(x - 3) + 7 = 5x - 8$$

$$2x - 6 + 7 = 5x - 8 \ (1)$$

$$2x + 1 = 5x - 8 \ (1)$$

$$\underline{-2x \qquad -2x}$$

$$1 = 3x - 8 \ (2)$$

$$\underline{+8 \qquad +8}$$

$$9 = 3x \qquad (2)$$

$$\frac{9}{3} = \frac{3x}{3} \qquad (4)$$

$$3 = x$$

Let us see if $x = 3$ makes the original equation true.

$$2(3 - 3) + 7 \overset{?}{=} 5(3) - 8$$

$$2(0) + 7 \overset{?}{=} 15 - 8$$

$$7 = 7 \qquad \text{This is a true statement, so } x = 3 \text{ is the solution.}$$

Solve the equation.

$$3(2x - 1) - 2 = 10 - (x + 1)$$

$$3(2x - 1) - 2 = 10 - (x + 1)$$

$$6x - 3 - 2 = 10 - x - 1 \ (1)$$

$$6x - 5 = 9 - x \qquad (1)$$

$$\underline{+x \qquad \quad +x} \qquad (2)$$

$$7x - 5 = 9$$

$$\underline{+5 \ +5} \qquad (2)$$

$$7x = 14$$

$$\frac{7}{7}x = \frac{14}{7} \qquad (4)$$

$$x = 2$$

PRACTICE

Solve the equation.

1. $2x + 16 = 10$

2. $\dfrac{3}{2}x - 1 = 5$

3. $6(8 - 2x) + 25 = 5(2 - 3x)$

4. $-4(8 - 3x) = 2x + 8$

5. $7(2x - 3) - 4(x + 5) = 8(x - 1) + 3$

6. $\dfrac{1}{2}(6x - 8) + 3(x + 2) = 4(2x - 1)$

7. $5x + 7 = 6(x - 2) - 4(2x - 3)$

8. $3(2x - 5) - 2(4x + 1) = -5(x + 3) - 2$

9. $-4(3x - 2) - (-x + 6) = -5x + 8$

SOLUTIONS

1. $2x + 16 = 10$

$ \underline{-16 \; -16}$

$ 2x = -6$

$\dfrac{2}{2}x = \dfrac{-6}{2}$

$ x = -3$

2. $\dfrac{3}{2}x - 1 = 5$

$\phantom{\dfrac{3}{2}x} \underline{+1 \; +1}$

$\dfrac{3}{2}x = 6$

$\dfrac{2}{3} \cdot \dfrac{3}{2}x = \dfrac{2}{3} \cdot 6$

$\phantom{\dfrac{2}{3} \cdot} x = 4$

3. $6(8-2x)+25=5(2-3x)$

$48-12x+25=10-15x$

$73-12x=10-15x$

$\underline{-73 \qquad\quad -73}$

$-12x=-63-15x$

$\underline{+15x \qquad\quad +15x}$

$3x=-63$

$\dfrac{3}{3}x=\dfrac{-63}{3}$

$x=-21$

4. $-4(8-3x)=2x+8$

$-32+12x=2x+8$

$\underline{-2x \quad\; -2x}$

$-32+10x=8$

$\underline{+32 \qquad\; +32}$

$10x=40$

$\dfrac{10}{10}x=\dfrac{40}{40}$

$x=4$

5. $7(2x-3)-4(x+5)=8(x-1)+3$

$14x-21-4x-20=8x-8+3$

$10x-41=8x-5$

$\underline{-8x \qquad\quad -8x}$

$2x-41=-5$

$\underline{\qquad\quad +41 \;\; +41}$

$2x=36$

$\dfrac{2}{2}x=\dfrac{36}{2}$

$x=18$

6. $\dfrac{1}{2}(6x - 8) + 3(x + 2) = 4(2x - 1)$

$$3x - 4 + 3x + 6 = 8x - 4$$

$$6x + 2 = 8x - 4$$

$$-6x \qquad -6x$$

$$2 = 2x - 4$$

$$+4 \qquad +4$$

$$6 = 2x$$

$$\dfrac{6}{2} = \dfrac{2}{2}x$$

$$3 = x$$

7. $\quad 5x + 7 = 6(x - 2) - 4(2x - 3)$

$$5x + 7 = 6x - 12 - 8x + 12$$

$$5x + 7 = -2x$$

$$+2x \qquad +2x$$

$$7x + 7 = 0$$

$$-7 = -7$$

$$7x = -7$$

$$\dfrac{7}{7}x = \dfrac{-7}{7}$$

$$x = -1$$

8. $\ 3(2x - 5) - 2(4x + 1) = -5(x + 3) - 2$

$$6x - 15 - 8x - 2 = -5x - 15 - 2$$

$$-2x - 17 = -5x - 17$$

$$+5x \qquad +5x$$

$$3x - 17 = -17$$

$$+17 \quad +17$$

$$3x = 0$$

$$\dfrac{3}{3}x = \dfrac{0}{3}$$

$$x = 0$$

9. $-4(3x - 2) - (-x + 6) = -5x + 8$

$-12x + 8 + x - 6 = -5x + 8$

$-11x + 2 = -5x + 8$

$+5x +5x$

$-6x + 2 = 8$

$-2 -2$

$-6x = 6$

$\dfrac{-6}{-6}x = \dfrac{6}{-6}$

$x = -1$

When we are given an equation having fractions in it, we can "clear the fractions" in the first step. Of course, the solution might be a fraction, but that fraction will not occur until the last step. To clear the fractions, we find the LCD of all fractions and multiply *both* sides of the equation by this number. Once we distribute this quantity on each side of the equation, every denominator will be 1.

EXAMPLES

Solve the equation by first clearing the fraction(s).

$$\frac{4}{5}x + 1 = -4$$

The LCD is 5, so we multiply each side of the equation by 5.

$$\frac{4}{5}x + 1 = -4$$

$$5\left[\frac{4}{5}x + 1\right] = 5(-4)$$

$$5 \cdot \frac{4}{5}x + 5(1) = -20$$

$$4x + 5 = -20$$

$$\underline{-5 \quad -5}$$

$$4x = -25$$

$$\frac{4}{4}x = \frac{-25}{4}$$

$$x = \frac{-25}{4}$$

Solve the equation.

$$\frac{3}{2}x - \frac{1}{6}x + \frac{2}{9} = \frac{1}{3}$$

We begin by identifying the LCD.

$$\frac{3}{2}x - \frac{1}{6}x + \frac{2}{9} = \frac{1}{3} \qquad \textbf{The LCD is 18.}$$

$$18\left[\frac{3}{2}x - \frac{1}{6}x + \frac{2}{9}\right] = 18 \cdot \frac{1}{3}$$

$$18 \cdot \frac{3}{2}x - 18 \cdot \frac{1}{6}x + 18 \cdot \frac{2}{9} = 6$$

$$27x - 3x + 4 = 6$$

$$24x + 4 = 6$$

$$\underline{-4 \quad -4}$$

$$24x = 2$$

$$\frac{24}{24}x = \frac{2}{24}$$

$$x = \frac{2}{24}$$

$$x = \frac{1}{12}$$

Still Struggling

A common mistake is to fail to distribute the LCD. Another is to multiply only one side of the equation by the LCD. In the first example, $\frac{4}{5}x+1=-4$, one common mistake is to multiply both sides by 5 but not to distribute 5 on the left-hand side.

$$5\left(\frac{4}{5}x + 1\right) = -4$$
$$4x + 1 = -4 \quad \text{(incorrect)}$$

Another common mistake is not to multiply both sides of the equation by the LCD.

$$5\left[\frac{4}{5}x + 1\right] = -4$$
$$4x + 5 = -4 \quad \text{(incorrect)}$$

In each case, the last line is *not* equivalent to the first line—that is, the solution to the last equation is not the solution to the first equation.

In some cases, we will need to use the associative property of multiplication with the LCD instead of the distributive property.

 EXAMPLE

Solve the equation by first clearing the fraction(s).

$$\frac{1}{3}(x + 4) = \frac{1}{2}(x - 1)$$

The LCD is 6.

$$6\left[\frac{1}{3}(x + 4)\right] = 6\left[\frac{1}{2}(x - 1)\right]$$

On each side, three quantities are being multiplied together. On the left, the quantities are 6, $\frac{1}{3}$, and $x + 4$. By the associative property of multiplication, the 6 and $\frac{1}{3}$ can be multiplied, then that product is multiplied by $x + 4$. Similarly, on the right, we first multiply 6 and $\frac{1}{2}$ then multiply that product by $x - 1$.

$$\left[6\left(\frac{1}{3}\right)\right](x + 4) = \left[6\left(\frac{1}{2}\right)\right](x - 1)$$

$$2(x + 4) = 3(x - 1)$$

$$2x + 8 = 3x - 3$$

$$\underline{-2x \qquad -2x}$$

$$8 = x - 3$$

$$\underline{+3 \qquad +3}$$

$$11 = x$$

PRACTICE

Solve the equation by first clearing the fraction(s).

1. $\dfrac{1}{2}x + 3 = \dfrac{3}{5}x - 1$

2. $\dfrac{1}{6}x - \dfrac{1}{3} = \dfrac{2}{3}x - \dfrac{5}{12}$

3. $\dfrac{1}{5}(2x - 4) = \dfrac{1}{3}(x + 2)$

4. $\dfrac{2}{3}(x - 1) - \dfrac{1}{6}(2x + 3) = \dfrac{1}{8}$

5. $\dfrac{3}{4}x - \dfrac{1}{3}x + 1 = \dfrac{4}{5}x - \dfrac{3}{20}$

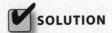 **SOLUTION** _____

1. $\dfrac{1}{2}x + 3 = \dfrac{3}{5}x - 1$

   The LCD is 10.

   $$10\left(\dfrac{1}{2}x + 3\right) = 10\left(\dfrac{3}{5}x - 1\right)$$

   $$10\left(\dfrac{1}{2}x\right) + 10(3) = 10\left(\dfrac{3}{5}x\right) - 10(1)$$

   $$5x + 30 = 6x - 10$$

   $$\underline{-5x \qquad\quad -5x}$$

   $$30 = x - 10$$

   $$\underline{+10 \qquad +10}$$

   $$40 = x$$

2. $\dfrac{1}{6}x - \dfrac{1}{3} = \dfrac{2}{3}x - \dfrac{5}{12}$

   The LCD is 12.

   $$12\left(\dfrac{1}{6}x - \dfrac{1}{3}\right) = 12\left(\dfrac{2}{3}x - \dfrac{5}{12}\right)$$

   $$12\left(\dfrac{1}{6}x\right) - 12\left(\dfrac{1}{3}\right) = 12\left(\dfrac{2}{3}x\right) - 12\left(\dfrac{5}{12}\right)$$

   $$2x - 4 = 8x - 5$$

   $$\underline{-2x \qquad -2x}$$

   $$-4 = 6x - 5$$

   $$\underline{+5 \qquad +5}$$

   $$1 = 6x$$

   $$\dfrac{1}{6} = \dfrac{6}{6}x$$

   $$\dfrac{1}{6} = x$$

3. $\dfrac{1}{5}(2x-4) = \dfrac{1}{3}(x+2)$

The LCD is 15.

$$15\left[\dfrac{1}{5}(2x-4)\right] = 15\left[\dfrac{1}{3}(x+2)\right]$$

$$\left[15\left(\dfrac{1}{5}\right)\right](2x-4) = \left[15\left(\dfrac{1}{3}\right)\right](x+2)$$

$$3(2x-4) = 5(x+2)$$

$$6x-12 = 5x+10$$

$$\underline{-5x \qquad\quad -5x}$$

$$x-12 = 10$$

$$\underline{+12 \quad +12}$$

$$x = 22$$

4. $\dfrac{2}{3}(x-1) - \dfrac{1}{6}(2x+3) = \dfrac{1}{8}$

The LCD Is 24.

$$24\left[\dfrac{2}{3}(x-1) - \dfrac{1}{6}(2x+3)\right] = 24\left(\dfrac{1}{8}\right)$$

$$\left[24\left(\dfrac{2}{3}\right)\right](x-1) - \left[24\left(\dfrac{1}{6}\right)\right](2x+3) = 3$$

$$16(x-1) - 4(2x+3) = 3$$

$$16x-16-8x-12 = 3$$

$$8x-28 = 3$$

$$\underline{+28 \ +28}$$

$$8x = 31$$

$$x = \dfrac{31}{8} \quad \text{or} \quad 3\tfrac{7}{8}$$

5. $\dfrac{3}{4}x - \dfrac{1}{3}x + 1 = \dfrac{4}{5}x - \dfrac{3}{20}$

   The LCD is 60.

$$60\left(\dfrac{3}{4}x - \dfrac{1}{3}x + 1\right) = 60\left(\dfrac{4}{5}x - \dfrac{3}{20}\right)$$

$$60\left(\dfrac{3}{4}x\right) - 60\left(\dfrac{1}{3}x\right) + 60(1) = 60\left(\dfrac{4}{5}x\right) - 60\left(\dfrac{3}{20}\right)$$

$$45x - 20x + 60 = 48x - 9$$

$$25x + 60 = 48x - 9$$

$$\begin{array}{r} -25x \qquad\qquad -25x \\ \hline 60 = 23x - 9 \\ +9 \qquad\qquad +9 \\ \hline 69 = 23x \end{array}$$

$$\dfrac{69}{23} = x$$

$$3 = x$$

Decimals

Because decimal numbers are fractions in disguise, the same trick can be used to "clear the decimal" in equations with decimal numbers. Count the largest number of digits behind each decimal point and multiply both sides of the equation by 10 raised to this power. This will change all the numbers in the equation to whole numbers.

 EXAMPLES

Solve the equation by first clearing the decimal.

$$0.25x + 0.6 = 0.1$$

Because there are two digits behind the decimal in 0.25, we need to multiply both sides of the equation by $10^2 = 100$. Remember to distribute the 100 inside the parentheses.

$$100(0.25x + 0.6) = 100(0.1)$$

$$100(0.25x) + 100(0.6) = 100(0.1)$$

$$25x + 60 = 10$$

$$\begin{array}{r} -60 \quad -60 \\ \hline 25x = -50 \end{array}$$

$$x = \dfrac{-50}{25}$$

$$x = -2$$

Solve the equation by first clearing the decimal.

$$x - 0.11 = 0.2x + 0.09$$

We begin by multiplying each side of the equation by 100 because there are at most two digits behind the decimal points.

$$x - 0.11 = 0.2x + 0.09$$

$$100(x - 0.11) = 100(0.2x + 0.09)$$

$$100x - 100(0.11) = 100(0.2x) + 100(0.09)$$

$$100x - 11 = 20x + 9$$

$$\underline{-20x \qquad -20x}$$

$$80x - 11 = 9$$

$$\underline{+11 \ +11}$$

$$80x = 20$$

$$x = \frac{20}{80}$$

$$x = \frac{1}{4} \ \text{ or } \ 0.25$$ (Normally, decimal solutions are given in equations that have decimals in them.)

PRACTICE

Solve for x after clearing the decimal. If the solution is a fraction, convert the fraction to a decimal.

1. $0.3(x - 2) + 0.1 = 0.4$

2. $0.12 - 0.4(x + 1) + x = 0.5x + 2$

3. $0.015x - 0.01 = 0.025x + 0.2$

4. $0.24(2x - 3) + 0.08 = 0.6(x + 8) - 1$

5. $0.01(2x + 3) - 0.003 = 0.11x$

 **SOLUTIONS** _____

1. $0.3(x - 2) + 0.1 = 0.4$
   Multiply both sides by $10^1 = 10$

   $$10[0.3(x - 2) + 0.1] = 10(0.4)$$

   $$10(0.3)(x - 2) + 10(0.1) = 4$$

   $$[10(0.3)](x - 2) + 1 = 4$$

   $$3(x - 2) + 1 = 4$$

   $$3x - 6 + 1 = 4$$

   $$3x - 5 = 4$$

   $$+5 \ +5$$

   $$3x = 9$$

   $$x = \frac{9}{3}$$

   $$x = 3$$

2. $0.12 - 0.4(x + 1) + x = 0.5x + 2$
   Multiply both sides by $10^2 = 100$.

   $$100[0.12 - 0.4(x + 1) + x] = 100(0.5x + 2)$$

   $$100(0.12) - 100[0.4(x + 1)] + 100x = 100(0.5x) + 100(2)$$

   $$12 - [100(0.4)](x + 1) + 100x = 50x + 200$$

   $$12 - 40(x + 1) + 100x = 50x + 200$$

   $$12 - 40x - 40 + 100x = 50x + 200$$

   $$60x - 28 = 50x + 200$$

   $$-50x -50x$$

   $$10x - 28 = 200$$

   $$+28 \ +28$$

   $$10x = 228$$

   $$x = \frac{228}{10} = 22.8$$

3. $0.015x - 0.01 = 0.025x + 0.2$

Multiply both sides by $10^3 = 1000$.

$$1000(0.015x - 0.01) = 1000(0.025x + 0.2)$$

$$1000(0.015x) - 1000(0.01) = 1000(0.025x) + 1000(0.2)$$

$$15x - 10 = 25x + 200$$

$$-15x \qquad -15x$$

$$-10 = 10x + 200$$

$$-200 \qquad -200$$

$$-210 = 10x$$

$$-\frac{210}{10} = x$$

$$-21 = x$$

4. $0.24(2x - 3) + 0.08 = 0.6(x + 8) - 1$

Multiply both sides by $10^2 = 100$.

$$100[0.24(2x - 3) + 0.08] = 100[0.6(x + 8) - 1]$$

$$100[0.24(2x - 3)] + 100(0.08) = 100[0.6(x + 8)] - 100(1)$$

$$[100(0.24)](2x - 3) + 8 = [100(0.6)](x + 8) - 100$$

$$24(2x - 3) + 8 = 60(x + 8) - 100$$

$$48x - 72 + 8 = 60x + 480 - 100$$

$$48x - 64 = 60x + 380$$

$$-48x \qquad -48x$$

$$-64 = 12x + 380$$

$$-380 \qquad -380$$

$$-444 = 12x$$

$$-\frac{444}{12} = x$$

$$-37 = x$$

5. $0.01(2x + 3) - 0.003 = 0.11x$

Multiply both sides by $10^3 = 1000$.

$$1000[0.01(2x + 3) - 0.003] = 1000(0.11x)$$
$$1000[0.01(2x + 3)] - 1000(0.003) = 110x$$
$$[1000(0.01)](2x + 3) - 3 = 110x$$
$$10(2x + 3) - 3 = 110x$$
$$20x + 30 - 3 = 110x$$
$$20x + 27 = 110x$$
$$\underline{-20x \qquad\qquad -20x}$$
$$27 = 90x$$
$$\frac{27}{90} = x$$
$$\frac{3}{10} = x$$
$$0.3 = x$$

Formulas

At times, math students are given a formula such as $I = Prt$ and asked to solve for one of the variables; that is, to isolate that particular variable on one side of the equation. In $I = Prt$, the equation is solved for I. The method used earlier for solving for x works on these, too. Many people are confused by the presence of multiple variables. The trick is to think of the variable for which you are trying to solve as x and all of the other variables as fixed numbers. For instance, if you were asked to solve for r in $I = Prt$, think of how you would solve something of the same form with numbers, say $100 = (500)(x)(2)$:

$$100 = (500)(x)(2)$$
$$100 = [(500)(2)]x$$
$$\frac{100}{(500)(2)} = x.$$

The steps for solving for r in $I = Prt$ are identical:

$$I = Prt$$

$$I = (Pt)r$$

$$\frac{I}{Pt} = r$$

All of the formulas in the following examples and practice problems are formulas used in business, science, and mathematics.

 EXAMPLES

Solve for q in the formula $P = pq - c$ (P and p are different variables).

$$P = pq - c$$

$$+c \qquad +c$$

$$P + c = pq$$

$$\frac{P+c}{p} = \frac{pq}{p}$$

$$\frac{P+c}{p} = q$$

Solve for C.

$$F = \frac{9}{5}C + 32$$

$$-32 \qquad -32$$

$$F - 32 = \frac{9}{5}C$$

$$\frac{5}{9}(F - 32) = \frac{5}{9} \cdot \frac{9}{5}C$$

$$\frac{5}{9}(F - 32) = C$$

Solve for b.

$$A = \frac{1}{2}(a+b)h \qquad \text{(A and a are different variables.)}$$

$$A = \left(\frac{1}{2}a + \frac{1}{2}b\right)h \qquad \left(\frac{1}{2} \text{ is distributed}\right)$$

$$A = \frac{1}{2}ah + \frac{1}{2}bh \qquad \text{(h is distributed)}$$

$$A = \frac{1}{2}ah + \left(\frac{1}{2}h\right)b$$

$$-\frac{1}{2}ah \quad -\frac{1}{2}ah$$

$$A - \frac{1}{2}ah = \frac{1}{2}hb$$

$$A - \frac{ah}{2} = \frac{h}{2}b \qquad \left(\frac{1}{2}h = \frac{h}{2} \text{ and } \frac{1}{2}ah = \frac{ah}{2}\right)$$

$$\frac{2}{h}\left(A - \frac{ah}{2}\right) = \frac{2}{h} \cdot \frac{h}{2}b$$

$$\frac{2}{h}\left(A - \frac{ah}{2}\right) = b$$

or

$$\frac{2A}{h} - \frac{2}{h} \cdot \frac{ah}{2} = b$$

$$\frac{2A}{h} - a = b$$

or

$$\frac{2A - ah}{h} = b$$

PRACTICE

Solve for indicated variable.

1. $A = \frac{1}{2}bh; h$

2. $C = 2\pi r; r$

3. $V = \dfrac{\pi r^2 h}{3}$; h

4. $P = 2L + 2W$; L

5. $L = L_0[1 + a(dt)]$; a

6. $S = C + RC$; C

7. $A = P + PRT$; R

8. $H = \dfrac{kA(t_1 - t_2)}{L}$; t_1

9. $L = a + (n-1)d$; n

10. $S = \dfrac{rL - a}{r - 1}$; r

 SOLUTIONS

1. $A = \dfrac{1}{2}bh$; h

$A = \left(\dfrac{1}{2}b\right)h$

$A = \dfrac{b}{2}h$

$\dfrac{2}{b}A = \dfrac{2}{b} \cdot \dfrac{b}{2}h$

$\dfrac{2A}{b} = h$

2. $C = 2\pi r$; r

$\dfrac{C}{2\pi} = \dfrac{2\pi}{2\pi}r$

$\dfrac{C}{2\pi} = r$

3. $V = \dfrac{\pi r^2 h}{3}$; h

$\dfrac{3}{\pi r^2}V = \dfrac{3}{\pi r^2} \cdot \dfrac{\pi r^2 h}{3}$

$\dfrac{3V}{\pi r^2} = h$

4. $P = 2L + 2W; L$

$$P = 2L + 2W$$

$$\underline{-2W \qquad -2W}$$

$$P - 2W = 2L$$

$$\frac{P - 2W}{2} = L$$

5. $L = L_0[1 + a(dt)]; a$

$$L = L_0[1 + a(dt)]$$

$$L = L_0(1) + L_0[a(dt)]$$

$$L = L_0 + (L_0 dt)a$$

$$\underline{-L_0 \; -L_0}$$

$$L - L_0 = (L_0 dt)a$$

$$\frac{L - L_0}{L_0 dt} = \frac{L_0 dt}{L_0 dt}a$$

$$\frac{L - L_0}{L_0 dt} = a$$

6. $S = C + RC; C$

$$S = C + RC$$

$$S = C(1 + R) \qquad \text{(Factor out } C \text{ since we are solving for } C.)$$

$$\frac{S}{1 + R} = C\frac{1 + R}{1 + R}$$

$$\frac{S}{1 + R} = C$$

7. $A = P + PRT; R$

$$A = P + PRT \qquad \text{(Do not factor out } P \text{ since we are not solving for } P.)$$

$$\underline{-P \; -P}$$

$$A - P = PRT$$

$$\frac{A - P}{PT} = \frac{PTR}{PT}$$

$$\frac{A - P}{PT} = R$$

8. $$H = \frac{kA(t_1 - t_2)}{L}; t_1$$

$$H = \frac{kA(t_1 - t_2)}{L}$$

$$HL = kA(t_1 - t_2)$$

$$HL = kAt_1 - kAt_2$$

$$+kAt_2 \qquad +kAt_2$$

$$HL + kAt_2 = kAt_1$$

$$\frac{HL + kAt_2}{kA} = \frac{kAt_1}{kA}$$

$$\frac{HL + kAt_2}{kA} = t_1$$

9. $$L = a + (n-1)d; n$$

$$L = a + (n-1)d$$

$$L = a + nd - d$$

$$-a \quad -a$$

$$L - a = nd - d$$

$$+d \qquad +d$$

$$L - a + d = nd$$

$$\frac{L - a + d}{d} = \frac{nd}{d}$$

$$\frac{L - a + d}{d} = n$$

10. $$S = \frac{rL - a}{r - 1}; r$$

$$S = \frac{rL - a}{r - 1}$$

$$(r-1)S = rL - a$$

$$rS - S = rL - a$$

$$-rL \qquad -rL$$

$$rS - rL - S = -a$$

$$+S \quad +S$$

$$rS - rL = S - a$$

$$r(S - L) = S - a$$

$$\frac{r(S - L)}{(S - L)} = \frac{S - a}{S - L}$$

$$r = \frac{S - a}{S - L}$$

Equations Leading to Linear Equations

Some equations are almost linear equations; after one or more steps, these equations become linear equations. In this section, we will convert rational equations (involving fractions containing variables) into linear equations and equations containing a square root into linear equations. After certain operations involving variables, though, we *must* check the solution(s) to the converted equation in the original equation.

To solve a rational equation, we begin by clearing the fraction. In this book, two approaches will be used. First, if the equation is in the form of "fraction = fraction," cross-multiply to eliminate the fraction. Second, if there is more than one fraction on one side of the equal sign, the LCD will be determined and each side of the equation will be multiplied by the LCD. These are not the only methods for solving rational equations, but they are usually the quickest.

The following is a rational equation in the form of one fraction equals another. We use the fact that for b and d nonzero, $\frac{a}{b} = \frac{c}{d}$ if and only if $ad = bc$. This method is called *cross-multiplication*.

EXAMPLE

Solve the equation.

$$\frac{1}{x-1} = \frac{1}{2}$$

We cross-multiply by multiplying the left side of the equation by 2, the denominator on the right side, and multiplying the right side of the equation by $x - 1$, the denominator on the left side.

$$\frac{1}{x-1} = \frac{1}{2}$$

$$2(1) = 1(x-1) \quad \text{(This is the cross-multiplication step.)}$$

$$2 = x - 1$$

$$\underline{+1 \qquad +1}$$

$$3 = x$$

We now check our solution.

$$\frac{1}{3-1} = \frac{1}{2} \text{ is a true statement, so } x = 3 \text{ is the solution.}$$

Anytime we multiply (or divide) both sides of the equation by an expression with a variable in it, we *must* check our solution(s) in the original equation. When we cross-multiply, we are implicitly multiplying both sides of the equations by the denominators of each fraction, so we must check our solution in this case as well. The reason that we must check our solution is that a solution to the converted equation might cause a zero to be in a denominator of the original equation. Such solutions are called *extraneous solutions*. Let us see what happens in the next example.

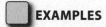 **EXAMPLES**

$$\frac{1}{x-2} = \frac{3}{x+2} - \frac{6x}{(x-2)(x+2)} \qquad \text{The LCD is } (x-2)(x+2).$$

$$(x-2)(x+2)\frac{1}{x-2} = (x-2)(x+2)\left[\frac{3}{x+2} - \frac{6x}{(x-2)(x+2)}\right]$$

Multiply each side by the LCD.

$$x+2 = (x-2)(x+2)\frac{3}{x+2} - (x-2)(x+2)\frac{6x}{(x-2)(x+2)}$$

Distribute the LCD.

$$x+2 = 3(x-2) - 6x$$
$$x+2 = 3x - 6 - 6x$$
$$x+2 = -3x - 6$$
$$\underline{+3x \qquad\quad +3x}$$
$$4x+2 = -6$$
$$\underline{-2 \quad -2}$$
$$4x = -8$$
$$x = -2$$

But $x = -2$ leads to a zero in a denominator of the original equation, so $x = -2$ is not a solution to the original equation. The original equation, then, has no solution.

Still Struggling

Have you ever wondered why expressions such as $\frac{2}{0}$ are not numbers? Let us see what complications arise when we try to see what $\frac{2}{0}$ might mean. Say $\frac{2}{0} = x$.

$$\frac{2}{0} = \frac{x}{1}$$

Now cross-multiply.

$$2(1) = 0(x)$$

Multiplication by zero *always* yields zero, so the right-hand side is zero.

$2 = 0$ No value for x can make this equation true.

We could, instead, try to "clear the fraction" by multiplying both sides of the equation by a common denominator. However, you will see that an absurd situation arises here, too.

$$0 \cdot \frac{2}{0} = 0x$$

So, $0 = 0x$, which is true for *any* x. Actually, the expression $\frac{0}{0}$ is not even defined (what could it mean, anyway?).

On some equations, we need to raise both sides of the equation to a power in order to solve for x. Be careful to raise both sides of the equation to the same power, not simply the side with the root. Raising both sides of an equation to an even power is another operation which can introduce extraneous solutions. To see how this can happen, let's look at the equation $x = 4$. If we square both sides of the equation, we have the equation $x^2 = 16$. This equation has *two* solutions: $x = 4$ and $x = -4$.

 EXAMPLE

Solve the equation.

$$\sqrt{x-1} = 6$$

Remember that $\left(\sqrt{a}\right)^2 = a$ if a is not negative. We will use this fact to pull $x - 1$ from the square root sign. To "undo" a square root, first isolate the

square root on one side of the equation (in this example, it already is) and then square both sides.

$$\left(\sqrt{x-1}\right)^2 = 6^2$$
$$x - 1 = 36$$
$$+1 \quad +1$$
$$x = 37$$

Because we squared both sides of the equation, we need to make sure $x = 37$ is a solution to the original equation.

$$\sqrt{37-1} = 6$$

is a true statement, so $x = 37$ is the solution.

In order for this method of square both sides of an equation to "undo" a square root, the radical *must* be isolated on one side of the equation.

EXAMPLE

Solve the equation.

$$\sqrt{4x+1}+7 = 10$$

Before we square both sides of this equation, we must isolate the root, $\sqrt{4x+1}$, so we subtract 7 from both sides.

$$\sqrt{4x+1}+7 = 10$$
$$\sqrt{4x+1} = 3$$
$$\left(\sqrt{4x+1}\right)^2 = 3^2$$
$$4x + 1 = 9$$
$$4x = 8$$
$$\frac{4x}{4} = \frac{8}{4}$$
$$x = 2$$

Because we squared both sides of the equation, we must check our solution.

$$\sqrt{4(2)+1}+7 = 10?$$
$$\sqrt{9}+7 = 10$$
$$3+7 = 10$$

Because $3 + 7 = 10$, is true, $x = 2$ is the solution.

Quadratic equations, to be studied in Chapter 10, have their variables squared—that is, the only powers on variables are one and two. Some quadratic equations are equivalent to linear equations. Once each side is simplified, the terms containing x^2 "cancel," that is, they add to 0.

EXAMPLE

Solve the equation.

$$(6x - 5)^2 = (4x + 3)(9x - 2)$$

$$(6x - 5)(6x - 5) = (4x + 3)(9x - 2) \quad \text{Begin with the FOIL method.}$$

$$36x^2 - 30x - 30x + 25 = 36x^2 - 8x + 27x - 6$$

$$36x^2 - 60x + 25 = 36x^2 + 19x - 6$$

Because $36x^2$ is on each side of the equation, they cancel each other, and we are left with $-60x + 25 = 19x - 6$, an ordinary linear equation.

$$-60x + 25 = 19x - 6$$
$$+60x \qquad\quad +60x$$
$$25 = 79x - 6$$
$$+6 \qquad\quad +6$$
$$31 = 79x$$
$$\frac{31}{79} = x$$

Because we neither multiplied (nor divided) both sides by an expression involving a variable nor raised both sides to a power, it is not necessary to check the solution. For the sake of accuracy, however, checking solutions is a good habit.

PRACTICE

Solve the equation.

1. $\dfrac{18 - 5x}{3x + 2} = \dfrac{7}{3}$

2. $\dfrac{6}{5x - 2} = \dfrac{9}{7x + 6}$

3. $(x-7)^2 - 4 = (x+1)^2$

4. $\dfrac{6x+7}{4x-1} = \dfrac{3x+8}{2x-4}$

5. $\dfrac{9x}{3x-1} = 4 + \dfrac{3}{3x-1}$

6. $\dfrac{1}{x-1} - \dfrac{2}{x+1} = \dfrac{3}{x^2-1}$

7. $(2x-1)^2 - 4x^2 = -4x+1$

8. $\sqrt{7x+1} = 13$

9. $\sqrt{x} - 6 = 10$

10. $\sqrt{2x-3} + 1 = 6$

11. $\sqrt{7-2x} = 3$

12. $\sqrt{3x+4} = \sqrt{2x+5}$

 SOLUTIONS

Unless a solution is extraneous, the check step is omitted.

1. $\dfrac{18-5x}{3x+2} = \dfrac{7}{3}$ This equation is in the form "Fraction = Fraction," so cross-multiply.

$$3(18-5x) = 7(3x+2)$$

$$54 - 15x = 21x + 14$$

$$ +15x +15x$$

$$54 = 36x + 14$$

$$-14 -14$$

$$40 = 36x$$

$$\frac{40}{36} = x$$

$$\frac{10}{9} = x$$

2. $\dfrac{6}{5x-2} = \dfrac{9}{7x+6}$ This equation is in the form "Fraction = Fraction," so cross-multiply.

$$6(7x + 6) = 9(5x - 2)$$

$$42x + 36 = 45x - 18$$

$$\underline{-42x \qquad -42x}$$

$$36 = 3x - 18$$

$$\underline{+18 \qquad +18}$$

$$54 = 3x$$

$$\dfrac{54}{3} = x$$

$$18 = x$$

3. $(x-7)^2 - 4 = (x+1)^2$

$$(x-7)(x-7) - 4 = (x+1)(x+1)$$

$$x^2 - 7x - 7x + 49 - 4 = x^2 + x + x + 1$$

$$x^2 - 14x + 45 = x^2 + 2x + 1 \qquad x^2\text{'s cancel}$$

$$-14x + 45 = 2x + 1$$

$$\underline{-2x \qquad -2x}$$

$$-16x + 45 = 1$$

$$\underline{-45 \quad -45}$$

$$-16x = -44$$

$$x = \dfrac{-44}{-16}$$

$$x = \dfrac{11}{4} \ \text{ or } \ 2\tfrac{3}{4}$$

4. $\dfrac{6x + 7}{4x - 1} = \dfrac{3x + 8}{2x - 4}$ This is in the form "Fraction = Fraction," so cross-multiply.

$$(6x+7)(2x-4)=(4x-1)(3x+8)$$

$$12x^2 -24x+14x-28=12x^2+32x-3x-8$$

$$12x^2 -10x-28=12x^2+29x-8 \quad 12x^2\text{'s cancel}$$

$$-10x-28=29x-8$$

$$+10x \qquad +10x$$

$$-28=39x-8$$

$$+8 \qquad +8$$

$$-20=39x$$

$$\dfrac{-20}{39}=x$$

5. $\dfrac{9x}{3x - 1} = 4 + \dfrac{3}{3x - 1}$ The LCD is $3x - 1$.

$$(3x - 1)\left[\dfrac{9x}{3x -1}\right] = (3x - 1)\left[4 + \dfrac{3}{3x -1}\right]$$

$$9x = (3x - 1)(4) + (3x - 1)\left[\dfrac{3}{3x - 1}\right]$$

$$9x = 12x - 4 + 3$$

$$9x = 12x - 1$$

$$-9x \quad -9x$$

$$0 = 3x - 1$$

$$+1 \qquad +1$$

$$1 = 3x$$

$$\dfrac{1}{3} = x$$

If we let $x = \dfrac{1}{3}$, then both denominators would be 0, so $x = \dfrac{1}{3}$ is *not* a solution to the original equation. The original equation has no solution.

6. $\dfrac{1}{x-1}-\dfrac{2}{x+1}=\dfrac{3}{x^2-1}$

$\dfrac{1}{x-1}-\dfrac{2}{x+1}=\dfrac{3}{(x-1)(x+1)}$ **The LCD is $(x-1)(x+1)$.**

$$(x-1)(x+1)\left[\dfrac{1}{x-1}-\dfrac{2}{x+1}\right]=(x-1)(x+1)\cdot\dfrac{3}{(x-1)(x+1)}$$

$$(x-1)(x+1)\cdot\dfrac{1}{(x-1)}-(x-1)(x+1)\cdot\dfrac{2}{x+1}=3$$

$$1(x+1)-2(x-1)=3$$
$$x+1-2x+2=3$$
$$-x+3=3$$
$$-3\ -3$$
$$-x=0$$
$$x=0$$

(0 and −0 are the same number)

7. $(2x-1)^2-4x^2=-4x+1$
$(2x-1)(2x-1)-4x^2=-4x+1$
$4x^2-2x-2x+1-4x^2=-4x+1$
$-4x+1=-4x+1$

The last equation is true for any real number x. An equation true for any real number x is called an *identity*. For example, $(2x-1)^2-4x^2=-4x+1$ is an identity.

8. $\sqrt{7x+1}=13$
$\left[\sqrt{7x+1}\right]^2=13^2$
$7x+1=169$
$-1\ -1$
$7x=168$
$x=\dfrac{168}{7}$
$x=24$

9. $\sqrt{x}-6=10$
$\sqrt{x}-6=10$ Isolate the square root *before* squaring both sides.
$+6\ +6$
$\sqrt{x}=16$
$\left[\sqrt{x}\right]^2=16^2$
$x=256$

10. $\sqrt{2x-3}+1=6$

    $\sqrt{2x-3}+1=6$ Isolate the square root *before* squaring both sides.

         $-1 \;\; -1$

       $\sqrt{2x-3}=5$

    $\left[\sqrt{2x-3}\right]^2 = 5^2$

        $2x-3=25$

        $+3 \;\; +3$

         $2x=28$

          $x=\dfrac{28}{2}$

          $x=14$

11. $\sqrt{7-2x}=3$

    $\left[\sqrt{7-2x}\right]^2 = 3^2$

      $7-2x=9$

    $-7 \qquad -7$

        $-2x=2$

          $x=\dfrac{2}{-2}$

          $x=-1$

12. $\sqrt{3x+4}=\sqrt{2x+5}$

    $\left[\sqrt{3x+4}\right]^2 = \left[\sqrt{2x+5}\right]^2$

      $3x+4=2x+5$

    $-2x \qquad -2x$

       $x+4=5$

       $-4 \;\; -4$

         $x=1$

Summary

In this chapter, we learned how to

- *Use PEMDAS to compute complex expressions.* The letters in PEMDAS indicate the order of operations. The "P" stands for "parentheses." We work inside parentheses first. The "E" stands for "exponents." We compute exponents second. The "M" stands for "multiplication" and "D" stands for "division." We multiply and divide third, working from left to right. Finally, "A" stands for "addition" and "S" stands for subtraction. We add and subtract last, working from left to right.

- *Solve linear equations.* To solve an equation for x (or some other variable) means to isolate x on one side of the equation. Our strategy is to simplify each side of the equation and then use addition/subtraction to collect terms having x in them on one side of the equation and terms without x on the other side. In the last step, we divide each side of the equation by the coefficient of x.

- *Solve equations having fractions/decimals in them by clearing the fractions/decimals.* If the equation has fractions, we identify the LCD and multiply each side of the equation by the LCD. This eliminates the fraction(s). If the equation has decimal numbers, we multiply both sides of the equation by a power of 10 large enough to eliminate any decimal. We then proceed as above.

- *Solve formulas containing multiple variables for one of the variables.* We use the same strategy as above to solve equations containing multiple variables—simplify each side of the equation and then collect the terms having the variable we want on one side and terms without this variable on the other side and then divide each side of the equation by the coefficient of this variable. The coefficient probably contains another variable.

- *Solve equations leading to linear equations.* If the equation contains a variable underneath a square root symbol (radical), we isolate the root symbol on one side of the equation and then square both sides of the equation. This eliminates the root. We then solve the remaining equation using the strategy outlined above. If the equation contains a rational expression (a fraction having a variable in the numerator/denominator), we use one of two strategies. For equations in the form "fraction = fraction," we cross-multiply. That is, we multiply each numerator by the denominator of the other fraction and then solve the equation. For other equations, we identify the LCD and multiply each side by the LCD, which eliminates the fractions.

QUIZ

1. Solve $5x - \dfrac{1}{2} = 4x + 3$.

   A. $x = \dfrac{1}{6}$

   B. $x = -\dfrac{7}{2}$

   C. $x = \dfrac{5}{2}$

   D. $x = \dfrac{7}{2}$

2. Solve $\dfrac{1}{2x+3} = \dfrac{5}{x+6}$.

   A. $x = -1$

   B. $x = -\dfrac{1}{9}$

   C. $x = 3$

   D. There is no solution.

3. $\dfrac{5(4^2-8)}{4^2-9} \cdot \left(\sqrt{49}\right) =$

   A. -20
   B. -49
   C. -56
   D. 40

4. Solve $5(x+2) - 8 = 3(4-2x)$.

   A. $x = \dfrac{11}{10}$

   B. $x = \dfrac{10}{11}$

   C. $x = \dfrac{18}{11}$

   D. $x = \dfrac{11}{18}$

5. Solve $\dfrac{4}{5}x + 2 = \dfrac{3}{4}x - \dfrac{5}{2}$.

   A. $x = -50$
   B. $x = 40$
   C. $x = -90$
   D. $x = 60$

6. Solve $\dfrac{3}{x+4} + \dfrac{2}{x-4} = \dfrac{16}{x^2-16}$.

   A. $x = 4$

   B. $x = \dfrac{4}{5}$

   C. $x = \dfrac{5}{4}$

   D. There is no solution.

7. Solve $\sqrt{1-2x} + 2 = 5$.

   A. $x = 0$

   B. $x = 4$

   C. $x = -4$

   D. There is no solution.

8. Solve $2.4x - 0.75 = 0.48x - 0.33$.

   A. $x = \dfrac{3}{8}$

   B. $x = \dfrac{7}{32}$

   C. $x = \dfrac{9}{16}$

   D. There is no solution.

9. Solve for W: $\dfrac{4}{W} + Z = 5$.

   A. $W = \dfrac{4+Z}{5}$

   B. $W = \dfrac{4}{5-Z}$

   C. $W = 4 + Z - 5$

   D. The equation cannot be solved for W.

10. Solve $(2x - 3)(x + 5) = (2x + 1)(x - 2)$.

    A. $x = \dfrac{13}{4}$

    B. $x = \dfrac{13}{16}$

    C. $x = -\dfrac{13}{4}$

    D. $x = \dfrac{13}{10}$

chapter 8

Linear Applications

Word problems (also called applications or applied problems) are usually not students' favorite math topic, but working with applied problems helps you to develop strategies for using mathematics to solve real problems. You will also need to be able to solve applied problems in other courses (science, accounting, business, and so on). The problems in this chapter will prepare you to solve such problems. In later chapters, we will build on the problem types that we will see in this chapter. We begin with problems that we see everyday—percents.

CHAPTER OBJECTIVES

In this chapter, you will

- Solve basic percent problems
- Work with formulas
- Solve number sense problems
- Solve age, coin, and grade problems
- Solve investment, mixture, and work problems
- Solve distance problems

Percents

A percent is a decimal number in disguise. In fact, the word "percent" literally means "per hundred." Remember that "per" means to divide and "cent" means hundred, so 16% means $16 \div 100$ or $16/100 = 0.16$ (notice that we move the decimal point to the left 2 places). Because the word "of" means to multiply, we see that the statement, "16% of 25" translates into $(0.16)(25) = 4$.

 EXAMPLE

Convert the percent to a decimal number and compute.

82% of 44

Writing 82% as a decimal number, we have 82% = 0.82, so 82% of 44 becomes $(0.82)(44) = 36.08$.

$$150\% \text{ of } 6 \text{ is } (1.50)(6) = 9$$
$$8\tfrac{3}{4}\% \text{ of } 24 \text{ is } (0.0875)(24) = 2.1$$
$$0.65\% \text{ of } 112 \text{ is } (0.0065)(112) = 0.728$$

 PRACTICE

Convert the percent to a decimal number and compute.

1. 64% of 50 is _____.
2. 126% of 38 is _____.
3. 0.42% of 16 is _____.
4. 18.5% of 48 is _____.
5. 213.6% of 90 is _____.

 SOLUTIONS

1. 64% of 50 is $(0.64)(50) = 32$
2. 126% of 38 is $(1.26)(38) = 47.88$
3. 0.42% of 16 is $(0.0042)(16) = 0.0672$
4. 18.5% of 48 is $(0.185)(48) = 8.88$
5. 213.6% of 90 is $(2.136)(90) = 192.24$

Increasing/Decreasing by a Percent

As consumers, we often see quantities being increased or decreased by some percentage. For instance, a cereal box claims "25% More." An item might be on sale, its tag saying "Reduced by 40%." We now learn how to interpret these percentage changes. If we increase the quantity x by percentage p (the percent written as a decimal number), the final quantity is $x + px$. If we decrease the quantity x by percentage p, the final quantity is $x - px$.

EXAMPLE
Compute.

80 increased by 20%

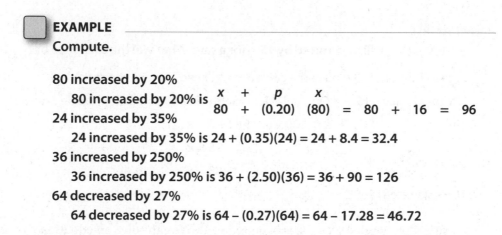

80 increased by 20% is
$$x + p \quad x$$
$$80 + (0.20)(80) = 80 + 16 = 96$$

24 increased by 35%

24 increased by 35% is $24 + (0.35)(24) = 24 + 8.4 = 32.4$

36 increased by 250%

36 increased by 250% is $36 + (2.50)(36) = 36 + 90 = 126$

64 decreased by 27%

64 decreased by 27% is $64 - (0.27)(64) = 64 - 17.28 = 46.72$

PRACTICE
Compute.

1. 46 increased by 60% is _____.

2. 78 increased by 125% is _____.

3. 16 decreased by 30% is _____.

4. 54 increased by 21.3% is _____.

5. 128 decreased by 8.16% is _____.

6. 15 increased by 0.03% is _____.

7. 24 decreased by 108.4% is _____.

SOLUTIONS

1. 46 increased by 60% is $46 + (0.60)(46) = 46 + 27.6 = 73.6$

2. 78 increased by 125% is $78 + (1.25)(78) = 78 + 97.5 = 175.5$

3. 16 decreased by 30% is $16 - (0.30)(16) = 16 - 4.8 = 11.2$

4. 54 increased by 21.3% is 54 + (0.213)(54) = 54 + 11.502 = 65.502

5. 128 decreased by 8.16% is 128 – (0.0816)(128) = 128 – 10.4448 = 117.5552

6. 15 increased by 0.03% is 15 + (0.0003)(15) = 15 + 0.0045 = 15.0045

7. 24 decreased by 108.4% is 24 – (1.084)(24) = 24 – 26.016 = –2.016

Many word problems involving percents fit the above model—that is, a quantity being increased or decreased by a percent of the original quantity.

EXAMPLE

A $100 jacket will be reduced by 15% for a sale. What will the sale price be?

Let x = sale price. Because the price is decreased, we use the model "Original Price – 15% × Original Price = Sale Price" to compute the sale price.

Then,
$$100 - (0.15)(100) = x$$
$$100 - 15 = x$$
$$85 = x$$

The sale price is $85.

We usually see a price *after* it has been reduced. We can solve an equation to find the original price. Again, the equation is

Sale Price = Original Price – Percent of the Original Price

EXAMPLE

The sale price for a computer is $1200, which represents a 20% markdown. What is the original price?

We let x represent the original price, so the sale price is $x - 0.20x = 0.80x$. The sale price is also 1200. These two facts give us the equation $0.80x = 1200$.

$$0.80x = 1200$$
$$x = \frac{1200}{0.80}$$
$$x = 1500$$

The original price is $1500.

Still Struggling

In the first example, the percent was multiplied by the number given; and in the second example, the percent was multiplied by the unknown. Be very careful when deciding from which quantity you take the percent. Suppose the first problem were worded, "An item is marked down 20% for a sale. The sale price is $60, what is the original price?" The equation to solve would be $x - 0.20x = 60$, where x represents the original price. A common mistake is to take 20% of $60 and not 20% of the *original price*.

The data used in the following examples and practice problems can be found in the 129th edition of *Statistical Abstract of the United States*. Many quantities and percentages are approximate.

EXAMPLE

In 2007, the average income for a U.S. adult, age 25 to 34, whose highest education level was completing high school was $28,982. For a U.S. adult of the same age having a bachelor's degree, the average income that same year was 66.5% more. What was the average income for a U.S. adult, age 25 to 34 in 2007 whose highest education level was bachelor's degree?

We want to compute the following:

HS Grad Income + 66.5% HS Grad Income = Bachelor's Degree Income

$$28,982 + (0.665)(28,982) = \text{Bachelor's Degree Income}$$
$$28,982 + 19,273 = 48,255$$

The average income in 2007 for U.S. adults, age 25 to 34, having a bachelor's degree was $48,255.

On January 1, 2010, the price of a first class postage stamp was $0.44, which is a 37.5% increase from the cost on January 1, 1995. What was the cost of a first class stamp on January 1, 1995?

Let x = 1st class price on January 1, 1995

$0.44 is 37.5% more than this quantity.

$$0.44 = x + 0.375x$$
$$0.44 = 1 \cdot x + 0.375x$$
$$0.44 = x(1 + 0.375) \quad \text{(factor } x\text{)}$$
$$0.44 = 1.375x$$
$$\frac{0.44}{1.375} = x$$
$$0.32 = x$$

The price of a first-class stamp on January 1, 1995 was $0.32.

PRACTICE

1. A local cable television company currently charges $75 per month. It plans an increase in its monthly charge by 15%. What will the new rate be?

2. In 1970, U.S. energy consumption was 67.84 quad BTUs. In the year 2008, the consumption had increased by 46.4%. What was energy consumption by the United States in the year 2008?

3. In 2008, the United States consumed 510 trillion BTUs of wind energy, which represents a 183% increase of wind energy consumed in the year 2005. How much wind energy was consumed in the United States in the year 2005?

4. A worker's take-home pay was $480 after deductions totaling 40%. What is the worker's gross pay?

5. A cereal company advertises that its 16-ounce cereal represents 25% more than before. What is the original amount?

6. A couple does not wish to spend more than $45 for dinner at their favorite restaurant. If a sales tax of $7\frac{1}{2}$% is added to the bill and they plan to tip 15% after the tax is added, what is the most they can spend for the meal?

7. A discount store prices some name brand toothpaste by raising the wholesale price by 40% and then adding $0.20. What must the toothpaste's wholesale price be if it sells for $3.00?

 **SOLUTIONS** _____

1. $75 will be increased by 15%:

   $$75 + (0.15)(75) = 75 + 11.25 = 86.25$$

   The new rate will be $86.25.

2. 67.84 increased by 46.4%:

   $$67.84 + (0.464)(67.84) = 67.84 + 31.48 = 99.32$$

   The energy consumption in the United States in the year 2008 was 99.3 quad BTUs.

3. Wind energy in the year 2005 increased by 183% (as a decimal, 1.83). Let x represent the wind energy (in trillion BTUs) in the year 2005.

   $$x + 1.83x = 510$$
   $$x(1 + 1.83) = 510$$
   $$2.83x = 510$$
   $$x = \frac{510}{2.83}$$
   $$x = 180$$

   The United States consumed 180 trillion BTUs of wind energy in the year 2005.

4. The gross pay is reduced by 40%. Let x represent gross pay.

   $$x - 0.40x = 480$$
   $$1x - 0.40x = 480$$
   $$x(1 - 0.40) = 480$$
   $$0.60x = 480$$
   $$x = \frac{480}{0.60}$$
   $$x = 800$$

   The worker's gross pay is $800.

5. The original amount is increased by 25%. Let x represent the original amount.

$$x + 0.25x = 16$$
$$1x + 0.25x = 16$$
$$x(1 + 0.25) = 16$$
$$1.25x = 16$$
$$x = \frac{16}{1.25}$$
$$x = 12.8$$

The original box of cereal contained 12.8 ounces.

6. The total bill is the cost of the meal plus the tax on the meal plus the tip.

Let x represent the cost of the meal, so the amount of the tax is $0.075x$.

The tip is 15% of the meal plus tax: ($x + 0.075x = 1.075x$ is the price of the meal including tax), so, the tip is $0.15(1.075x) = 0.16125x$.

The total bill is $x + 0.075x + 0.16125x$. We want this to equal 45:

meal   tax   tip   total
$$x + 0.075x + 0.16125x = 45$$
$$1x + 0.075x + 0.16125x = 45$$
$$x(1 + 0.075 + 0.16125) = 45$$
$$1.23625x = 45$$
$$x = \frac{45}{1.23625}$$
$$x \approx 36.40$$

The couple can spend $36.40 on their meal.

7. Let x represent the wholesale price. 40% of the wholesale price is $0.40x$. The retail price is the wholesale price plus 40% of the wholesale price plus $0.20:

$$x + 0.40x + 0.20 = 3.00$$
$$1x + 0.40x + 0.20 = 3.00$$
$$x(1 + 0.40) + 0.20 = 3.00$$
$$1.40x + 0.20 = 3.00$$
$$\underline{-0.20 \quad -0.20}$$
$$1.40x = 2.80$$
$$x = \frac{2.80}{1.40}$$
$$x = 2$$

The wholesale price is $2.

In the previous problems, we knew the percent and wanted the original quantity. We will now be given two numbers and asked what percent one number is of the other. We let x represent the unknown percent as a decimal number.

 EXAMPLE

Find the percent.

5 is what percent of 8?

This sentence translates into $5 = \qquad x \qquad \cdot \ 8$

$\qquad\qquad\qquad$ 5 is what percent of 8

The equation to solve is $8x = 5$.

$$8x = 5$$
$$x = \frac{5}{8} = 0.625 = 62.5\%.$$

So, 5 is 62.5% of 8.

8 is what percent of 5?

This question translates into the equation

$$8 = x \cdot 5$$
$$5x = 8$$
$$x = \frac{8}{5} = 1.6 = 160\%$$

8 is 160% of 5.

PRACTICE

1. 2 is what percent of 5?

2. 5 is what percent of 2?

3. 3 is what percent of 15?

4. 15 is what percent of 3?

5. 1.8 is what percent of 18?

6. 18 is what percent of 1.8?

7. $\frac{1}{4}$ is what percent of 2?

8. 2 is what percent of $\frac{1}{4}$?

SOLUTIONS

1. $5x = 2$

$$x = \frac{2}{5} = 0.40 = 40\%$$

2 is 40% of 5.

2. $2x = 5$

$$x = \frac{5}{2} = 2.5 = 250\%$$

5 is 250% of 2.

3. $15x = 3$

$$x = \frac{3}{15} = \frac{1}{5} = 0.20 = 20\%$$

3 is 20% of 15.

4. $3x = 15$

$$x = \frac{15}{3} = 5 = 500\%$$

15 is 500% of 3.

5. $18x = 1.8$

$$x = \frac{1.8}{18}$$

$$x = \frac{1.8(10)}{18(10)}$$

$$x = \frac{18}{180} = 0.10 = 10\%$$

1.8 is 10% of 18.

6. $1.8x = 18$

$$x = \frac{18}{1.8}$$

$$x = \frac{18(10)}{1.8(10)}$$

$$x = \frac{180}{18} = 10 = 1000\%$$

18 is 1000% of 1.8.

7. $2x = \dfrac{1}{4}$

$$x = \frac{1}{2} \cdot \frac{1}{4}$$

$$x = \frac{1}{8} = 0.125 = 12.5\%$$

$\frac{1}{4}$ **is 12.5% of 2.**

8. $\dfrac{1}{4}x = 2$

$$x = 4(2)$$

$$x = 8 = 800\%$$

2 is 800% of $\frac{1}{4}$.

Working with Formulas

For some word problems, nothing more is required other than to plug a given value into a formula, which is either given or is readily available. The most difficult part of these problems is to decide which variable the given quantity should represent. For example, the formula might look like $R = 8q$ and the value given to you is 440. Is $R = 440$ or is $q = 440$? The answer lies in the way the variables are described. In $R = 8q$, suppose R represents revenue (in dollars) and q represents quantity (in units) sold of some item. In the question, "If 440 units were sold, what is the revenue?" q is 440. We would then solve $R = 8(440)$. In the question, "If the revenue is \$440, how many units were sold?" 440 is R, and we would solve $440 = 8q$.

EXAMPLE

The cost formula for a manufacturer's product is $C = 5000 + 2x$, where C is the cost (in dollars) and x is the number of units manufactured.

(a) If no units are produced, what is the cost?

(b) If the manufacturer produces 3000 units, what is the cost?

(c) If the manufacturer has spent $16,000 on production, how many units were produced?

We answer these questions by substituting the numbers into the formula.

(a) If no units are produced, then $x = 0$, and $C = 5000 + 2x$ becomes $C = 5000 + 2(0) = 5000$. The cost is $5,000.

(b) If the manufacturer produces 3000 units, then $x = 3000$, and $C = 5000 + 2x$ becomes $C = 5000 + 2(3000) = 5000 + 6000 = 11,000$. The manufacturer's cost is $11,000.

(c) The manufacturer's cost is $16,000, so $C = 16,000$. Substitute 16,000 for C in the equation $C = 5000 + 2x$, giving us $16,000 = 5000 + 2x$.

$$16,000 = 5000 + 2x$$
$$ -5000 \quad -5000$$
$$11,000 = 2x$$
$$\frac{11,000}{2} = x$$
$$5500 = x$$

There were 5500 units produced.

The profit formula for a manufacturer's product is $P = 2x - 4000$ where x is the number of units sold and P is the profit (in dollars).

(a) What is the profit when 12,000 units were sold?

(b) What is the loss when 1500 units were sold?

(c) How many units must be sold for the manufacturer to have a profit of $3000?

(d) How many units must be sold for the manufacturer to break even?

(This question could equivalently be phrased, "How many units must be sold in order for the manufacturer to cover its costs?")

We begin by substituting numbers given in the problem for variables in the formula.

(a) If 12,000 units are sold, then $x = 12{,}000$. The profit equation then becomes $P = 2(12{,}000) - 4000 = 24{,}000 - 4000 = 20{,}000$. The profit is $20,000.

(b) Think of a loss as a negative profit. When 1500 units are sold, $P = 2x - 4000$ becomes $P = 2(1500) - 4000 = 3000 - 4000 = -1000$. The manufacturer loses $1000 when 1500 units are sold.

(c) If the profit is $3000, then $P = 3000$; $P = 2x - 4000$ becomes $3000 = 2x - 4000$.

$$3000 = 2x - 4000$$
$$\underline{+4000 \qquad +4000}$$
$$7000 = 2x$$
$$\frac{7000}{2} = x$$
$$3500 = x$$

A total of 3500 units were sold.

(d) The break-even point occurs when the profit is zero, that is when $P = 0$, so $P = 2x - 4000$ becomes $0 = 2x - 4000$.

$$0 = 2x - 4000$$
$$\underline{+4000 \qquad +4000}$$
$$4000 = 2x$$
$$\frac{4000}{2} = x$$
$$2000 = x$$

The manufacturer must sell 2000 units in order to break even.

A box has a square bottom. The height has not yet been determined, but the bottom is 10" by 10". The volume formula is $V = lwh$, because each of the length and width is 10, lw becomes $10 \cdot 10 = 100$. The formula for the box's volume is $V = 100h$.

(a) If the height of the box is to be 6 inches, what is its volume?

(b) If the volume is to be 450 cubic inches, what should its height be?

(c) If the volume is to be 825 cubic inches, what should its height be?

We begin by substituting numbers given in the question for variables in the formula.

(a) The height is 6 inches, so $h = 6$. Then $V = 100h$ becomes $V = 100(6) = 600$.

The box's volume is 600 cubic inches.

(b) The volume is 450 cubic inches, so $V = 450$, and $V = 100h$ becomes $450 = 100h$.

$$450 = 100h$$

$$\frac{450}{100} = h$$

$$4.5 = h$$

The box's height would need to be 4.5 inches.

(c) The volume is 825, so $V = 100h$ becomes $825 = 100h$.

$$825 = 100h$$

$$\frac{825}{100} = h$$

$$8.25 = h$$

The height should be 8.25 inches.

A square has a perimeter of 18 cm. What is the length of each of its sides?

We recall the formula for the perimeter of a square: $P = 4l$ where l is the length of each of its sides.

$P = 18$; so $P = 4l$ becomes $18 = 4l$.

$$18 = 4l$$

$$\frac{18}{4} = l$$

$$4.5 = l$$

The length of each of its sides is 4.5 cm.

The relationship between degrees Fahrenheit and degrees Celsius is given by the formula $C = \frac{5}{9}(F - 32)$. At what temperature will degrees Fahrenheit and degrees Celsius be the same?

If degrees Fahrenheit and degrees Celsius are the same, then we begin with $F = C$. We can replace C with $\frac{5}{9}(F - 32)$ because they are equal.

$F = C$

$F = \frac{5}{9}(F - 32)$

The LCD is 9.

$$9F = 9\left[\frac{5}{9}(F - 32)\right]$$

$$9F = \left[9\left(\frac{5}{9}\right)\right](F - 32)$$

$$9F = 5(F - 32)$$

$$9F = 5F - 5(32)$$

$$9F = 5F - 160$$

$$-5F \quad -5F$$

$$4F = -160$$

$$F = \frac{-160}{4}$$

$$F = -40$$

At −40 degrees Fahrenheit and degrees Celsius are the same.

PRACTICE

1. The daily charge for a small rental car is $C = 18 + 0.35x$, where x is the number of miles driven.

   (a) If the car was driven 80 miles, what was the charge for the day?

   (b) Suppose that a day's bill was $39. How many miles were driven?

2. The profit obtained for a company's product is given by $P = 25x - 8150$, where P is the profit in dollars and x is the number of units sold. How many units must be sold in order for the company to have a profit of $5000 from this product?

3. A salesman's weekly salary is based on the formula $S = 200 + 0.10s$, where S is the week's salary in dollars and s is the week's sales level in dollars. One week, his salary was $410. What was the sales level for that week?

4. The volume of a box with a rectangular bottom is given by $V = 120h$, where V is the volume in cubic inches and h is the height in inches. If the volume of the box is to be 1140 cubic inches, what should its height be?

5. The volume of a certain cylinder with radius 2.8 cm is given by $V = 7.84\pi h$, where h is the height of the cylinder in centimeters. If the volume needs to be 25.088π cubic centimeters, what does the height need to be?

6. At what temperature will the Celsius reading be twice as high as the Fahrenheit reading?

7. At what temperature will degrees Fahrenheit be twice degrees Celsius?

 SOLUTIONS

1. (a) Here $x = 80$, so $C = 18 + 0.35x$ becomes $C = 18 + 0.35(80)$.

$$C = 18 + 0.35(80)$$
$$C = 18 + 28$$
$$C = 46$$

The charge is $46.

(b) The cost is $39, so $C = 18 + 0.35x$ becomes $39 = 18 + 0.35x$.

$$39 = 18 + 0.35x$$
$$\underline{-18 \quad -18}$$
$$21 = 0.35x$$
$$\frac{21}{0.35} = x$$
$$60 = x$$

Sixty miles were driven.

2. The profit is $5000, so $P = 25x - 8150$ becomes $5000 = 25x - 8150$.

$$5000 = 25x - 8150$$
$$\underline{+8150 \qquad +8150}$$
$$13{,}150 = 25x$$
$$\frac{13{,}150}{25} = x$$
$$526 = x$$

The company must sell 526 units.

3. The salary is \$410, so $S = 200 + 0.10s$ becomes $410 = 200 + 0.10s$.

$$410 = 200 + 0.10s$$

$$-200 \quad -200$$

$$210 = 0.10s$$

$$\frac{210}{0.10} = s$$

$$2100 = s$$

The week's sales level was \$2100.

4. The volume is 1140 cubic inches, so $V = 120h$ becomes $1140 = 120h$.

$$1140 = 120h$$

$$\frac{1140}{120} = h$$

$$9.5 = h$$

The box needs to be 9.5 inches tall.

5. The volume is 25.088π cm³, so $V = 7.84\pi h$ becomes $25.088\pi = 7.84\pi h$.

$$25.088\pi = 7.84\pi h$$

$$\frac{25.088\pi}{7.84\pi} = h$$

$$3.2 = h$$

The height of the cylinder needs to be 3.2 cm.

6. $\frac{5}{9}(F - 32)$ represents degrees Celsius, and $2F$ represents twice degrees Fahrenheit. We want these two quantities to be equal.

$$\frac{5}{9}(F - 32) = 2F$$

The LCD is 9.

$$9\left[\left(\frac{5}{9}\right)(F-32)\right]=9(2F)$$

$$\left[9\left(\frac{5}{9}\right)\right](F-32)=18F$$

$$5(F-32)=18F$$

$$5F-160=18F$$

$$\underline{-5F\qquad\quad-5F}$$

$$-160=13F$$

$$\frac{-160}{13}=F$$

When $F=-\dfrac{160}{13}$, $C=\dfrac{5}{9}\left(-\dfrac{160}{13}-32\right)=-\dfrac{320}{13}$ or $-24\frac{8}{13}$.

7. Degrees Celsius is represented by $\dfrac{5}{9}(F-32)$ so twice degrees Celsius is represented by $2\left[\dfrac{5}{9}(F-32)\right]$. We want for this to equal degrees Fahrenheit.

$$F=2\left[\frac{5}{9}(F-32)\right]$$

$$F=\left[2\left(\frac{5}{9}\right)\right](F-32)$$

$$F=\frac{10}{9}(F-32)$$

$$9F=9\left[\frac{10}{9}(F-32)\right]$$

$$9F=\left[9\left(\frac{10}{9}\right)\right](F-32)$$

$$9F=10(F-32)$$

$$9F=10F-320$$

$$\underline{-10F\quad-10F}$$

$$-F=-320$$

$$(-1)(-F)=(-1)(-320)$$

$$F=320$$

When $F=320$, $C=\dfrac{5}{9}(320-32)=160.$

Number Sense

Many problems require us to use "common sense" to solve them—that is, basic mathematical reasoning. For instance, when a problem refers to consecutive integers, we are expected to realize that any two consecutive integers differ by one. If two numbers are consecutive, we normally let x equal the first number and $x + 1$, the second.

 EXAMPLE

The sum of two consecutive integers is 25. What are the numbers?

Let x = first number, and $x + 1$ = second number.

Their sum is 25, so $x + (x + 1) = 25$.

$$x + (x + 1) = 25$$
$$2x + 1 = 25$$
$$\underline{-1 \quad -1}$$
$$2x = 24$$
$$x = \frac{24}{2}$$
$$x = 12$$

The first number is 12 and the second number is $x + 1 = 12 + 1 = 13$.

The sum of three consecutive integers is 27. What are the numbers?

Let x = first number, $x + 1$ = second number, and $x + 2$ = third number.

Their sum is 27, so $x + (x + 1) + (x + 2) = 27$.

$$x + (x + 1) + (x + 2) = 27$$
$$3x + 3 = 27$$
$$\underline{-3 \quad -3}$$
$$3x = 24$$
$$x = \frac{24}{3}$$
$$x = 8$$

The first number is 8; the second is $x + 1 = 8 + 1 = 9$; the third is $x + 2 = 8 + 2 = 10$.

PRACTICE _____

1. Find two consecutive numbers whose sum is 57.

2. Find three consecutive numbers whose sum is 48.

3. Find four consecutive numbers whose sum is 90.

SOLUTIONS _____

1. Let x = first number, and $x + 1$ = second number.

   Their sum is 57, so $x + (x + 1) = 57$.

$$x + (x + 1) = 57$$
$$2x + 1 = 57$$
$$\underline{-1 \quad -1}$$
$$2x = 56$$
$$x = \frac{56}{2}$$
$$x = 28$$

   The first number is 28 and the second is $x + 1 = 28 + 1 = 29$.

2. Let x = first number, $x + 1$ = second number, and $x + 2$ = third number.

   Their sum is 48, so $x + (x + 1) + (x + 2) = 48$.

$$x + (x + 1) + (x + 2) = 48$$
$$3x + 3 = 48$$
$$\underline{-3 \quad -3}$$
$$3x = 45$$
$$x = \frac{45}{3}$$
$$x = 15$$

   The first number is 15; the second, $x + 1 = 15 + 1 = 16$; and the third, $x + 2 = 15 + 2 = 17$.

3. Let x = first number, $x + 1$ = second number, $x + 2$ = third number, and $x + 3$ = fourth number.

    Their sum is 90, so $x + (x + 1) + (x + 2) + (x + 3) = 90$.

$$x + (x + 1) + (x + 2) + (x + 3) = 90$$
$$4x + 6 = 90$$
$$\underline{-6 \quad -6}$$
$$4x = 84$$
$$x = \frac{84}{4}$$
$$x = 21$$

The first number is 21; the second, $x + 1 = 21 + 1 = 22$; the third, $x + 2 = 21 + 2 = 23$; and the fourth, $x + 3 = 21 + 3 = 24$.

We can solve other kinds of "number sense" problems when numbers are not consecutive. In the following problems, we will look for two numbers and will be told their sum and how much larger one is than the other. The information on the sum gives us the equation to solve. We use the other information to find a relationship between the numbers so that we can represent all of the numbers in the equation using a single variable, as we did above with consecutive numbers.

EXAMPLE

The sum of two numbers is 70. One number is eight more than the other. What are the numbers?

Let x = first number. (The term "first" is used because it is the first number we are looking for; it is not necessarily the "first" in order.) The other number is eight more than this, so $x + 8$ represents the other number. Their sum is 70, giving us $x + (x + 8) = 70$.

$$x + (x + 8) = 70$$

$$2x + 8 = 70$$

$$-8 \quad -8$$

$$2x = 62$$

$$x = \frac{62}{2}$$

$$x = 31$$

The numbers are 31 and $x + 8 = 31 + 8 = 39$.

The sum of two numbers is 63. One of the numbers is twice the other. Find the numbers.

Let $x =$ first number, and $2x =$ other number.

Their sum is 63, so $x + 2x = 63$.

$$x + 2x = 63$$

$$3x = 63$$

$$x = \frac{63}{3}$$

$$x = 21$$

The numbers are 21 and $2x = 2(21) = 42$.

PRACTICE

1. The sum of two numbers is 85. One number is 15 more than the other. What are the two numbers?

2. The sum of two numbers is 48. One number is three times the other. What are the numbers?

 **SOLUTIONS**

1. Let x = first number, and $x + 15$ = second number.

    Their sum is 85, so $x + (x + 15) = 85$.

    $$x + (x + 15) = 85$$

    $$2x + 15 = 85$$

    $$-15 \quad -15$$

    $$2x = 70$$

    $$x = \frac{70}{2}$$

    $$x = 35$$

    The numbers are 35 and $x + 15 = 35 + 15 = 50$.

2. Let x = first number and $3x$ = second number.

    Their sum is 48, so $x + 3x = 48$.

    $$x + 3x = 48$$

    $$4x = 48$$

    $$x = \frac{48}{4}$$

    $$x = 12$$

    The numbers are 12 and $3(12) = 36$.

The relationship between the unknown numbers in the previous problems was fairly simple. The relationship between the two numbers in the following problems is less simple, but the strategy is the same.

 EXAMPLE

The difference between two numbers is 13. Twice the smaller plus three times the larger is 129. What are the numbers?

If the difference between two numbers is 13, then one of the numbers is 13 more than the other. As before, we let x represent the first number. Then, $x + 13$ represents the other. "Twice the smaller" means "$2x$" (x is the smaller quantity because the other quantity is 13 more than x). Three times the larger number is $3(x + 13)$. "Twice the smaller plus three times the larger is 129" becomes the equation $2x + 3(x + 13) = 129$.

$$2x + 3(x + 13) = 129$$
$$2x + 3x + 39 = 129$$
$$5x + 39 = 129$$
$$\underline{-39 \quad -39}$$
$$5x = 90$$
$$x = \frac{90}{5}$$
$$x = 18$$

The numbers are 18 and $x + 13 = 18 + 13 = 31$.

The sum of two numbers is 14. Three times the smaller plus twice the larger is 33. What are the two numbers?

We let x represent the smaller number. How can we represent the larger number? We know that the sum of the smaller number and larger number is 14. Let "?" represent the larger number and we'll find "?" in terms of x.

The smaller number plus the larger number is 14.

$$x \quad + \quad ? \quad = 14$$

We solve this "equation" for the "?" symbol.

$$x + ? = 14$$
$$\underline{-x \qquad -x}$$
$$? = 14 - x$$

So, $14 - x$ is the larger number. Three times the smaller is $3x$. Twice the larger is $2(14 - x)$. Their sum is 33, giving us the equation $3x + 2(14 - x) = 33$.

$$3x + 2(14 - x) = 33$$
$$3x + 28 - 2x = 33$$
$$x + 28 = 33$$
$$\underline{-28 \quad -28}$$
$$x = 5$$

The smaller number is 5 and the larger is $14 - x = 14 - 5 = 9$.

PRACTICE

1. The sum of two numbers is 10. Three times the smaller plus five times the larger number is 42. What are the numbers?

2. The difference between two numbers is 12. Twice the smaller plus four times the larger is 108. What are the two numbers?

3. The difference between two numbers is 8. The sum of one and a half times the smaller and four times the larger is 54. What are the numbers?

4. The sum of two numbers is 11. When twice the larger is subtracted from five times the smaller, the difference is 6. What are the numbers?

 SOLUTIONS

1. Let x represent the smaller number. The larger number is then $10 - x$.

$$3x + 5(10 - x) = 42$$
$$3x + 50 - 5x = 42$$
$$-2x + 50 = 42$$
$$\underline{-50 \quad -50}$$
$$-2x = -8$$
$$x = \frac{-8}{-2}$$
$$x = 4$$

The numbers are 4 and $10 - x = 10 - 4 = 6$.

2. The difference between the numbers is 12, so one number is 12 more than the other. Let x represent the smaller number. Then $x + 12$ is the larger. Twice the smaller is $2x$, and four times the larger is $4(x + 12)$.

$$2x + 4(x + 12) = 108$$
$$2x + 4x + 48 = 108$$
$$6x + 48 = 108$$
$$\underline{-48 \quad -48}$$
$$6x = 60$$
$$x = \frac{60}{6}$$
$$x = 10$$

The smaller number is 10 and the larger is $x + 12 = 10 + 12 = 22$.

3. The difference between the numbers is 8, so one of the numbers is 8 more than the other. Let x represent smaller number. The larger number

is $x + 8$. One and a half of the smaller number is $1\frac{1}{2}x$; four times the larger is $4(x + 8)$.

$$1\frac{1}{2}x + 4(x + 8) = 54$$

$$\frac{3}{2}x + 4x + 32 = 54$$

$$2\left(\frac{3}{2}x + 4x + 32\right) = 2(54)$$

$$\left[2\left(\frac{3}{2}x\right)\right] + 2(4x) + 2(32) = 108$$

$$3x + 8x + 64 = 108$$

$$11x + 64 = 108$$

$$\underline{-64 \quad -64}$$

$$11x = 44$$

$$x = \frac{44}{11}$$

$$x = 4$$

The smaller number is 4 and the larger, $x + 8 = 4 + 8 = 12$.

4. Let x = smaller number. Then $11 - x$ is the larger. Five times the smaller is $5x$, and twice the larger is $2(11 - x)$. "Twice the larger subtracted from five times the smaller" becomes "$5x - 2(11 - x)$."

$$5x - 2(11 - x) = 6$$

$$5x - 22 + 2x = 6$$

$$7x - 22 = 6$$

$$\underline{+22 \quad +22}$$

$$7x = 28$$

$$x = \frac{28}{7}$$

$$x = 4$$

The smaller number is 4 and the larger is $11 - x = 11 - 4 = 7$.

Problems Involving Three Unknowns

The problems in this section are a variation of number sense problems. What is new in this section is that we are given information about *three* quantities instead of two. Usually, two of the quantities are compared to a third. We let x represent the third quantity and write the other two in terms of x.

EXAMPLE

Jill is twice as old as Jim and Jim is 3 years older than Ken. The sum of their ages is 61. What are their ages?

Three quantities are being compared, so we find one age and relate the other two ages to it. This allows us to represent the sum using only one variable. Ken's age is being compared to Jim's and Jim's to Jill's. One route to take is to let x represent Jim's age. We can then write Jill's age in terms of Jim's age: $2x$. Jim is 3 years older than Ken, so Ken is 3 years younger than Jim. This makes Ken's age as $x - 3$. The sum of their ages is 61.

$$x + 2x + (x - 3) = 61$$
$$4x - 3 = 61$$
$$ +3 \quad +3$$
$$4x = 64$$
$$x = \frac{64}{4}$$
$$x = 16$$

Jim's age is 16; Jill's age is $2x = 2(16) = 32$; and Ken's age is $x - 3 = 16 - 3 = 13$.

Karen is 4 years older than Robert, and Jerri is half as old as Robert. The sum of their ages is 44. Find Karen's, Robert's, and Jerri's ages.

Both Karen's and Jerri's ages are being compared to Robert's age, so we let x represent Robert's age. Karen is four years older than Robert, so Karen's age is $x + 4$. Jerri is half as old as Robert, so Jerri's age is $\frac{1}{2}x$.

$$x + (x + 4) + \frac{x}{2} = 44$$
$$2x + 4 + \frac{x}{2} = 44$$
$$2\left(2x + 4 + \frac{x}{2}\right) = 2(44)$$
$$2(2x) + 2(4) + 2\left(\frac{x}{2}\right) = 88$$
$$4x + 8 + x = 88$$
$$5x + 8 = 88$$
$$ -8 \quad -8$$
$$5x = 80$$
$$x = \frac{80}{5}$$
$$x = 16$$

Robert's age is 16; Karen's age is $x + 4 = 16 + 4 = 20$; and Jerri's, $\frac{1}{2}x = \frac{1}{2}(16) = 8$.

PRACTICE

1. Andy is 3 years older than Bea and Bea is 5 years younger than Rose. If Rose is 28, how old are Andy and Bea?

2. Michele is 4 years younger than Steve and three times older than Sean. If the sum of their ages is 74, how old are they?

3. Monica earns three times per hour as John. John earns $2 more per hour than Alicia. Together they earn $43 per hour. How much is each one's hourly wage?

 SOLUTIONS

1. Because Rose is 28 and Bea is 5 years younger than Rose, Bea is $28 - 5 = 23$ years old. Andy is 3 years older than Bea, so Andy is $23 + 3 = 26$ years old.

2. Let x = Michele's age. Steve is 4 years older, so his age is $x + 4$. Sean is one-third Michele's age, so his age is $\frac{1}{3}x = \frac{x}{3}$.

$$x + (x + 4) + \frac{x}{3} = 74$$

$$2x + 4 + \frac{x}{3} = 74$$

$$3\left(2x + 4 + \frac{x}{3}\right) = 3(74)$$

$$3(2x) + 3(4) + 3\left(\frac{x}{3}\right) = 222$$

$$6x + 12 + x = 222$$

$$7x + 12 = 222$$

$$\underline{-12 \quad -12}$$

$$7x = 210$$

$$x = \frac{210}{7}$$

$$x = 30$$

Michele is 30 years old; Steve is $x + 4 = 34$; and Sean is $\frac{x}{2} = \frac{30}{3} = 10$.

We can avoid the fraction $\frac{x}{3}$ in this problem if we let x represent Sean's age. Then Michele's age would be $3x$; and Steve's, $3x + 4$.

3. Monica's earnings are being compared to John's, and John's to Alicia's. The easiest thing to do is to let x represent Alicia's hourly wage, so John's hourly wage is $x + 2$. Monica earns three times as much as John, so her hourly wage is $3(x + 2)$.

$$x + (x + 2) + 3(x + 2) = 43$$
$$x + x + 2 + 3x + 6 = 43$$
$$5x + 8 = 43$$
$$\underline{-8 \quad -8}$$
$$5x = 35$$
$$x = \frac{35}{5}$$
$$x = 7$$

Alicia earns $7 per hour; John, $x + 2 = 7 + 2 = \$9$; and Monica $3(x + 2) = 3(7 + 2) = \$27$.

Grade Problems

Grade computation problems are probably the most useful problem types to students. In these problems, the formula for the course grade and all but one grade are given. We are asked to compute the unknown grade in order to ensure a particular course average.

 EXAMPLE

A student has grades of 72, 74, 82, and 90. What does the next grade have to be to obtain an average of 80?

We take the average of five numbers: 72, 74, 82, 90 and the next grade. Call this next grade x. We want this average to be 80.

$$\frac{72 + 74 + 82 + 90 + x}{5} = 80$$
$$\frac{318 + x}{5} = 80$$
$$5\left(\frac{318 + x}{5}\right) = 5(80)$$
$$318 + x = 400$$
$$\underline{-318 \qquad -318}$$
$$x = 82$$

The student needs an 82 to raise his/her average to 80.

Angela has grades of 78, 83, 86, 82, and 88. If the next grade counts twice as much as each of the others, what does this grade need to yield an average of 85?

Even though there are a total of six grades, the last one counts twice as much as the others, so it is like having a total of seven grades; that is, the divisor needs to be seven. Let x represent the next grade.

$$\frac{78+83+86+82+88+2x}{7} = 85$$

$$\frac{417+2x}{7} = 85$$

$$7\left(\frac{417+2x}{7}\right) = 7(85)$$

$$417+2x = 595$$

$$\begin{array}{cc} -417 & -417 \end{array}$$

$$2x = 178$$

$$x = \frac{178}{2}$$

$$x = 89$$

Angela needs a grade of 89 to raise her average to 85.

A major project accounts for one-third of the course grade. The rest of the course grade is determined by the quiz average. Allan has quiz grades of 82, 80, 99, and 87, each counting equally. What does the project grade need to be to raise his average to 90?

The quiz average accounts for two-thirds of the grade and the project, one-third.

The equation to use, then, is $\frac{2}{3}$ quiz average $+ \frac{1}{3}$ project grade $= 90$. The quiz average is

$$\frac{82+80+99+87}{4} = 87$$

Let x represent the project grade.

$$\frac{2}{3}(87) + \frac{1}{3}x = 90$$

$$58 + \frac{x}{3} = 90$$

$$3\left(58 + \frac{x}{3}\right) = 3(90)$$

$$3(58) + 3\left(\frac{x}{3}\right) = 270$$

$$174 + x = 270$$

$$\underline{-174 \qquad -174}$$

$$x = 96$$

Allan needs a 96 on his project to raise his average to 90.

PRACTICE

1. Crystal's grades are 93, 89, 96, and 98. What does the next grade have to be to raise her average to 95?

2. Nichelle's grades are 79, 82, 77, 81, and 78. What does the next grade have to be to raise her average to 80?

3. A presentation grade counts toward one-fourth of the course grade. The average of the four tests counts toward the remaining three-fourths of the course grade. If a student's test scores are 61, 63, 65, and 83, what does he need to make on the presentation grade to raise his average to 70?

4. The final exam accounts for one-third of the course grade. The average of the four tests accounts for another third, and a presentation accounts for the final third. Ava's tests scores are 68, 73, 80, and 95. Her presentation grade is 75. What does the final exam grade need to be to raise her average to 80?

5. A book report counts toward one-fifth of a student's course grade. The remaining four-fifths of the courses' average is determined by the average of six quizzes. One student's book report grade is 90 and has quiz grades of 72, 66, 69, 80, and 85. What does she need to earn on her sixth quiz to raise her average to 80?

 **SOLUTIONS**

1. Let x = the next grade.

$$\frac{93+89+96+98+x}{5}=95$$

$$\frac{376+x}{5}=95$$

$$5\left(\frac{376+x}{5}\right)=5(95)$$

$$376+x=475$$

$$\begin{array}{r} -376 \quad -376 \\ x=99 \end{array}$$

Crystal's last grade needs to be 99 in order to raise her average to 95.

2. Let x = the next grade.

$$\frac{79+82+77+81+78+x}{6}=80$$

$$\frac{397+x}{6}=80$$

$$6\left(\frac{397+x}{6}\right)=6(80)$$

$$397+x=480$$

$$\begin{array}{r} -397 \quad -397 \\ x=83 \end{array}$$

Nichelle's next grade needs to be 83 to raise her average to 80.

3. Let x represent the presentation grade. The test average is $(61 + 63 + 65 + 83)/4 = 68$. Then $\frac{3}{4}$ test average + $\frac{1}{4}$ presentation grade = 70 becomes $\frac{3}{4}(68) + \frac{1}{4}x = 70$.

$$\frac{3}{4}(68)+\frac{1}{4}x=70$$

$$51+\frac{x}{4}=70$$

$$\begin{array}{r} -51 \quad -51 \\ \frac{x}{4}=19 \end{array}$$

$$x=4(19)$$

$$x=76$$

The student needs a 76 on his presentation to have a course grade of 70.

4. Let x represent Ava's final exam grade. Her test average is $(68 + 73 + 80 + 95)/4 = 79$. Then $\frac{1}{3}$ test average $+ \frac{1}{3}$ presentation grade $+ \frac{1}{3}$ final exam grade is 80 becomes

$$\frac{1}{3}(79) + \frac{1}{3}(75) + \frac{1}{3}(x) = 80.$$

$$\frac{79}{3} + 25 + \frac{x}{3} = 80$$

$$3\left(\frac{79}{3} + 25 + \frac{x}{3}\right) = 3(80)$$

$$3\left(\frac{79}{3}\right) + 3(25) + 3 \cdot \frac{x}{3} = 240$$

$$79 + 75 + x = 240$$

$$154 + x = 240$$

$$\underline{\quad -154 \qquad -154 \quad}$$

$$x = 86$$

Ava's final exam grade needs to be 86 to obtain an average of 80.

5. Let x = sixth quiz grade. The course grade $\frac{4}{5}$ quiz grade $+ \frac{1}{5}$ book report becomes

$$\frac{4}{5}\left(\frac{72 + 66 + 69 + 80 + 85 + x}{6}\right) + \frac{1}{5}(90).$$

Simplified, the above is

$$\frac{4}{5}\left(\frac{372 + x}{6}\right) + 18 = \frac{4(372) + 4x}{30} + 18 = \frac{1488 + 4x}{30} + 18.$$

We want this quantity to equal 80.

$$\frac{1488 + 4x}{30} + 18 = 80$$

$$30\left(\frac{1488 + 4x}{30} + 18\right) = 30(80)$$

$$30\left(\frac{1488 + 4x}{30}\right) + 30(18) = 2400$$

$$1488 + 4x + 540 = 2400$$

$$2028 + 4x = 2400$$

$$\underline{-2028 \qquad\quad -2028}$$

$$4x = 372$$

$$x = \frac{372}{4}$$

$$x = 93$$

The student needs a 93 on her quiz to raise her average to 80.

Coin Problems

Coin problems are also common algebra applications. Usually the total number of coins is given as well as the total dollar value. The question is normally "How many of each coin is there?" When there is more than one coin involved, we let x represent the number of one specific coin and the number of other coins in terms of x. The steps involved are:

1. Let x represent the number of a specific coin.
2. Write the number of other coins in terms of x.
3. Multiply the value of the coin by its number; this gives the total amount of money represented by each coin.
4. Add all of the terms obtained in Step 3 and set equal to the total money value.
5. Solve for x.
6. Answer the question. Don't forget this step! It is easy to feel like you are done when you've solved for x, but the answer to the question could be one step later.

As in all word problems, units of measure must be consistent. In the following problems, this means that all money will need to be in terms of dollars or in terms of cents. We use dollars in the following examples.

EXAMPLE

Terri has $13.45 in dimes and quarters. If there are 70 coins in all, how many of each type does she have?

Let x represent the number of dimes. Because the number of dimes and quarters is 70, $70 - x$ represents the number of quarters. Terri has x dimes, so she has $0.10x$ in dimes. She has $70 - x$ quarters, so she has $0.25(70 - x)$ in quarters. These two amounts must sum to $13.45.

$$0.10x \quad + 0.25(70 - x) = 13.45$$

$$\text{(amount in} \quad \text{(amount in}$$
$$\text{dimes)} \quad \text{quarters)}$$

$$0.10x + 0.25(70 - x) = 13.45$$
$$0.10x + 17.5 - 0.25x = 13.45$$
$$-0.15x + 17.5 = 13.45$$
$$\underline{-17.5 \ -17.50}$$
$$-0.15x = -4.05$$
$$x = \frac{-4.05}{-0.15}$$
$$x = 27$$

Terri has 27 dimes and $70 - x = 70 - 27 = 43$ quarters.

Bobbie has $1.54 in quarters, dimes, nickels, and pennies. He has twice as many dimes as quarters and three times as many nickels as dimes. The number of pennies is the same as the number of dimes. How many of each type of coin does he have?

Nickels are being compared to dimes, and dimes are being compared to quarters, so we will let x represent the number of quarters. Bobbie has twice as many dimes as quarters, so $2x$ is the number of dimes he has. He has three times as many nickels as dimes, namely triple $2x$: $3(2x) = 6x$. He has the same number of pennies as dimes, so he has $2x$ pennies.

How much of the total $1.54 does Bobbie have in each coin? He has x quarters, each worth $0.25, so he has a total of $0.25x$ (dollars) in quarters. He has $2x$ dimes, each worth $0.10; this gives him $0.10(2x) = 0.20x$ (dollars) in dimes. Bobbie has $6x$ nickels, each worth $0.05. The total amount of money in nickels, then, is $0.05(6x) = 0.30x$ (dollars). Finally, he has $2x$ pennies, each worth $0.01. The pennies count as $0.01(2x) = 0.02x$ (dollars).

The total amount of money is $1.54, so

$$
\underset{\substack{\text{(amount in}\\\text{quarters)}}}{0.25x} + \underset{\substack{\text{(amount in}\\\text{dimes)}}}{0.20x} + \underset{\substack{\text{(amount in}\\\text{nickels)}}}{0.30x} + \underset{\substack{\text{(amount in}\\\text{pennies)}}}{0.02x} = 1.54
$$

$$0.25x + 0.20x + 0.30x + 0.02x = 1.54$$
$$0.77x = 1.54$$
$$x = \frac{1.54}{0.77}$$
$$x = 2$$

Bobbie has 2 quarters; $2x = 2(2) = 4$ dimes; $6x = 6(2) = 12$ nickels; and $2x = 2(2) = 4$ pennies.

PRACTICE

1. A vending machine as $19.75 in dimes and quarters. There are 100 coins in all. How many dimes and quarters are in the machine?

2. Ann has $2.25 in coins. She has the same number of quarters as dimes. She has half as many nickels as quarters. How many of each coin does she have?

3. Sue has twice as many quarters as nickels and half as many dimes as nickels. If she has a total of $4.80, how many of each coin does she have?

SOLUTIONS

1. Let x represent the number of dimes. Then $100 - x$ is the number of quarters. There is $0.10x$ dollars in dimes and $0.25(100 - x)$ dollars in quarters.

$$0.10x + 0.25(100 - x) = 19.75$$
$$0.10x + 25 - 0.25x = 19.75$$
$$-0.15x + 25 = 19.75$$
$$\underline{-25 \quad -25.00}$$
$$-0.15x = -5.25$$
$$x = \frac{-5.25}{-0.15}$$
$$x = 35$$

There are 35 dimes and $100 - x = 100 - 35 = 65$ quarters.

2. Let x represent the number of quarters. There are as many dimes as quarters, so x also represents the number of dimes. There are half as many nickels as quarters, so $\frac{1}{2}x$ (or $0.50x$) is the number of nickels.

$$0.25x = \text{amount in quarters}$$
$$0.10x = \text{amount in dimes}$$
$$0.05(0.50x) = \text{amount in nickels}$$

$$0.25x + 0.10x + 0.05(0.50x) = 2.25$$
$$0.25x + 0.10x + 0.025x = 2.25$$
$$0.375x = 2.25$$
$$x = \frac{2.25}{0.375}$$
$$x = 6$$

There are 6 quarters, 6 dimes, $0.50x = 0.50(6) = 3$ nickels.

3. As both the number of quarters and dimes are being compared to the number of nickels, let x represent the number of nickels. Then $2x$ represents the number of quarters and $\frac{1}{2}x$ (or $0.50x$) is the number of dimes.

$$0.05x = \text{amount of money in nickels}$$
$$0.10(0.50x) = \text{amount of money in dimes}$$
$$0.25(2x) = \text{amount of money in quarters}$$

$$0.05x + 0.10(0.50x) + 0.25(2x) = 4.80$$
$$0.05x + 0.05x + 0.50x = 4.80$$
$$0.60x = 4.80$$
$$x = \frac{4.80}{0.60}$$
$$x = 8$$

There are 8 nickels, $0.50x = 0.50(8) = 4$ dimes, and $2x = 2(8) = 16$ quarters.

Investment Problems

We now learn how to solve basic investment problems. An amount of money is divided into two investments earning different interest rates. We are told how much interest is earned in a year, information that we will use to determine how much was invested at each rate.

HINT *If we let x represent the amount earned at interest rate A, then the rest is invested at interest rate B. "The rest" is the total dollars less the amount invested at interest rate A. If x dollars is invested at interest rate A, then the amount of interest earned in a year is Ax. If "Total – x" dollars is invested at interest rate B, then B(Total – x) is the amount of interest earned in a year at interest rate B. We use the following equation (model) to solve these problems.*

$$Ax + B(\text{Total} - x) = \text{Interest earned}$$

EXAMPLE

Dora had $10,000 to invest. She deposited her money into two accounts—one paying 6% interest and the other $7\frac{1}{2}$% interest. If at the end of the year the total interest earned was $682.50, how much was originally deposited in each account?

We could either let x represent the amount deposited at 6% or at $7\frac{1}{2}$%. Here, we will let x represent the amount deposited into the 6% account. Because the two amounts must sum to 10,000, 10,000 – x is the amount deposited at $7\frac{1}{2}$%. The amount of interest earned at 6% is 0.06x, and the amount of interest earned at $7\frac{1}{2}$% is 0.075(10,000 – x). The total amount of interest is $682.50, so we want to solve 0.06x + 0.075(10,000 – x) = 682.50. The solution to this equation gives us the amount to invest at 6%, and putting this solution into 10,000 – x gives us the amount to invest at $7\frac{1}{2}$%.

$$0.06x + 0.075(10,000 - x) = 682.50$$
$$0.06x + 750 - 0.075x = 682.50$$
$$-0.015x + 750 = 682.50$$
$$\underline{-750 \quad -750.00}$$
$$-0.015x = -67.50$$
$$x = \frac{-67.50}{-0.015}$$
$$x = 4500$$

Dora deposited $4500 in the 6% account and 10,000 – x = 10,000 – 4500 = $5500 in the $7\frac{1}{2}$% account.

PRACTICE

1. A businessman invested $50,000 into two funds which yielded profits of $16\frac{1}{2}$% and 18%. If the total profit was $8520, how much was invested in each fund?

2. A college student deposited $3500 into two savings accounts, one with an annual yield of $4\frac{3}{4}$% and the other with an annual yield of $5\frac{1}{4}$%. If he earned $171.75 total interest the first year, how much was deposited in each account?

3. A banker plans to lend $48,000 at a simple interest rate of 16% and the remainder at 19%. How should she allot the loans in order to obtain a return of $18\frac{1}{2}$%?

SOLUTIONS

1. Let x represent the amount invested at $16\frac{1}{2}$%. Then $50,000 - x$ represents the amount invested at 18%. The profit from the $16\frac{1}{2}$% account is $0.165x$, and the profit from the 18% investment is $0.18(50,000 - x)$. The sum of the profits is $8520.

$$0.165x + 0.18(50,000 - x) = 8520$$

$$0.165x + 9000 - 0.180x = 8520$$

$$-0.015x + 9000 = 8520$$

$$\underline{-9000 \quad -9000}$$

$$-0.015x = -480$$

$$x = \frac{-480}{-0.015}$$

$$x = 32,000$$

The amount invested at $16\frac{1}{2}$% is $32,000, and the amount invested at 18% is $50,000 - x = 50,000 - 32,000 = \$18,000$.

2. Let x represent the amount deposited at $4\frac{3}{4}\%$. Then the amount deposited at $5\frac{1}{4}\%$ is $3500 - x$. The interest earned at $4\frac{3}{4}\%$ is $0.0475x$; the interest earned at $5\frac{1}{4}\%$ is $0.0525(3500 - x)$. The sum of these two quantities is 171.75.

$$0.0475x + 0.0525(3500 - x) = 171.75$$
$$0.0475x + 183.75 - 0.0525x = 171.75$$
$$183.75 - 0.005x = 171.75$$
$$\underline{-183.75 \qquad\qquad -183.75}$$
$$-0.005x = -12$$
$$x = \frac{-12}{-0.005}$$
$$x = 2400$$

$2400 was deposited in the $4\frac{3}{4}\%$ account, and $3500 - x = 3500 - 2400 = \$1100$ was deposited in the $5\frac{1}{4}\%$ account.

3. Let x represent the amount to be loaned at 16%, so $48{,}000 - x$ represents the amount to be loaned at 19%. The total amount of return should be $18\frac{1}{2}\%$ of 48,000 which is $0.185(48{,}000) = 8880$.

$$0.16x + 0.19(48{,}000 - x) = 8880$$
$$0.16x + 9120 - 0.19x = 8880$$
$$9120 - 0.03x = 8880$$
$$\underline{-9120 \qquad\qquad -9120}$$
$$-0.03x = -240$$
$$x = \frac{-240}{-0.03}$$
$$x = 8000$$

$8000 should be loaned at 16%, and $48{,}000 - x = 48{,}000 - 8000 = \$40{,}000$ should be loaned at 19%.

Mixture Problems

Mixture problems involve mixing two different concentrations of a substance to obtain some concentration in between. Often these problems are stated as alcohol or acid solutions, but there are many more types. For example, you might want to know how many pure peanuts should be mixed with a 40%

peanut mixture to obtain a 50% peanut mixture. You might have a two-cycle engine requiring a particular oil and gas mixture. Or you might have a recipe calling for 1% fat milk and all you have on hand is 2% fat milk and $\frac{1}{2}$% fat milk. We solve these problems with the following method.

Mixture problems involve three quantities—the two concentrations being mixed together and the final concentration. Of these three quantities, one of them is a fixed number. We let the variable represent one of the two concentrations being mixed. The other unknown quantity is written as some combination of the variable and the fixed quantity. If one of the quantities being mixed is known, then we let x represent the other quantity being mixed and the final solution would be "x + known quantity." If the final solution is known, we again let x represent one of the quantities being mixed, the other quantity being mixed would be of the form "final solution quantity $-x$." For example, in the following problem, the amount of one of the two concentrations being mixed is known: "How many liters of 10% acid solution should be mixed with 75 liters of 30% acid solution to yield a 25% acid solution?"

If we let x represent the number of liters of 10% acid solution, then $x + 75$ represents the number of liters of the final solution. If the problem were stated, "How many liters of 10% acid solution and 30% solution should be mixed together with to produce 100 liters of 25% solution?" We can let x represent either the number of liters of 10% solution or the 30% solution. Here, we let x represent the number of liters of 10% solution. How do we represent the number of liters of 30% solution? For the moment, let "?" represent the number of liters of 30% solution. We know that the final solution must be 100 liters, so the two amounts must sum to 100:

$$x + ? = 100.$$
$$x + ? = 100$$
$$\underline{-x \qquad\quad -x}$$
$$? = 100 - x$$

Now we see that $100 - x$ represents the number of liters of 30% solution.

Many mixture problems can be represented by drawing three boxes. We write the percentages given above the boxes and the volume inside the boxes. Below the boxes, we multiply the percentages (converted to decimal numbers) and the volume below the boxes; this gives us the equation to solve. Incidentally, the product of the percent and the volume, is the amount of pure acid/alcohol/milk-fat, etc. in that particular concentration.

EXAMPLE

How much 10% acid solution should be added to 30 liters of 25% acid solution to achieve a 15% solution?

Let x represent the amount of 10% solution. Then the total amount of solution is $30 + x$.

10%		25%		15%
x liters	$+$	30 liters	$=$	$30 + x$ liters
$0.10x$	$+$	$0.25(30)$	$=$	$0.15(30 + x)$

There are $0.10x$ liters of pure acid in the 10% mixture, $0.25(30)$ liters of pure acid in the 25% mixture, and $0.15(x + 30)$ liters of pure acid in the 15% mixture.

We solve the equation that is on the bottom row.

$$0.10x + 0.25(30) = 0.15(x + 30)$$
$$0.10x + 7.5 = 0.15x + 4.5$$
$$\underline{-0.10x \qquad\quad -0.10x}$$
$$7.5 = 0.05x + 4.5$$
$$\underline{-4.5 \qquad\qquad -4.5}$$
$$3 = 0.05x$$
$$\frac{3}{0.05} = x$$
$$60 = x$$

Add 60 liters of 10% acid solution to 30 liters of 25% acid solution to achieve a 15% acid solution.

How much 10% acid solution and 30% acid solution should be mixed together to yield 100 liters of a 25% acid solution?

Let x represent the amount of 10% acid solution, so $100 - x$ represents the amount of 30% acid solution.

10%		30%		25%
x liters	+	$100 - x$ liters	=	100 liters
0.10x	+	0.30(100 − x)	=	0.25(100)

$$0.10x + 0.30(100 - x) = 0.25(100)$$
$$0.10x + 30 - 0.30x = 25$$
$$30 - 0.20x = 25$$
$$\underline{-30 \qquad\qquad -30}$$
$$-0.20x = -5$$
$$x = \frac{-5}{-0.20}$$
$$x = 25$$

Add 25 liters of 10% solution to $100 - x = 100 - 25 = 75$ liters of 30% solution to obtain 100 liters of 25% solution.

How much pure alcohol should be added to 6 liters of 30% alcohol solution to obtain a 40% alcohol solution?

We represent pure alcohol as a 100% mixture. Let x represent the amount of pure alcohol.

100%		30%		40%
x liters	+	6 liters	=	$x + 6$ liters
1.00x	+	0.30(6)	=	0.40(x + 6)

$$1.00x + 0.30(6) = 0.40(x + 6)$$
$$1.00x + 1.80 = 0.40x + 2.40$$
$$0.60x + 1.80 = 2.40$$
$$0.60x = 0.60$$
$$x = 1$$

Add one liter of pure alcohol to 6 liters of 30% alcohol solution to obtain a 40% alcohol solution.

How much water should be added to 9 liters of 45% solution to weaken it to a 30% solution?

Think of water as a "0% mixture."

0%		45%		30%
x liters	+	9 liters	=	$x + 9$ liters
$0x$	+	$0.45(9)$	=	$0.30(x + 9)$

$$0x + 0.45(9) = 0.30(x + 9)$$
$$0 + 4.05 = 0.30x + 2.70$$
$$-2.70 \qquad -2.70$$
$$1.35 = 0.30x$$
$$\frac{1.35}{0.30} = x$$
$$4.5 = x$$

Add 4.5 liters of water to weaken 9 liters of 45% solution to a 30% solution.

How much pure acid and 30% acid solution should be mixed together to obtain 28 quarts of 40% acid solution?

Think of pure acid as a 100% mixture. We let x represent the amount of pure acid.

100%		30%		40%
x quarts	+	$28 - x$ quarts	=	28 quarts
$1.00x$	+	$0.30(28 - x)$	=	$0.40(28)$

$$1.00x + 0.30(28 - x) = 0.40(28)$$
$$x + 8.40 - 0.30x = 11.2$$
$$0.70x + 8.40 = 11.2$$
$$-8.40 \quad -8.4$$
$$0.70x = 2.8$$
$$x = \frac{2.8}{0.70}$$
$$x = 4$$

Add 4 quarts of pure acid to $28 - x = 28 - 4 = 24$ quarts of 30% acid solution to yield 28 quarts of a 40% solution.

PRACTICE

1. How much 60% acid solution should be added to 8 liters of 25% acid solution to produce a 40% acid solution?

2. How many quarts of $\frac{1}{2}$% fat milk should be added to 4 quarts of 2% fat milk to produce 1% fat milk?

3. How much 30% alcohol solution should be mixed with 70% alcohol solution to produce 12 liters of 60% alcohol solution?

4. How much 65% acid solution and 25% acid solution should be mixed together to produce 180 ml of 40% acid solution?

5. How much water should be added to 10 liters of 45% alcohol solution to produce a 30% solution?

6. How much decaffeinated coffee (assume this means 0% caffeine) and 50% caffeine coffee should be mixed to produce 25 cups of 40% caffeine coffee?

7. How much pure acid should be added to 18 ounces of 35% acid solution to produce 50% acid solution?

8. How many peanuts should be mixed with a nut mixture that is 40% peanuts to produce 36 ounces of a 60% peanut mixture?

SOLUTIONS

1.

60%		25%		40%
x liters	$+$	8 liters	$=$	$x + 28$ liters
0.60x	$+$	0.25(8)	$=$	0.40($x + 28$)

$$0.60x + 0.25(8) = 0.40(x + 8)$$
$$0.60x + 2 = 0.40x + 3.2$$
$$\underline{-0.40x \qquad -0.40x}$$
$$0.20x + 2.0 = 3.2$$
$$\underline{-2.0 \qquad -2.0}$$
$$0.20x = 1.2$$
$$x = \frac{1.2}{0.20}$$
$$x = 6$$

Add 6 liters of 60% solution to 8 liters of 25% solution to produce a 40% solution.

2.   0.5%   2%   1%

| x quarts | $+$ | 4 quarts | $=$ | $x + 4$ quarts |

    $0.005x$ $+$ $0.02(4)$ $=$ $0.01(x + 4)$

$$0.005x + 0.02(4) = 0.01(x + 4)$$
$$0.005x + 0.08 = 0.010x + 0.04$$
$$-0.005x \qquad\qquad -0.005x$$
$$0.08 = 0.005x + 0.04$$
$$-0.04 \qquad\qquad -0.04$$
$$0.04 = 0.005x$$
$$\frac{0.04}{0.005} = x$$
$$8 = x$$

Add 8 quarts of $\frac{1}{2}$% fat milk to 4 quarts of 2% milk to produce 1% milk.

3.   30%   70%   60%

| x liters | $+$ | $12 - x$ liters | $=$ | 12 liters |

    $0.30x$ $+$ $0.70(12 - x)$ $=$ $0.60(12)$

$$0.30x + 0.70(12 - x) = 0.60(12)$$
$$0.30x + 8.4 - 0.70x = 7.2$$
$$-0.40x + 8.4 = 7.2$$
$$-8.4 \quad -8.4$$
$$-0.40x = -1.2$$
$$x = \frac{-1.2}{-0.40}$$
$$x = 3$$

Add 3 liters of 30% alcohol solution to $12 - x = 12 - 3 = 9$ liters of 70% alcohol solution to produce 12 liters of 60% alcohol solution.

4.

65%		25%		40%
x ml	+	$180 - x$ ml	=	180 ml

$$0.65x \quad + \quad 0.25(180 - x) \quad = \quad 0.40(180)$$

$$0.65x + 0.25(180 - x) = 0.40(180)$$

$$0.65x + 45 - 0.25x = 72$$

$$0.40x + 45 = 72$$

$$\underline{-45 \quad -45}$$

$$0.40x = 27$$

$$x = \frac{27}{0.40}$$

$$x = 67.5$$

Add 67.5 ml of 65% acid solution to 180 – x = 180 – 67.5 = 112.5 ml of 25% solution to produce 180 ml of 40% acid solution.

5.

0%		45%		30%
x liters	+	10 liters	=	$x + 10$ liters

$$0x \quad + \quad 0.45(10) \quad = \quad 0.30(x + 10)$$

$$0x + 0.45(10) = 0.30(x + 10)$$

$$0 + 4.5 = 0.30x + 3$$

$$\underline{-3.0 \qquad -3}$$

$$1.5 = 0.30x$$

$$\frac{1.5}{0.30} = x$$

$$5 = x$$

Add 5 liters of water to 10 liters of 45% alcohol solution to produce a 30% alcohol solution.

6.

0%		50%		40%
x cups	$+$	$25 - x$ cups	$=$	25 cups

$$0x \quad + \quad 0.50(25 - x) \quad = \quad 0.40(25)$$

$$0x + 0.50(25 - x) = 0.40(25)$$

$$0 + 12.5 - 0.50x = 10.0$$

$$\underline{-12.5 \qquad\qquad -12.5}$$

$$-0.50x = -2.5$$

$$x = \frac{-2.5}{-0.50}$$

$$x = 5$$

Mix 5 cups of decaffeinated coffee with $25 - x = 25 - 5 = 20$ cups of 50% caffeine coffee to produce 25 cups of 40% caffeine coffee.

7.

100%		35%		50%
x ounces	$+$	18 ounces	$=$	$x + 18$ ounces

$$1.00x \quad + \quad 0.35(18) \quad = \quad 0.50(x + 18)$$

$$1.00x + 0.35(18) = 0.50(x + 18)$$

$$1.00x + 6.3 = 0.50x + 9$$

$$\underline{-0.50x \qquad\qquad -0.50x}$$

$$0.50x + 6.3 = 9.0$$

$$\underline{-6.3 \quad -6.3}$$

$$0.50x = 2.7$$

$$x = \frac{2.7}{0.50}$$

$$x = 5.4$$

Add 5.4 ounces of pure acid to 18 ounces of 35% acid solution to produce a 50% acid solution.

8.

100%		40%		60%
x ounces	$+$	$36 - x$ ounces	$=$	36 ounces

$$1.00x \quad + \quad 0.40(36 - x) \quad = \quad 0.60(36)$$

$$1.00x + 0.40(36 - x) = 0.60(36)$$

$$1.00x + 14.4 - 0.40x = 21.6$$

$$0.60x + 14.4 = 21.6$$

$$-14.4 \quad -14.4$$

$$0.60x = 7.2$$

$$x = \frac{7.2}{0.6}$$

$$x = 12$$

Add 12 ounces of peanuts to $36 - x = 36 - 12 = 24$ ounces of a 40% peanut mixture to produce 36 ounces of a 60% peanut mixture.

Work Problems

Work problems are another staple of algebra courses. A work problem is normally stated as two workers (two people, machines, hoses, drains, etc.) working together and working separately to complete a task. Usually one worker performs faster than the other. Sometimes the problem states how fast each can complete the task alone and we are asked to find how long it takes for them to complete the task together. At other times, we are told how long one worker takes to complete the task alone and how long it takes for both working together to complete it; we are asked how long the second worker would take to complete the task alone. The formula is quantity (work done—usually 1) = rate times time: $Q = rt$. The method outlined below will help you solve most, if not all, work problems. The chart shown is useful in solving these problems.

Worker	Quantity	Rate	Time
1st Worker			
2nd Worker			
Together			

There are four equations in this chart. One of them is the one we use to solve for the unknown. Each horizontal line in the chart represents the equation $Q = rt$ for that particular line. The fourth equation comes from the sum of each worker's rate equaling the together rate. Often, the fourth equation is the one we need to solve. Remember, as in all word problems, all units of measure must be consistent.

EXAMPLE

Joe takes 45 minutes to mow a lawn. His older brother Jerry takes 30 minutes to mow the lawn. If they work together, how long will it take for them to mow the lawn?

The quantity in each of the three cases is 1—there is one yard to be mowed. Use the formula $Q = rt$ and the data given in the problem to fill in all nine boxes. Because we are looking for the time (in minutes) it takes for them to mow the lawn together, let t represent the number of minutes needed to mow the lawn together.

Worker	Quantity	Rate	Time
Joe	1		45
Jerry	1		30
Together	1		t

Because $Q = rt$, $r = \dfrac{Q}{t}$. But $Q = 1$, so $r = \dfrac{1}{t}$. This makes Joe's rate $\frac{1}{45}$ and Jerry's rate $\frac{1}{30}$. The together rate is $\dfrac{1}{t}$.

Worker	Quantity	Rate	Time
Joe	1	1/45	45
Jerry	1	1/30	30
Together	1	1/t	t

Of the four equations on the chart, only "Joe's rate + Jerry's rate = Together rate" has enough information in it to solve for t.

The equation to solve is $\dfrac{1}{45} + \dfrac{1}{30} = \dfrac{1}{t}$. The LCD is $90t$.

$$\frac{1}{45} + \frac{1}{30} = \frac{1}{t}$$

$$90t\left(\frac{1}{45} + \frac{1}{30}\right) = 90t\left(\frac{1}{t}\right)$$

$$90t\left(\frac{1}{45}\right) + 90t\left(\frac{1}{30}\right) = 90$$

$$2t + 3t = 90$$

$$5t = 90$$

$$t = \frac{90}{5}$$

$$t = 18$$

Joe and Jerry can mow the lawn in 18 minutes.

Tammy can wash a car in 40 minutes. When working with Jim, they can wash the same car in 15 minutes. How long would Jim need to wash the car by himself?

Let t represent the number of minutes Jim needs to wash the car by himself.

Worker	Quantity	Rate	Time
Tammy	1	1/40	40
Jim	1	1/t	t
Together	1	1/15	15

The equation to solve is $\dfrac{1}{40} + \dfrac{1}{t} = \dfrac{1}{15}$. The LCD is $120t$.

$$\frac{1}{40} + \frac{1}{t} = \frac{1}{15}$$

$$120t\left(\frac{1}{40} + \frac{1}{t}\right) = 120t\left(\frac{1}{15}\right)$$

$$120t\left(\frac{1}{40}\right) + 120t\left(\frac{1}{t}\right) = 8t$$

$$3t + 120 = 8t$$

$$-3t \qquad\qquad -3t$$

$$120 = 5t$$

$$\frac{120}{5} = t$$

$$24 = t$$

Jim needs 24 minutes to wash the car by himself.

Kellie can mow the campus yard in $2\frac{1}{2}$ hours. When Bobby helps, they can mow the yard in $1\frac{1}{2}$ hours. How long would Bobby need to mow the yard by himself?

Let t represent the number of hours Bobby needs to mow the yard himself.

Kellie's time is $2\frac{1}{2}$ or $\frac{5}{2}$. Then her rate is $\frac{1}{\frac{5}{2}} = \frac{2}{5}$.

The together time is $1\frac{1}{2}$ or $\frac{3}{2}$, so the together rate is $\frac{1}{\frac{3}{2}} = \frac{2}{3}$.

Worker	Quantity	Rate	Time
Kellie	1	2/5	$2\frac{1}{2}$
Bobby	1	1/t	t
Together	1	2/3	$1\frac{1}{2}$

The equation to solve is $\dfrac{2}{5} + \dfrac{1}{t} = \dfrac{2}{3}$. The LCD is $15t$.

$$\frac{2}{5} + \frac{1}{t} = \frac{2}{3}$$

$$15t\left(\frac{2}{5} + \frac{1}{t}\right) = 15t\left(\frac{2}{3}\right)$$

$$15t\left(\frac{2}{5}\right) + 15t\left(\frac{1}{t}\right) = 10t$$

$$6t + 15 = 10t$$

$$-6t \qquad\quad -6t$$

$$15 = 4t$$

$$\frac{15}{4} = t$$

Bobby needs $\dfrac{15}{4} = 3\frac{3}{4}$ hours or 3 hours 45 minutes to mow the yard by himself.

PRACTICE

1. Sherry and Denise together can mow a yard in 20 minutes. Alone, Denise can mow the yard in 30 minutes. How long would Sherry need to mow the yard by herself?

2. Together, Ben and Brandon can split a pile of wood in 2 hours. If Ben could split the same pile of wood in 3 hours, how long would it take Brandon to split the pile alone?

3. A boy can weed the family garden in 90 minutes. His sister can weed it in 60 minutes. How long will they need to weed the garden if they work together?

4. Robert needs 40 minutes to assemble a bookcase. Paul needs 20 minutes to assemble the same bookcase. How long will it take them to assemble the bookcase if they work together?

5. Together, two pipes can fill a reservoir in $\frac{3}{4}$ of an hour. Pipe I requires 1 hour 10 minutes ($1\frac{1}{6}$ hours) to fill the reservoir by itself. How long would Pipe II need to fill the reservoir by itself?

6. Pipe I can drain a reservoir in 6 hours 30 minutes ($6\frac{1}{2}$ hours). Pipe II can drain the same reservoir in 4 hours 20 minutes ($4\frac{1}{3}$ hours). How long will it take to drain the reservoir if both pipes are used?

 SOLUTIONS

In the following, t will represent the unknown time.

1.

Worker	Quantity	Rate	Time
Sherry	1	1/t	t
Denise	1	1/30	30
Together	1	1/20	20

The equation to solve is $\dfrac{1}{t} + \dfrac{1}{30} = \dfrac{1}{20}$. The LCD is $60t$.

$$\frac{1}{t} + \frac{1}{30} = \frac{1}{20}$$

$$60t\left(\frac{1}{t} + \frac{1}{30}\right) = 60t\left(\frac{1}{20}\right)$$

$$60t\left(\frac{1}{t}\right) + 60t\left(\frac{1}{30}\right) = 3t$$

$$60 + 2t = 3t$$

$$-2t \quad -2t$$

$$60 = t$$

Alone, Denise can mow the yard in 60 minutes.

2.

Worker	Quantity	Rate	Time
Ben	1	1/3	3
Brandon	1	1/t	t
Together	1	1/2	2

The equation to solve is $\dfrac{1}{3} + \dfrac{1}{t} = \dfrac{1}{2}$. The LCD is $6t$.

$$\frac{1}{3} + \frac{1}{t} = \frac{1}{2}$$

$$6t\left(\frac{1}{3} + \frac{1}{t}\right) = 6t\left(\frac{1}{2}\right)$$

$$6t\left(\frac{1}{3}\right) + 6t\left(\frac{1}{t}\right) = 3t$$

$$2t + 6 = 3t$$

$$\underline{-2t \qquad -2t}$$

$$6 = t$$

Brandon can split the wood-pile by himself in 6 hours.

3.

Worker	Quantity	Rate	Time
Boy	1	1/90	90
Girl	1	1/60	60
Together	1	1/t	t

The equation to solve is $\dfrac{1}{90} + \dfrac{1}{60} = \dfrac{1}{t}$. The LCD is $180t$.

$$\frac{1}{90} + \frac{1}{60} = \frac{1}{t}$$

$$180t\left(\frac{1}{90} + \frac{1}{60}\right) = 180t\left(\frac{1}{t}\right)$$

$$180t\left(\frac{1}{90}\right) + 180t\left(\frac{1}{60}\right) = 180$$

$$2t + 3t = 180$$

$$5t = 180$$

$$t = \frac{180}{5}$$

$$t = 36$$

Working together, the boy and girl need 36 minutes to weed the garden.

4.

Worker	Quantity	Rate	Time
Robert	1	1/40	40
Paul	1	1/20	20
Together	1	1/t	t

The equation to solve is $\dfrac{1}{40} + \dfrac{1}{20} = \dfrac{1}{t}$. **The LCD is 40$t$.**

$$\frac{1}{40} + \frac{1}{20} = \frac{1}{t}$$

$$40t\left(\frac{1}{40} + \frac{1}{20}\right) = 40t\left(\frac{1}{t}\right)$$

$$40t\left(\frac{1}{40}\right) + 40t\left(\frac{1}{20}\right) = 40$$

$$t + 2t = 40$$

$$3t = 40$$

$$t = \frac{40}{3} = 13\tfrac{1}{3}$$

Together Robert and Paul can assemble the bookcase in $13\tfrac{1}{3}$ minutes or 13 minutes 20 seconds.

5.

Worker	Quantity	Rate	Time
Pipe I	1	6/7 $\dfrac{1}{7/6} = 6/7$	7/6
Pipe II	1	1/t	t
Together	1	4/3 $\dfrac{1}{3/4} = 4/3$	3/4

The equation to solve is $\dfrac{6}{7} + \dfrac{1}{t} = \dfrac{4}{3}$. **The LCD is 21$t$.**

$$\frac{6}{7} + \frac{1}{t} = \frac{4}{3}$$

$$21t\left(\frac{6}{7} + \frac{1}{t}\right) = 21t\left(\frac{4}{3}\right)$$

$$21t\left(\frac{6}{7}\right) + 21t\left(\frac{1}{t}\right) = 28t$$

$$18t + 21 = 28t$$

$$\underline{-18t \qquad\qquad -18t}$$

$$21 = 10t$$

$$\frac{21}{10} = t$$

Alone, Pipe II can fill the reservoir in $2\tfrac{1}{10}$ hours or 2 hours 6 minutes.
($\tfrac{1}{10}$ of an hour is $\tfrac{1}{10}$ of 60 minutes and $\dfrac{1}{10} \cdot 60 = 6$.)

6.

Worker	Quantity	Rate	Time
Pipe I	1	$2/13$ $\dfrac{1}{13/2} = 2/13$	$6\frac{1}{2} = 13/2$
Pipe II	1	$3/13$ $\dfrac{1}{13/3} = 3/13$	$4\frac{1}{3} = \dfrac{13}{3}$
Together	1	$1/t$	t

The equation to solve is $\dfrac{2}{13} + \dfrac{3}{13} = \dfrac{1}{t}$. The LCD is $13t$.

$$\frac{2}{13} + \frac{3}{13} = \frac{1}{t}$$

$$13t\left(\frac{2}{13} + \frac{3}{13}\right) = 13t\left(\frac{1}{t}\right)$$

$$13t\left(\frac{2}{13}\right) + 13t\left(\frac{3}{13}\right) = 13$$

$$2t + 3t = 13$$

$$5t = 13$$

$$t = \frac{13}{5}$$

Together the pipes can drain the reservoir in $2\frac{3}{5}$ hours or 2 hours 36 minutes. ($\frac{3}{5}$ of hour is $\frac{3}{5}$ of 60 minutes and $\frac{3}{5} \cdot 60 = 36$.)

Some work problems require part of the work being performed by one worker before the other worker joins in, or both start the job and one finishes the job. In these cases, the together quantity will not be 1. Take the time the one worker works alone divided by the time that worker requires to do the entire job by himself and then subtract from this 1. This number is the proportion left over for both to work together. For example, if a pipe can empty a tank in 3 hours and works alone for 1 hour, then $\frac{1}{3}$ of the job is complete and $\frac{2}{3}$ of the job remains when the second begins work. The proportion of the work remaining becomes the quantity for the "together" line.

EXAMPLE

Jerry needs 40 minutes to mow the lawn. Lou can mow the same lawn in 30 minutes. If Jerry works alone for 10 minutes then Lou joins in, how long will it take for them to finish the job?

Because Jerry worked for 10 minutes, he did $\frac{10}{40} = \frac{1}{4}$ of the job alone. So, there is $1 - \frac{1}{4} = \frac{3}{4}$ of the job remaining when Lou started working. Let t represent the number of minutes they worked together—after Lou joins in. Even though Lou does not work the entire job, his rate is still $\frac{1}{30}$.

Worker	Quantity	Rate	Time
Jerry	1	1/40	40
Lou	1	1/30	30
Together	3/4	$\dfrac{3/4}{t} = \dfrac{3}{4t}$	t

The equation to solve is $\dfrac{1}{40} + \dfrac{1}{30} = \dfrac{3}{4t}$. The LCD is $120t$.

$$\frac{1}{40} + \frac{1}{30} = \frac{3}{4t}$$

$$120t\left(\frac{1}{40} + \frac{1}{30}\right) = 120t\left(\frac{3}{4t}\right)$$

$$120t\left(\frac{1}{40}\right) + 120t\left(\frac{1}{30}\right) = 90$$

$$3t + 4t = 90$$

$$7t = 90$$

$$t = \frac{90}{7}$$

Together, they will work $\frac{90}{7} = 12\frac{6}{7}$ minutes.

Pipe I can fill a reservoir in 6 hours. Pipe II can fill the same reservoir in 4 hours. If Pipe II is used alone for $2\frac{1}{2}$ hours, then Pipe I joins in to finish the job, how long will the Pipe I be used?

The amount of time Pipe I is used is the same as the amount of time both pipes work together. Let t represent the number of hours both pipes are used. Alone, Pipe II performed $2\frac{1}{2}$ parts of a 4-part job:

$$\frac{2\frac{1}{2}}{4} = \frac{5}{2} \div 4 = \frac{5}{2} \cdot \frac{1}{4} = \frac{5}{8}, \text{ so } 1 - \frac{5}{8} = \frac{3}{8}$$

of the job remains.

Worker	Quantity	Rate	Time
Pipe I	1	1/6	6
Pipe II	1	1/4	4
Together	3/8	$\dfrac{3/8}{t} = \dfrac{3}{8t}$	t

The equation to solve is $\dfrac{1}{6} + \dfrac{1}{4} = \dfrac{3}{8t}$. The LCD is 24$t$.

$$\frac{1}{6} + \frac{1}{4} = \frac{3}{8t}$$

$$24t\left(\frac{1}{6} + \frac{1}{4}\right) = 24t\left(\frac{3}{8t}\right)$$

$$24t\left(\frac{1}{6}\right) + 24t\left(\frac{1}{4}\right) = 9$$

$$4t + 6t = 9$$

$$10t = 9$$

$$t = \frac{9}{10}$$

Both pipes together will be used for $\frac{9}{10}$ hours or $\frac{9}{10} \cdot 60 = 54$ minutes. Hence, Pipe I will be used for 54 minutes.

Press A can print 100 fliers per minute. Press B can print 150 fliers per minute. The presses will be used to print 150,000 fliers.

(a) How long will it take for both presses to complete the run if they work together?

(b) If Press A works alone for 24 minutes then Press B joins in, how long will it take both presses to complete the job?

These problems are different from the previous work problems because the *rates* are given, not the times. Before, we used $Q = rt$ implies $r = Q/t$. Here, we will use $Q = rt$ to fill in the Quantity boxes.

(a) Press A's rate is 100, and Press B's rate is 150. The together quantity is 150,000. Let t represent the number of minutes both presses work together; this is also how much time each individual press will run. Press A's quantity is 100t, and Press B's quantity is 150t. The together rate is $r = Q/t = 150{,}000/t$.

Worker	Quantity	Rate	Time
Press A	$100t$	100	t
Press B	$150t$	150	t
Together	150,000	$150{,}000/t$	t

In this problem, the quantity produced by Press A plus the quantity produced by Press B will equal the quantity produced together. This gives the equation $100t + 150t = 150{,}000$. (Another equation that works is $100 + 150 = 150{,}000/t$.)

$$100t + 150t = 150{,}000$$
$$250t = 150{,}000$$
$$t = \frac{150{,}000}{250}$$
$$t = 600$$

The presses will run for 600 minutes or 10 hours.

(b) Because Press A works alone for 24 minutes, it has run $24 \times 100 = 2400$ fliers. When Press B begins its run, there are $150{,}000 - 2400 = 147{,}600$ fliers left to run. Let t represent the number of minutes both presses are running. This is also how much time Press B spends on the run. The boxes will represent work done together.

Worker	Quantity	Rate	Time
Press A	$100t$	100	t
Press B	$150t$	150	t
Together	147,600	$147{,}600/t$	t

The equation to solve is $100t + 150t = 147{,}600$. (Another equation that works is $100 + 150 = 147{,}600/t$.)

$$100t + 150t = 147{,}600$$
$$250t = 147{,}600$$
$$t = \frac{147{,}600}{250} = 590\tfrac{2}{5} = 590.4 \text{ minutes}$$

The presses will work together for 590.4 minutes or 9 hours 50 minutes 24 seconds. (This is 590 minutes and $0.4(60) = 24$ seconds.)

PRACTICE _____

1. Neil can paint a wall in 45 minutes; Scott, in 30 minutes. If Neil begins painting the wall and Scott joins in after 15 minutes, how long will it take both to finish the job?

2. Two hoses are used to fill a water trough. Hose 1 can fill it in 20 minutes while Hose 2 needs only 16 minutes. If the Hose 2 is used for the first 4 minutes and then Hose 1 is also used, how long will Hose 1 be used?

3. Jeremy can mow a lawn in one hour. Sarah can mow the same lawn in one and a half hours. If Jeremy works alone for 20 minutes then Sarah starts to help, how long will it take for them to finish the lawn?

4. A mold press can produce 1200 buttons an hour. Another mold press can produce 1500 buttons an hour. They need to produce 45,000 buttons.

    (a) How long will be needed if both presses are used to run the job?

    (b) If the first press runs for 3 hours then the second press joins in, how long will it take for them to finish the run?

 **SOLUTIONS** _____

1. Neil worked alone for $\frac{15}{45} = \frac{1}{3}$ of the job, so $1 - \frac{1}{3} = \frac{2}{3}$ of the job remains. Let t represent the number of minutes both will work together.

Worker	Quantity	Rate	Time
Neil	1	1/45	45
Scott	1	1/30	30
Together	2/3	$\frac{2}{3t}$ $\frac{2/3}{t} = \frac{2}{3t}$	t

The equation to solve is $\dfrac{1}{45} + \dfrac{1}{30} = \dfrac{2}{3t}$. The LCD is $90t$.

$$\frac{1}{45} + \frac{1}{30} = \frac{2}{3t}$$

$$90t\left(\frac{1}{45} + \frac{1}{30}\right) = 90t\left(\frac{2}{3t}\right)$$

$$90t\left(\frac{1}{45}\right) + 90t\left(\frac{1}{30}\right) = 60$$

$$2t + 3t = 60$$

$$5t = 60$$

$$t = \frac{60}{5} = 12$$

It will take Scott and Neil 12 minutes to finish painting the wall.

2. Hose 2 is used alone for $\frac{4}{16} = \frac{1}{4}$ of the job, so $1 - \frac{1}{4} = \frac{3}{4}$ of the job remains. Let t represent the number of minutes both hoses will be used.

Worker	Quantity	Rate	Time
Hose 1	1	1/20	20
Hose 2	1	1/16	16
Together	3/4	$\dfrac{3}{4t}$ $\dfrac{3/4}{t} = \dfrac{3}{4t}$	t

The equation to solve is $\dfrac{1}{20} + \dfrac{1}{16} = \dfrac{3}{4t}$. The LCD is $80t$.

$$\frac{1}{20} + \frac{1}{16} = \frac{3}{4t}$$

$$80t\left(\frac{1}{20} + \frac{1}{16}\right) = 80t\left(\frac{3}{4t}\right)$$

$$80t\left(\frac{1}{20}\right) + 80t\left(\frac{1}{16}\right) = 60$$

$$4t + 5t = 60$$

$$9t = 60$$

$$t = \frac{60}{9} = 6\tfrac{2}{3}$$

Both hoses will be used for $6\tfrac{2}{3}$ minutes or 6 minutes 40 seconds. Therefore, Hose 1 will be used for $6\tfrac{2}{3}$ minutes.

3. Some of the information given in this problem is given in hours and other information in minutes. We must use only one unit of measure. Using minutes as the unit of measure will make the computations a little less messy. Let t represent the number of minutes both Sarah and Jeremy work together. Alone, Jeremy completed $\frac{20}{60} = \frac{1}{3}$ of the job, so $1 - \frac{1}{3} = \frac{2}{3}$ of the job remains to be done.

Worker	Quantity	Rate	Time
Jeremy	1	1/60	60
Sarah	1	1/90	90
Together	2/3	$\dfrac{2}{3t}$ $\dfrac{2/3}{t} = \dfrac{2}{3t}$	t

The equation to solve is $\dfrac{1}{60} + \dfrac{1}{90} = \dfrac{2}{3t}$. The LCD is $180t$.

$$\frac{1}{60} + \frac{1}{90} = \frac{2}{3t}$$

$$180t\left(\frac{1}{60} + \frac{1}{90}\right) = 180t \cdot \frac{2}{3t}$$

$$180t\left(\frac{1}{60}\right) + 180t\left(\frac{1}{90}\right) = 120$$

$$3t + 2t = 120$$

$$5t = 120$$

$$t = \frac{120}{5}$$

$$t = 24$$

They will need 24 minutes to finish the lawn.

4. (a) Let t represent the number of hours the presses need, working together, to complete the job.

Worker	Quantity	Rate	Time
Press 1	$1200t$	1200	t
Press 2	$1500t$	1500	t
Together	45,000	45,000/t	t

The equation to solve is $1200t + 1500t = 45,000$. (Another equation that works is $1200 + 1500 = 45,000/t$.)

$$1200t + 1500t = 45,000$$

$$2700t = 45,000$$

$$t = \frac{45,000}{2700}$$

$$t = 16\tfrac{2}{3}$$

They will need $16\frac{2}{3}$ hours or 16 hours 40 minutes ($\frac{2}{3}$ of an hour is $\frac{2}{3}$ of 60 minutes—$\frac{2}{3} \cdot 60 = 40$) to complete the run.

(b) Press 1 has produced $3(1200) = 3600$ buttons alone, so there remains $45{,}000 - 3600 = 41{,}400$ buttons to be produced. Let t represent the number of hours the presses, running together, need to complete the job.

Worker	Quantity	Rate	Time
Press 1	1200t	1200	t
Press 2	1500t	1500	t
Together	41,400	41,400/t	t

The equation to solve is $1200t + 1500t = 41{,}400$. (Another equation that works is $1200 + 1500 = 41{,}400/t$.)

$$1200t + 1500t = 41{,}400$$
$$2700t = 41{,}400$$
$$t = \frac{41{,}400}{2700}$$
$$t = 15\frac{1}{3}$$

The presses will need $15\frac{1}{3}$ hours or 15 hours 20 minutes to complete the run.

Distance Problems

Another common word problem is the distance problem, sometimes called the uniform rate problem. The underlying formula is $d = rt$ (distance equals rate times time). From $d = rt$, we have two other equations: $r = d/t$ and $t = d/r$. These problems come in many forms: two bodies traveling in opposite directions, two bodies traveling in the same direction, two bodies traveling away from each other or toward each other at right angles. Sometimes the bodies leave at the same time, sometimes one gets a head start. Usually they are traveling at different rates, or speeds. As in all applied problems, the units of measure must be consistent throughout the problem. For instance, if rates are given in miles per hour and time is given in minutes, we convert minutes to hours. We could convert miles per hour into miles per minute, but this can be awkward.

We begin with two cars/cyclists/runners, etc. moving in the same direction. The rate at which the distance between them is changing is the difference in their rates. For example, if a car is traveling northward on a highway at 60 mph and another car behind the first traveling at 70 mph, then the rate at which the distance between them is decreasing is 10 mph.

EXAMPLE _____

A bicyclist starts at a certain point and rides at a rate of 10 mph. Twelve minutes later, another bicyclist starts from the same point in the same direction and rides 16 mph. How long will it take for the second cyclist to catch up with the first?

When the second cyclist begins, the first has traveled:

$$
\begin{array}{ccc}
r & t & = d \\
10 & \left(\dfrac{12}{60}\right) & = 2
\end{array} \quad \text{miles}
$$

(Converting 12 minutes to hours gives us $\frac{12}{60}$ hours.) Because the cyclists are moving in the same direction, the rate at which the distance between them is decreasing is $16 - 10 = 6$ mph. Then, the question boils down to "How long will it take for someone traveling 6 mph to cover 2 miles?"

Let t represent the number of hours the second cyclist is traveling.

$$d = rt$$
$$2 = 6t$$
$$\frac{2}{6} = t$$
$$\frac{1}{3} = t$$

It will take the second cyclist $\frac{1}{3}$ of an hour, or 20 minutes, to catch up the first cyclist.

A car passes an intersection heading north at 40 mph. Another car passes the same intersection 15 minutes later heading north traveling at 45 mph. How long will it take for the second car to overtake the first?

In 15 minutes, the first car has traveled $40\left(\dfrac{15}{60}\right) = 10$ miles. The second car is gaining on the first at a rate of $45 - 40 = 5$ mph. So the question becomes "How long will it someone traveling 5 mph to cover 10 miles?"

Let *t* represent the number of hours the second car has traveled after passing the intersection.

$$d = rt$$
$$10 = 5t$$
$$\frac{10}{5} = t$$
$$2 = t$$

It will take the second car 2 hours to overtake the first.

 PRACTICE

1. Lori starts jogging from a certain point and runs 5 mph. Jeffrey jogs from the same point 15 minutes later at a rate of 8 mph. How long will it take Jeffrey to catch up to Lori?

2. A truck driving east at 50 mph passes a certain mile marker. A motorcyclist also driving east passes that same mile marker 45 minutes later. If the motorcyclist is driving 65 mph, how long will it take for the motorcyclist to pass the truck?

SOLUTIONS

1. Lori has jogged

$$
\begin{array}{ccc}
r & t & = d \\
5 & \left(\dfrac{15}{60}\right) & = \dfrac{5}{4}
\end{array}
$$

miles before Jeffrey began. Jeffrey is catching up to Lori at the rate of $8 - 5 = 3$ mph. How long will it take someone traveling 3 mph to cover $\frac{5}{4}$ miles?

Let *t* represent the number of hours Jeffrey jogs.

$$3t = \frac{5}{4}$$
$$t = \frac{1}{3} \cdot \frac{5}{4}$$
$$t = \frac{5}{12}$$

Jeffrey will catch up to Lori in $\frac{5}{12}$ hours or $\left(\frac{5}{12}\right)(60) = 25$ minutes.

2. The truck traveled $50\left(\dfrac{45}{60}\right) = \dfrac{75}{2}$ miles. The motorcyclist is catching up to the truck at a rate of $65 - 50 = 15$ mph. How long will it take somebody moving at a rate of 15 mph to cover $\frac{75}{2}$ miles?

Let t represent the number of hours the motorcyclist has been driving since passing the mile marker.

$$\dfrac{75}{2} = 15t$$

$$\dfrac{1}{15} \cdot \dfrac{75}{2} = t$$

$$\dfrac{5}{2} = t$$

$$2\tfrac{1}{2} = t$$

The motorcyclist will overtake the truck in $2\frac{1}{2}$ hours.

When two bodies are moving in opposite directions, whether toward each other or away from each other, the rate at which the distance between them is changing, whether growing larger or smaller, is the sum of their individual rates.

 EXAMPLE

Two cars meet at an intersection, one heading north; the other, south. If the northbound driver drives at a rate of 30 mph and the southbound driver at the rate of 40 mph, when will they be 35 miles apart?

The distance between them is growing at the rate of $30 + 40 = 70$ mph. The question then becomes, "how long will it take for someone moving 70 mph to travel 35 miles?"

Let t represent the number of hours the cars travel after leaving the intersection.

$$70t = 35$$

$$t = \dfrac{35}{70}$$

$$t = \dfrac{1}{2}$$

In half an hour, the cars will be 35 miles apart.

Katy left her house on bicycle heading north at 8 mph. At the same time, her sister Molly headed south at 12 mph. How long will it take for them to be 24 miles apart?

The distance between them is increasing at the rate of $8 + 12 = 20$ mph. The question then becomes "How long will it take someone moving 20 mph to travel 24 miles?"

Let t represent the number of hours each girl is traveling.

$$20t = 24$$
$$t = \frac{24}{20}$$
$$t = \frac{6}{5} = 1\tfrac{1}{5}$$

The girls will be 24 miles apart after $1\tfrac{1}{5}$ hours or 1 hour 12 minutes.

 PRACTICE

1. Two airplanes leave an airport simultaneously—one heading east, the other west. The eastbound plane travels at 140 mph and the westbound plane travels at 160 mph. How long will it take for the planes to be 750 miles apart?

2. Mary began walking home from school, heading south at a rate of 4 mph. Sharon left school at the same time heading north at 6 mph. How long will it take for them to be 3 miles apart?

3. Two freight trains pass each other on parallel tracks. One train is traveling west, going 40 mph. The other is traveling east, going 60 mph. When will the trains be 325 miles apart?

 **SOLUTIONS**

1. The planes are moving apart at a rate of $140 + 160 = 300$ mph. Let t represent the number of hours the planes are flying.

$$300t = 750$$
$$t = \frac{750}{300}$$
$$t = 2\tfrac{1}{2}$$

In $2\tfrac{1}{2}$ hours, or 2 hours 30 minutes, the planes will be 750 miles apart.

2. The distance between the girls is increasing at the rate of 4 + 6 = 10 mph. Let t represent the number of hours the girls are walking.

$$10t = 3$$
$$t = \frac{3}{10}$$

Mary and Sharon will be 3 miles apart in $\frac{3}{10}$ of an hour or $60\left(\frac{3}{10}\right) = 18$ minutes.

3. The distance between the trains is increasing at the rate of 40 + 60 = 100 mph. Let t represent the number of hours the trains travel after passing each other.

$$100t = 325$$
$$t = \frac{325}{100}$$
$$t = 3\tfrac{1}{4}$$

The trains will be 325 miles apart after $3\tfrac{1}{4}$ hours or 3 hours 15 minutes.

In the following problems, two bodies leave different locations at the same time, and they travel in opposite directions (either away from each other or toward each other). The rate at which the distance between them changes is the sum of their individual rates.

EXAMPLE

Dale left his high school at 3:45 and walked toward his brother's school at 5 mph. His brother, Jason, left his elementary school at the same time and walked toward Dale's high school at 3 mph. If their schools are 2 miles apart, when will they meet?

The rate at which the brothers are moving toward each other is 3 + 5 = 8 mph. Let t represent the number of hours the boys walk.

$$8t = 2$$
$$t = \frac{2}{8}$$
$$t = \frac{1}{4}$$

The boys will meet after $\tfrac{1}{4}$ an hour or 15 minutes; that is, at 4:00.

A jet leaves Dallas going to Houston, averaging at 400 mph. At the same time, another jet leaves Houston, flying to Dallas, at the same rate. How long will it take for the two planes to meet? (Dallas and Houston are 250 miles apart.)

The distance between the jets is decreasing at the rate of 400 + 400 = 800 mph. Let t represent the number of hours they are flying.

$$800t = 250$$

$$t = \frac{250}{800}$$

$$t = \frac{5}{16}$$

The planes will meet after $\frac{5}{16}$ hours or $60\left(\frac{5}{16}\right) = 18\frac{3}{4}$ minutes or 18 minutes 45 seconds.

 PRACTICE

1. Jessie leaves her house on bicycle, traveling at 8 mph. She is going to her friend Kerrie's house. Coincidentally, Kerrie leaves her house at the same time and rides her bicycle at 7 mph to Jessie's house. If they live 5 miles apart, how long will it take for the girls to meet?

2. Two cars 270 miles apart enter an interstate highway traveling toward one another. One car travels at 65 mph and the other at 55 mph. When will they meet?

3. At one end of town, a jogger jogs southward at the rate of 6 mph. At the opposite end of town, at the same time, another jogger heads northward at the rate of 9 mph. If the joggers are 9 miles apart, how long will it take for them to meet?

SOLUTIONS

1. The distance between the girls is decreasing at the rate of 8 + 7 = 15 mph. Let t represent the number of hours they are on their bicycles.

$$15t = 5$$

$$t = \frac{5}{15}$$

$$t = \frac{1}{3}$$

The girls will meet in $\frac{1}{3}$ of an hour or 20 minutes.

2. The distance between the cars is decreasing at the rate of 65 + 55 = 120 mph. Let *t* represent the number of hours the cars have traveled since entering the highway.

$$120t = 270$$
$$t = \frac{270}{120}$$
$$t = 2\tfrac{1}{4}$$

The cars will meet after $2\tfrac{1}{4}$ hours or 2 hours 15 minutes.

3. The distance between the joggers is decreasing at the rate of 6 + 9 = 15 mph. Let *t* represent the number of the hours they are jogging.

$$15t = 9$$
$$t = \frac{9}{15}$$
$$t = \frac{3}{5}$$

The joggers will meet after $\tfrac{3}{5}$ of an hour or $60\left(\dfrac{3}{5}\right) = 36$ minutes.

Some distance problems involve the complication of the two bodies starting at different times. For these, we compute the head start of the first one and let *t* represent the time they are both moving (which is the same as the amount of time the second is moving). We then subtract the head start from the distance in question then proceed as if they started at the same time.

EXAMPLE

A car driving eastbound passes through an intersection at 6:00 at the rate of 30 mph. Another car driving westbound passes through the same intersection ten minutes later at the rate of 35 mph. When will the cars be 18 miles apart?

The eastbound driver has a 10-minute head start. In 10 minutes ($\tfrac{10}{60}$ hours), that driver has traveled $30\left(\dfrac{10}{60}\right) = 5$ miles. So when the westbound driver passes the intersection, there is already 5 miles between them, so the question is now "How long will it take for there to be 18 − 5 = 13 miles between two bodies moving away from each other at the rate of 30 + 35 = 65 mph?"

Let *t* represent the number of hours after the second car has passed the intersection.

$$65t = 13$$

$$t = \frac{13}{65}$$

$$t = \frac{1}{5}$$

In $\frac{1}{5}$ of an hour or $60\left(\frac{1}{5}\right) = 12$ minutes, an additional 13 miles is between them. Twelve minutes after the second car passes the intersection, there is a total of 18 miles between the cars. That is, at 6:22 the cars will be 18 miles apart.

Two employees ride their bikes to work. At 10:00 one leaves work and rides southward home at 9 mph. At 10:05 the other leaves work and rides home northward at 8 mph. When will they be 5 miles apart?

The first employee has ridden $9\left(\frac{5}{60}\right) = \frac{3}{4}$ miles by the time the second employee has left. So we now need to see how long, after 10:05, it takes for an additional $5 - \frac{3}{4} = 4\frac{1}{4} = \frac{17}{4}$ miles to be between them. Let *t* represent the number of hours after 10:05. When both employees are riding, the distance between them is increasing at the rate of $9 + 8 = 17$ mph.

$$(9+8)t = \frac{17}{4}$$

$$17t = \frac{17}{4}$$

$$t = \frac{1}{17} \cdot \frac{17}{4}$$

$$t = \frac{1}{4}$$

After $\frac{1}{4}$ hour, or 15 minutes, they will be an additional $4\frac{1}{4}$ miles apart. That is, at 10:20, the employees will be 5 miles apart.

Two boys are 1250 meters apart when one begins walking toward the other. If one walks at a rate of 2 meters per second and the other, who starts walking toward the first boy four minutes later, walks at the rate of 1.5 meters per second, how long will it take for them to meet?

The boy with the head start has walked for $4(60) = 240$ seconds. (Because the rate is given in meters per second, we convert all times to seconds.) So, he has traveled $240(2) = 480$ meters. At the time the other boy begins walking, there remains $1250 - 480 = 770$ meters to cover. When the second boy begins to walk, they are moving toward one another at the rate of $2 + 1.5 = 3.5$ meters per second.

Let t represent the number of seconds the second boy walks.

$$3.5t = 770$$
$$t = \frac{770}{3.5}$$
$$t = 220$$

The boys will meet 220 seconds, or 3 minutes 40 seconds, after the second boy starts walking.

A plane leaves City A toward City B at 9:10, flying at 200 mph. Another plane leaves City B towards City A at 9:19, flying at 180 mph. If the cities are 790 miles apart, when will the planes pass each other?

In 9 minutes the first plane has flown $200\left(\frac{9}{60}\right) = 30$ miles, so when the second plane takes off, there are $790 - 30 = 760$ miles between them. The planes are traveling toward each other at $200 + 180 = 380$ mph. Let t represent the number of hours the second plane flies.

$$380t = 760$$
$$t = \frac{760}{380}$$
$$t = 2$$

Two hours after the second plane has left the planes will pass each other; that is, at 11:19 the planes will pass each other.

PRACTICE

1. Two joggers start jogging on a trail. One jogger heads north at the rate of 7 mph. Eight minutes later, the other jogger begins at the same point and heads south at the rate of 9 mph. When will they be 2 miles apart?

2. Two boats head toward each other from opposite ends of a lake, which is 6 miles wide. One boat left at 2:05 going 12 mph. The other boat left at 2:09 at a rate of 14 mph. What time will they meet?

3. The Smiths leave the Tulsa city limits, heading toward Dallas, at 6:05 driving 55 mph. The Hewitts leave Dallas and drive to Tulsa at 6:17, driving 65 mph. If Dallas and Tulsa are 257 miles apart, when will they pass each other?

✓ **SOLUTIONS**

1. The first jogger had jogged $7\left(\dfrac{8}{60}\right) = \dfrac{56}{60} = \dfrac{14}{15}$ miles when the other jogger began. So there is $2 - \dfrac{14}{15} = 1\tfrac{1}{5} = \dfrac{16}{15}$ miles left to cover. The distance between them is growing at a rate of $7 + 9 = 16$ mph. Let t represent the number of hours the second jogger jogs.

$$16t = \dfrac{16}{15}$$
$$t = \dfrac{1}{16} \cdot \dfrac{16}{15}$$
$$t = \dfrac{1}{15}$$

In $\tfrac{1}{15}$ of an hour, or $60\left(\dfrac{1}{15}\right) = 4$ minutes after the second jogger began, the joggers will be two miles apart.

2. The first boat got a $12\left(\dfrac{4}{60}\right) = \dfrac{4}{5}$ mile head start. When the second boat leaves, there remains $6 - \dfrac{4}{5} = 5\tfrac{1}{5} = \dfrac{26}{5}$ miles between them. When the second boat leaves the distance between them is decreasing at a rate of $12 + 14 = 26$ mph. Let t represent the number of hours the second boat travels.

$$26t = \dfrac{26}{5}$$
$$t = \dfrac{1}{26} \cdot \dfrac{26}{5}$$
$$t = \dfrac{1}{5}$$

In $\tfrac{1}{5}$ hour, or $\dfrac{1}{5}(60) = 20$ minutes, after the second boat leaves, the boats will meet. That is, at 2:29 both boats will meet.

3. The Smiths have driven $55\left(\dfrac{12}{60}\right) = 11$ miles outside of Tulsa by the time the Hewitts have left Dallas. So, when the Hewitts leave Dallas, there are $257 - 11 = 246$ miles between the Smiths and Hewitts. When the Hewitts leave Dallas, the distance between them is decreasing at the rate of $55 + 65 = 120$ mph. Let t represent the number of hours after the Hewitts have left Dallas.

$$120t = 246$$
$$t = \dfrac{246}{120}$$
$$t = 2\tfrac{1}{20}$$

$2\frac{1}{20}$ hours, or 2 hours 3 minutes $\left(60 \cdot \frac{1}{20} = 3\right)$, after the Hewitts leave Dallas, the Smiths and Hewitts will pass each other. In other words, at 8:20, the Smiths and Hewitts will pass each other.

In the following problems, we have information on how long a trip and return trip are and use this information to find an unknown distance.

EXAMPLE

A semi-truck traveled from City A to City B at 50 mph. On the return trip, it averaged only 45 mph and took 15 minutes longer. How far is it from City A to City B?

We have three unknowns—the distance between City A and City B, the time spent traveling from City A to City B, and the time spent traveling from City B to City A. We must eliminate two of these unknowns. Let t represent the number of hours spent on the trip from City A to City B. We know that it took 15 minutes longer traveling from City B to City A (the return trip), so $t + \frac{15}{60}$ represents the amount of time traveling from City B to City A. We also know that the distance from City A to City B is the same as from City B to City A. Let d represent the distance between the two cities. We now have the following two equations.

From City A to City B: From City B to City A:

$d = 50t$ $d = 45\left(t + \frac{15}{60}\right)$

But if the distance between them is the same, then $50t =$ Distance from City A to City B is equal to the distance from City B to City A $= 45\left(t + \frac{15}{60}\right)$. Therefore,

$$50t = 45\left(t + \frac{15}{60}\right)$$

$$50t = 45\left(t + \frac{1}{4}\right)$$

$$50t = 45t + \frac{45}{4}$$

$$\underline{-45t \quad -45t}$$

$$5t = \frac{45}{4}$$

$$t = \frac{1}{5} \cdot \frac{45}{4}$$

$$t = \frac{9}{4}$$

We now know the time, but the problem asked for the distance. The distance from City A to City B is given by $d = 50t$, so $d = 50\left(\frac{9}{4}\right) = \frac{225}{2} = 112\frac{1}{2}$. The cities are $112\frac{1}{2}$ miles apart.

Another approach to this problem would be to let t represent the number of hours the semi spent traveling from City B to City A. Then $t - \frac{15}{60}$ would represent the number of hours the semi spent traveling from City A to City B, and the equation to solve is $50\left(t - \frac{15}{60}\right) = 45t$.

Kaye rode her bike to the library. The return trip took 5 minutes less. If she rode to the library at the rate of 10 mph and home from the library at the rate of 12 mph, how far is her house from the library?

Again there are three unknowns—the distance between Kaye's house and the library, the time spent riding to the library and the time spent riding home. Let t represent the number of hours spent riding to the library. She spent 5 minutes less riding home, so $t - \frac{5}{60}$ represents the number of hours spent riding home. Let d represent the distance between Kaye's house and the library. The trip to the library is given by $d = 10t$, and the trip home is given by $d = 12\left(t - \frac{5}{60}\right)$. As these distances are equal, we have that $10t = d = 12\left(t - \frac{5}{60}\right)$.

$$10t = 12\left(t - \frac{5}{60}\right)$$

$$10t = 12\left(t - \frac{1}{12}\right)$$

$$10t = 12t - 1$$

$$\underline{-12t \quad -12t}$$

$$-2t = -1$$

$$t = \frac{-1}{-2}$$

$$t = \frac{1}{2}$$

The distance from home to the library is $d = 10t = 10\left(\frac{1}{2}\right) = 5$ miles.

PRACTICE

1. Terry, a marathon runner, ran from her house to the high school then back. The return trip took 5 minutes longer. If her speed was 10 mph to the high school and 9 mph back, how far is Terry's house from the high school?

2. Because of heavy morning traffic, Toni spent 18 minutes more driving to work than driving home. If she averaged 30 mph on her drive to work and 45 mph on her drive home, how far is her home from her work?

3. Leo walked his grandson to school. If he averaged 3 mph on the way to school and 5 mph on his way home, and if it took 16 minutes long to get to school, how far is it between his home and his grandson's school?

SOLUTIONS

1. Let t represent the number of hours spent running from home to school. Then $t + \frac{5}{60}$ represents the number of hours spent running from school to home. The distance to school is given by $d = 10t$, and the distance home is given by $d = 9\left(t + \frac{5}{60}\right)$.

$$10t = 9\left(t + \frac{5}{60}\right)$$

$$10t = 9\left(t + \frac{1}{12}\right)$$

$$10t = 9t + \frac{9}{12}$$

$$10t = 9t + \frac{3}{4}$$

$$\underline{-9t \quad -9t}$$

$$t = \frac{3}{4}$$

The distance between home and school is $10t = 10\left(\frac{3}{4}\right) = \frac{15}{2} = 7\frac{1}{2}$ miles.

2. Let t represent the number of hours Toni spent driving to work. Then $t - \frac{18}{60}$ represents the number of hours driving home. The distance from

home to work is given by $d = 30t$, and the distance from work to home is given by $d = 45\left(t - \frac{18}{60}\right)$ miles.

$$30t = 45\left(t - \frac{18}{60}\right)$$

$$30t = 45\left(t - \frac{3}{10}\right)$$

$$30t = 45t - \frac{135}{10}$$

$$30t = 45t - \frac{27}{2}$$

$$\underline{-45t \quad -45t}$$

$$-15t = \frac{-27}{2}$$

$$t = \frac{1}{-15} \cdot \frac{-27}{2}$$

$$t = \frac{9}{10}$$

The distance from Toni's home and work is $30t = 30\left(\frac{9}{10}\right) = 27$ miles.

3. Let t represent the number of hours Leo spent walking his grandson to school. Then $t - \frac{16}{60}$ represents the number of hours Leo spent walking home. The distance from home to school is given by $d = 3t$ and the distance from school to home is given by $d = 5\left(t - \frac{16}{60}\right)$.

$$3t = 5\left(t - \frac{16}{60}\right)$$

$$3t = 5\left(t - \frac{4}{15}\right)$$

$$3t = 5t - \frac{20}{15}$$

$$3t = 5t - \frac{4}{3}$$

$$\underline{-5t \quad -5t}$$

$$-2t = \frac{-4}{3}$$

$$t = \frac{1}{-2} \cdot \frac{-4}{3}$$

$$t = \frac{2}{3}$$

The distance from home to school is $3t = 3\left(\frac{2}{3}\right) = 2$ miles.

In the above examples and practice problems, we substituted the number for *t* in the first distance equation to get *d*. It does not matter which equation we us to find *d*, we would get the same value.

For distance problems in which the bodies are moving away from each other or toward each other at right angles (for example, one heading east, the other north), we use the Pythagorean theorem. This topic will be covered in the last chapter.

Geometric Figures

Algebra problems involving geometric figures are very common. In algebra, we normally work with rectangles, triangles, and circles. On occasion, we are asked to solve problems involving other shapes such as right circular cylinders and right circular cones. By mastering the more common types of geometric problems, you will find that the more exotic shapes are just as easy.

In many of these problems, we will have several unknowns which we again must reduce to a single unknown. In the problems above, we reduced a problem of three unknowns to two unknowns by relating one quantity to another (the time on one direction related to the time on the return trip) and by setting the equal distances equal to each other. We use similar techniques here.

 PROBLEM

A rectangle is $1\frac{1}{2}$ times as long as it is wide. The perimeter is 100 cm. Find the dimensions of the rectangle.

The formula for the perimeter of a rectangle is given by $P = 2l + 2w$. We are told the perimeter is 100, so the equation becomes $100 = 2l + 2w$. We are also told that the length is $1\frac{1}{2}$ times the width, so $l = 1.5w$. We can substitute 1.5w for *l* into the equation:

$100 = 2l + 2w = 2(1.5w) + 2w$. We have reduced an equation with three unknowns to an equation with a single unknown.

$$100 = 2(1.5w) + 2w$$
$$100 = 3w + 2w$$
$$100 = 5w$$
$$\frac{100}{5} = w$$
$$20 = w$$

The width is 20 cm and the length is $1.5w = 1.5(20) = 30$ cm.

PRACTICE

1. A box's width is two-thirds its length. The perimeter of the box is 40 inches. What are the box's length and width?

2. A rectangular yard is twice as long as it is wide. The perimeter is 120 feet. What are the yard's dimensions?

SOLUTIONS

1. The perimeter of the box is 40 inches, so $P = 2l + 2w$ becomes $40 = 2l + 2w$. The width is $\frac{2}{3}$ its length, and $w = \left(\frac{2l}{3}\right)$, so $40 = 2l + 2w$ becomes $40 = 2l + 2\left(\frac{2l}{3}\right) = 2l + \left(\frac{4l}{3}\right)$.

$$40 = 2l + \frac{4l}{3}$$

$$40 = \frac{10l}{3} \qquad \left(2 + \frac{4}{3} = \frac{6}{3} + \frac{4}{3} = \frac{10}{3}\right)$$

$$\frac{3}{10} \cdot 40 = \frac{3}{10} \cdot \frac{10l}{3}$$

$$12 = l$$

The length of the box is 12 inches and its width is $\frac{2}{3}l = \frac{2}{3}(12) = 8$ inches.

2. The perimeter of the yard is 120 feet, so $P = 2l + 2w$ becomes $120 = 2l + 2w$. The length is twice the width, so $l = 2w$, and $120 = 2l + 2w$ becomes $120 = 2(2w) + 2w$.

$$120 = 2(2w) + 2w$$

$$120 = 4w + 2w$$

$$120 = 6w$$

$$\frac{120}{6} = w$$

$$20 = w$$

The yard's width is 20 feet and its length is $2l = 2(20) = 40$ feet.

In the following problems, one or more dimensions are changed and we are given information about how this change has affected the figure's area. We can then decide how the two areas are related so that we can reduce the problem from several unknowns to just one.

EXAMPLE

A rectangle is twice as long as it is wide. If the length is decreased by 4 inches and its width is decreased by 3 inches, the area is decreased by 88 square inches. Find the original dimensions.

The area formula for a rectangle is $A = lw$. Let A represent the original area; l, the original length; and w, the original width. We know that the original length is twice the original width, so $l = 2w$ and $A = lw$ becomes $A = (2w)w = 2w^2$. The new length is $l - 4 = 2w - 4$ and the new width is $w - 3$, so the new area is $(2w - 4)(w - 3)$. But the new area is also 88 square inches less than the old area, so $A - 88$ represents the new area, also. We then have for the new area, $A - 88 = (2w - 4)(w - 3)$. But the A can be replaced with $2w^2$. We now have the equation $2w^2 - 88 = (2w - 4)(w - 3)$, an equation with one unknown.

$$2w^2 - 88 = (2w - 4)(w - 3)$$

$$2w^2 - 88 = 2w^2 - 6w - 4w + 12 \quad \text{(Use the FOIL method.)}$$

$$2w^2 - 88 = 2w^2 - 10w + 12 \quad \text{(}2w^2\text{on each side cancels.)}$$

$$\begin{aligned} -12 \qquad\qquad\quad -12 \\ -100 = -10w \end{aligned}$$

$$\frac{-100}{-10} = w$$

$$10 = w$$

The width of the original rectangle is 10 inches and its length is $2w = 2(10) = 20$ inches.

A square's length is increased by 3 cm, which causes the area of the new square to increase by 33 cm². What is the length of the original square?

A square's length and width are the same, so the area formula for the square is $A = l \cdot l = l^2$. Let l represent the original length. The new length is $l + 3$. The original area is $A = l^2$ and its new area is $(l + 3)^2$. The new area is also the original area plus 33, so $(l + 3)^2 = $ new area $= A + 33 = l^2 + 33$.

So we now have the equation, with one unknown: $(l + 3)^2 = l^2 + 33$.

$$(l+3)^2 = l^2 + 33$$
$$(l+3)(l+3) = l^2 + 33$$
$$l^2 + 6l + 9 = l^2 + 33 \quad (l^2 \text{ on each side cancels.})$$
$$\underline{ -9 \qquad -9}$$
$$6l = 24$$
$$l = \frac{24}{6}$$
$$l = 4$$

The original length is 4 cm.

The radius of a circle is increased by 3 cm. As a result, the area is increased by 45π cm². What is the original radius?

Remember that the area of a circle is $A = \pi r^2$, where r represents the radius. So, let r represent the original radius. The new radius is then represented by $r + 3$. The new area is represented by $\pi(r + 3)^2$. But the new area is also the original area plus 45π cm². This gives us $A + 45\pi = \pi(r + 3)^2$. Because $A = \pi r^2$, $A + 45\pi$ becomes $\pi r^2 + 45\pi$. Our equation, then, is $\pi r^2 + 45\pi = \pi(r + 3)^2$.

$$\pi r^2 + 45\pi = \pi(r + 3)^2$$
$$\pi r^2 + 45\pi = \pi(r + 3)(r + 3)$$
$$\pi r^2 + 45\pi = \pi(r^2 + 3r + 3r + 9)$$
$$\pi r^2 + 45\pi = \pi(r^2 + 6r + 9)$$
$$\pi r^2 + 45\pi = \pi r^2 + 6r\pi + 9\pi \quad (\pi r^2 \text{ on each side cancels.})$$
$$45\pi = 6r\pi + 9\pi$$
$$\underline{-9\pi \qquad\qquad -9\pi}$$
$$36\pi = 6r\pi$$
$$\frac{36\pi}{6\pi} = r$$
$$6 = r$$

The original radius is 6 cm.

PRACTICE

1. A rectangular piece of cardboard starts out with its width being three-fourths its length. Four inches are cut off its length and 2 inches from its

width. The area of the cardboard is 72 square inches smaller than before it was trimmed. What was its original length and width?

2. A rectangle's length is $1\frac{1}{2}$ times its width. The length is increased by 4 inches and its width by 3 inches. The resulting area is 97 square inches more than the original rectangle. What were the original dimensions?

3. A circle's radius is increased by 5 inches and as a result, its area is increased by 155π square inches. What is the original radius?

 SOLUTIONS

1. Let l represent the original length and w, the original width. The original area is $A = lw$. The new length is $l - 4$ and the new width is $w - 2$. The new area is then $(l - 4)(w - 2)$. But the new area is 72 square inches smaller than the original area, so $(l - 4)(w - 2) = A - 72 = lw - 72$. So far, we have $(l - 4)(w - 2) = lw - 72$.

The original width is three-fourths its length, so $w = \frac{3}{4}l$. We will now replace w with $\left(\frac{3}{4}l\right) = \frac{3l}{4}$.

$$(l - 4)\left(\frac{3l}{4} - 2\right) = l\left(\frac{3l}{4}\right) - 72$$

$$\frac{3l^2}{4} - 2l - 4\left(\frac{3l}{4}\right) + 8 = \frac{3l^2}{4} - 72 \quad \left(\frac{3l^2}{4} \text{ on each side cancels.}\right)$$

$$-2l - 3l + 8 = -72$$

$$-5l + 8 = -72$$

$$\underline{-8 \quad -8}$$

$$-5l = -80$$

$$l = \frac{-80}{-5}$$

$$l = 16$$

The original length was 16 inches and the original width was $\frac{3}{4}l = \frac{3}{4}(16) = 12$ inches.

2. Let l represent the original length and w, the original width. The original area is then given by $A = lw$. The new length is $l + 4$ and the new width is $w + 3$. The new area is now $(l + 4)(w + 3)$. But the new area is also the old area plus 97 square inches, so $A + 97 = (l + 4)(w + 3)$. But $A = lw$, so $A + 97$ becomes $lw + 97$. We now have

$$lw + 97 = (l + 4)(w + 3).$$

Since the original length is $1\frac{1}{2} = \frac{3}{2}$ of the original width, $l = \frac{3}{2}w$. Replace each l by $\frac{3}{2}w$.

$$\frac{3}{2}ww + 97 = \left(\frac{3}{2}w + 4\right)(w + 3)$$

$$\frac{3}{2}w^2 + 97 = \left(\frac{3}{2}w + 4\right)(w + 3)$$

$$\frac{3}{2}w^2 + 97 = \frac{3}{2}w^2 + \frac{9}{2}w + 4w + 12 \qquad \left(\frac{3}{2}w^2 \text{ on each side cancels.}\right)$$

$$97 = \frac{9}{2}w + 4w + 12$$

$$97 = \frac{17}{2}w + 12 \qquad \left(\frac{9}{2} + 4 = \frac{9}{2} + \frac{8}{2} = \frac{17}{2}\right)$$

$$\underline{-12 \qquad\qquad -12}$$

$$85 = \frac{17}{2}w$$

$$\frac{2}{17} \cdot 85 = \frac{2}{17} \cdot \frac{17}{2}w$$

$$10 = w$$

The original width is 10 inches and the original length is $\frac{3}{2}w = \frac{3}{2}(10) = 15$ inches.

3. Let r represent the original radius. Then $r + 5$ represents the new radius, and $A = \pi r^2$ represents the original area. The new area is 155π square inches more than the original area, so $155\pi + A = \pi(r + 5)^2 = 155\pi + \pi r^2$.

$$\pi(r + 5)^2 = 155\pi + \pi r^2$$

$$\pi(r + 5)(r + 5) = 155\pi + \pi r^2$$

$$\pi(r^2 + 10r + 25) = 155\pi + \pi r^2$$

$$\pi r^2 + 10r\pi + 25\pi = 155\pi + \pi r^2 \qquad (\pi r^2 \text{ on each side cancels.})$$

$$10r\pi + 25\pi = 155\pi$$

$$\underline{-25\pi \quad -25\pi}$$

$$10r\pi = 130\pi$$

$$r = \frac{130\pi}{10\pi}$$

$$r = 13$$

The original radius is 13 inches.

Summary

In this chapter, we learned how to:

- *Solve percent problems.* If we increase x by $p \times 100\%$ (with p written as a decimal number), the new amount is $x + px$. If we decrease x by $p \times 100\%$, the new amount is $x - px$.

- *Solve problems with a known formula.* If a problem involves a known formula, such as a formula from geometry, we use information in the problem to eliminate all but one variable, giving us an equation to solve.

- *Solve number sense problems having two unknowns.* We are given two sets of information about a pair of numbers. One of these sets of information involves a sum. We use the information in the other set to eliminate one of the variables in the sum, giving us an equation to solve.

- *Solve number sense problems having three unknowns.* We are told the sum of the three numbers. The numbers are compared to each other. We use these comparisons to write the sum using a single variable, giving us an equation to solve. If two variables are compared to a third, we let the variable represent this number and write the other two numbers in terms of this number.

- *Solve problems involving coins.* Coin problems are similar to number sense problems. Instead of the being told the sum of the numbers, we are told a total amount of money. We let the variable represent the number of one of the coins and write the number of the other coins in terms of the same variable. We then multiply the number of each coin by the value of the coin. The sum of these values is the total amount of money.

- *Solve investment problems.* Money is divided into two different investments, paying different interest rates. We are told the interest earned. We represent the amount of one of the investments with the variable. If A and B are the interest rates, written as decimal numbers and x the amount invested at interest A, then we solve the following equation: $Ax + B(\text{Total} - x) = \text{Interest earned}$.

- *Solve mixture problems.* Mixture problems involve mixing together two different concentrations of a solution to produce the final concentration. We let x represent the amount of one of the concentrations. We know the amount of either the other concentration or the final concentration. If we know the amount of the second concentration, the amount of the final concentration is "$x +$ Amount of the second concentration." If we know the amount of the total concentration, the amount of the second concentration is "Amount of final concentration $- x$." We then solve the following equation:

$$(\text{Amt of Mixture \#1})(\%) + (\text{Amt of Mixture \#2})(\%)$$
$$= (\text{Amt of Final Mixture})(\%)$$

- *Solve work problems*. Two workers can complete a task. We either know how long each worker needs to complete the task working alone and are asked how long it would take them to complete the task if they work together, or we know how long it would take for them to complete the task working together and are asked how long it would take one of them to complete the task alone (with some information on how much faster one works than the other). We let the variable represent the time we want to find. We then solve the following equation:

$$\frac{1}{\text{Worker \#1 Time}} + \frac{1}{\text{Worker \#2 Time}} = \frac{1}{\text{Together Time}}$$

If one of the workers works alone for a while and then the other joins in to complete the task, the numerator of the last fraction is the proportion of the task that remains when the second worker joins in.

- *Solve distance problems*. Distance problems come in many forms, but all of them are based on the formula $d = rt$. If two cars/runners/etc. are traveling in the same direction, the rate at which the distance between them is either increasing or decreasing is the difference of their two rates. If they are traveling in opposite directions, the rate at which the distance between them is either increasing or decreasing is the sum of their rates. We compute this changing rate for r. We usually know d. We use the equation $d = rt$ to answer the question. For some distance problems, a trip is divided into two parts and we know the total distance and the total time. If this is the case, we solve one of the following:

$$\frac{d_1}{r_1} + \frac{d_2}{r_2} = \text{Total time} \quad \text{or} \quad r_1 t_1 = r_2 t_2$$

where $d_1 = r_1 t_1$ represents one part of the trip and $d_2 = r_2 t_2$ represents the other part.

QUIZ

1. How much 9% acid solution should be mixed with 21% acid solution to produce 16 ml of $15\frac{3}{4}$% acid solution?

   A. 6 ml
   B. 7 ml
   C. 8 ml
   D. 9 ml

2. The perimeter of a rectangle is 80 inches, and its width is two-thirds its length. What is the area of the rectangle? (Hint: After finding the length and width, multiply them.)

   A. 384 in²
   B. 864 in²
   C. 256 in²
   D. 80 in²

3. What is 35% of 14?

   A. 4.9
   B. 40
   C. 4
   D. 49

4. The weekly profit, P, for producing q units is $P = 26q - 1950$. How many units must be produced each week in order to break even?

   A. 25
   B. 50
   C. 75
   D. 100

5. At noon, one car heading north on a freeway passes an exit ramp, averaging 70 mph. Another northbound car passes the same exit ramp 5 minutes later, averaging 77 mph. When will the second car to catch up to the first?

   A. 12:45
   B. 12:50
   C. 12:55
   D. 1:00

6. The difference between two numbers is 15 and three times the larger is six times the smaller. What is the smaller number?

   A. 5
   B. 10
   C. 15
   D. 30

7. A pair of boots is on sale for $112, which is 30% off the original price. What is the original price?

    A. $160.00

    B. $145.60

    C. $373.30

    D. $142.00

8. When the radius of a circle is increased by 3 inches, its area is increased by 75π in^2. What is the radius of the original circle?

    A. 10 inches

    B. 11 inches

    C. 12 inches

    D. 14 inches

9. Crystal is 3 years older than Melissa but 2 years younger than Carolina. The sum of their ages is 62. How old is Crystal?

    A. 21 years

    B. 22 years

    C. 23 years

    D. 24 years

10. Nichelle can wash the windows of a small store in 5 hours 15 minutes. Maxwell can wash the windows of the same store in 2 hours 6 minutes. If they work together, how long would it take for them to wash the store's windows?

    A. 60 minutes

    B. 80 minutes

    C. 90 minutes

    D. 110 minutes

11. Troy has $1.55 in coins in his pocket. He has twice as many nickels as dimes and one more dime than he has quarters. How many quarters does he have in his pocket?

    A. 2

    B. 3

    C. 4

    D. 5

12. Ryan's course grade is based on three tests, the homework average, and the final exam. Each test is worth 20%, the homework average is worth 15%, and the final exam is worth 25%. His homework average is 84, and his test grades are 68, 79, and 75. He must make at least a B (80) in order to keep his scholarship. What is the lowest score he can make on the final exam and keep his scholarship?

    A. 92

    B. 93

    C. 94

    D. 95

13. A school district's printing services department owns two printing presses. The department must produce an order of brochures. Press A would take 10 hours to produce the order, and Press B would take 8 hours. Press A is used alone for 1 hour before Press B starts producing the brochures. After Press B joins in, how long will the presses need to complete the order?

   A. 3 hours
   B. 3.5 hours
   C. 4 hours
   D. 4.5 hours

14. 80 is what percent of 125?

   A. 64%
   B. 75%
   C. 92%
   D. 156.25%

15. The relationship between degrees Fahrenheit and degrees Celsius is given by the formula $C = \frac{5}{9}(F - 32)$. For what temperature will degrees Celsius be 40 more than degrees Fahrenheit?

   A. $-90°C$
   B. $-100°C$
   C. $-120°C$
   D. $-130°C$

chapter 9

Linear Inequalities

Because linear inequalities are similar to linear equations, many of the problems in this chapter are similar to the problems in Chapters 7 and 8. Instead of having a single number as a solution, the solution to an equality usually involves a range (or interval) of numbers. We will learn how to solve linear inequalities and how to represent these solutions in interval notation and as a shaded region on the number line. We will also use inequalities to solve applied problems. For example, suppose the directions on a bottle of medicine recommends storing the bottle between 20°C and 30°C, we can use linear inequalities to find the appropriate range of temperatures on the Fahrenheit scale.

CHAPTER OBJECTIVES

In this chapter, you will

- Represent an inequality by shading a region on the number line
- Write the interval notation for an inequality
- Solve linear inequalities
- Solve double inequalities
- Solve applied problems with linear inequalities

Inequalities and the Number Line

The solution to an algebraic inequality usually consists of a range (or ranges) of numbers, which we write as an inequality in the form $x < a$, $x \leq a$, $x > a$, or $x \geq a$, where a is a number. The inequality $x < a$ means all numbers smaller than a but not including a, and the inequality $x \leq a$ means all numbers smaller than a including a itself. For example, the inequality $x \leq 2$ means all numbers smaller than 2, *including* 2 itself. The inequality $x < 2$ means all numbers smaller than 2 but does *not* include 2. Similarly the inequality $x > a$ means all numbers larger than a but not a itself, and $x \geq a$ means all numbers larger than a including a itself.

Every interval on the number line is represented by an inequality, and every inequality is represented by an interval on the number line. We begin by representing inequalities with shaded regions on the number line. Later we will represent inequalities with intervals.

The inequality $x < a$ is represented on the number line by shading to the left of the number a with an open dot at a.

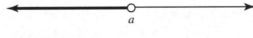

a

FIGURE 9-1

A closed dot at a is used for $x \leq a$.

a

FIGURE 9-2

We shade to the right of a for $x > a$.

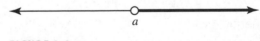

a

FIGURE 9-3

Use a closed dot at a for $x \geq a$.

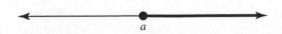

a

FIGURE 9-4

Before we solve inequalities, we will learn how to shade the region on the number line represented by an inequality.

EXAMPLE

Shade the region on the number line.

$x < 0$

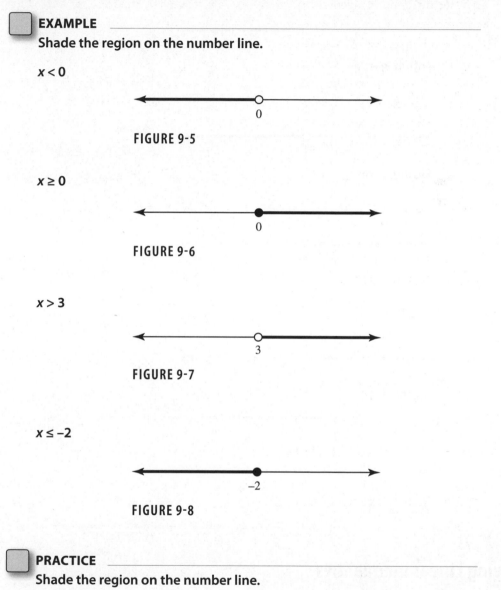

FIGURE 9-5

$x \geq 0$

FIGURE 9-6

$x > 3$

FIGURE 9-7

$x \leq -2$

FIGURE 9-8

PRACTICE

Shade the region on the number line.

1. $x > 4$
2. $x > -5$
3. $x \leq 1$
4. $x < -3$
5. $x \geq 10$

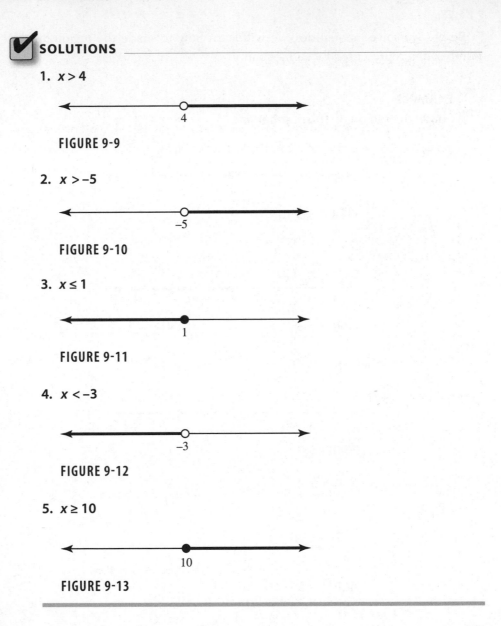

✔ **SOLUTIONS**

1. $x > 4$

FIGURE 9-9

2. $x > -5$

FIGURE 9-10

3. $x \leq 1$

FIGURE 9-11

4. $x < -3$

FIGURE 9-12

5. $x \geq 10$

FIGURE 9-13

Solving Linear Inequalities

Linear inequalities are solved much the same way as linear equations are solved, with one important exception. When multiplying or dividing both sides of an inequality by a negative number, the inequality sign must be reversed. For example $2 < 3$ but $-2 > -3$. Adding and subtracting the same quantity to both sides of an inequality never changes the direction of the inequality sign.

EXAMPLE

Solve the inequality.

$$2x - 7 > 5x + 2$$

We begin by collecting terms with x in them on the left side of the inequality and terms without x in them on the right side.

$$2x - 7 > 5x + 2$$
$$ +7 +7$$
$$2x > 5x + 9$$
$$-5x -5x$$
$$-3x > 9$$
$$\frac{-3}{-3}x < \frac{9}{-3} \quad \text{The sign changed at this step.}$$
$$x < -3$$

Solve the inequality.

$$-\frac{1}{2}(4x - 6) + 2 > x - 7$$

$$-2x + 3 + 2 > x - 7 \qquad \text{Distribute } -\frac{1}{2}.$$

$$-2x + 5 > x - 7$$
$$-x -x$$
$$-3x + 5 > -7$$
$$ -5 -5$$
$$-3x > -12$$
$$\frac{-3}{-3}x < \frac{-12}{-3} \qquad \text{The sign changed at this step.}$$
$$x < 4$$

Solve the inequality.

$$2x + 1 < 5$$
$$-1 -1$$
$$2x < 4$$
$$\frac{2}{2}x < \frac{4}{2} \qquad \text{Because 2 is positive, we do not change the sign.}$$
$$x < 2$$

PRACTICE

Solve the inequality and graph the solution on the number line.

1. $7x - 4 \le 2x + 8$

2. $\dfrac{2}{3}(6x - 9) + 4 > 5x + 1$

3. $0.2(x - 5) + 1 \ge 0.12$

4. $10x - 3(2x + 1) \ge 8x + 1$

5. $3(x - 1) + 2(x - 1) \le 7x + 7$

 **SOLUTIONS**

1. $7x - 4 \le 2x + 8$
    $\underline{-2x \qquad -2x}$
    $5x - 4 \le 8$
    $\underline{\quad +4 \ +4}$
    $5x \le 12$
    $x \le \dfrac{12}{5}$

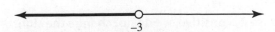

FIGURE 9-14

2. $\dfrac{2}{3}(6x - 9) + 4 > 5x + 1$
    $4x - 6 + 4 > 5x + 1$
    $4x - 2 > 5x + 1$
    $\underline{\quad +2 \qquad +2}$
    $4x > 5x + 3$
    $\underline{-5x \ -5x}$
    $-x > 3$
    $x < -3$

FIGURE 9-15

3. $0.2(x - 5) + 1 \geq 0.12$

$0.2x - 1 + 1 \geq 0.12$

$0.2x \geq 0.12$

$x \geq \dfrac{0.12}{0.2}$

$x \geq \dfrac{3}{5}$

$x \geq 0.60$

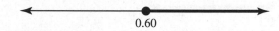

0.60

FIGURE 9-16

4. $10x - 3(2x + 1) \geq 8x + 1$

$10x - 6x - 3 \geq 8x + 1$

$4x - 3 \geq 8x + 1$

$+3 \qquad +3$

$4x \geq 8x + 4$

$-8x \quad -8x$

$-4x \geq 4$

$\dfrac{-4}{-4}x \leq \dfrac{4}{-4}$

$x \leq -1$

−1

FIGURE 9-17

5. $3(x-1)+2(x-1) \le 7x+7$

$\quad\quad 3x-3+2x-2 \le 7x+7$

$\quad\quad\quad\quad 5x-5 \le 7x+7$

$\quad\quad\quad\quad\quad +5 \quad\quad\quad +5$

$\quad\quad\quad\quad\quad 5x \le 7x+12$

$\quad\quad\quad\quad -7x \quad -7x$

$\quad\quad\quad\quad -2x \le 12$

$\quad\quad\quad\quad \dfrac{-2}{-2}x \ge \dfrac{12}{-2}$

$\quad\quad\quad\quad\quad x \ge -6$

FIGURE 9-18

Interval Notation

Sometimes, we use interval notation to represent inequalities. The notation for an interval tells us how far the solution extends to the left and to the right. From the notation, we can tell whether or not the endpoint(s) belong to the interval. We begin with unbounded intervals. These intervals involve infinity.

The symbol for positive infinity is "∞," and "$-\infty$" is the symbol for negative infinity. These symbols mean that the numbers in the interval are getting larger in the positive or negative direction. The intervals for the previous problems are examples of *infinite* intervals, or *unbounded* intervals.

An interval consists of, in order, an open parenthesis "(" or open bracket "[," a number or "$-\infty$," a comma, a number or "∞," and a closing parenthesis ")" or closing bracket "]." A parenthesis is used for strict inequalities ($x < a$ and $x > a$) and a bracket is used for an "or equal to" inequality ($x \le a$ and $x \ge a$). A parenthesis is always used next to an infinity symbol.

TABLE 9-1

Inequality	Interval
x < number	$(-\infty, \text{number})$
x > number	(number, ∞)
$x \le$ number	$(-\infty, \text{number}]$
$x \ge$ number	$[\text{number}, \infty)$

EXAMPLE

$$x < 3 \qquad (-\infty, 3)$$
$$x > -6 \qquad (-6, \infty)$$
$$x \le 100 \qquad (-\infty, 100]$$
$$x \ge 4 \qquad [4, \infty)$$

The relationship between an unbounded inequality, its interval notation, and its region on the number line is summarized in Figure 9-19.

Inequality	Number Line Region	Interval Notation
$x < a$	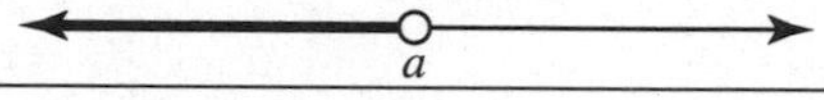 a	$(-\infty, a)$
$x \le a$	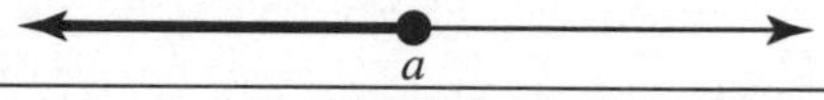 a	$(-\infty, a]$
$x > a$	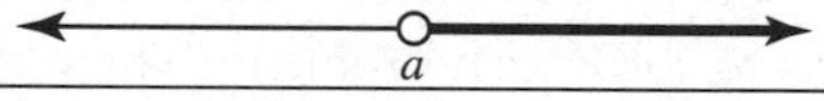 a	(a, ∞)
$x \ge a$	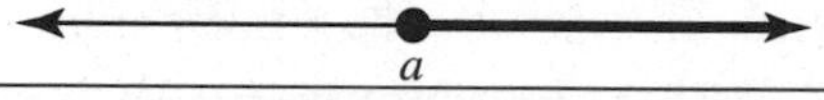 a	$[a, \infty)$

FIGURE 9-19

Ordinarily the variable is written on the left in an inequality but not always. For instance to say that x is less than 3 ($x < 3$) is the same as saying 3 is greater than x ($3 > x$).

PRACTICE

Give the interval notation for the inequality.

1. $x \ge 5$

2. $x < 1$

3. $x \le 4$

4. $x \ge -10$

5. $x \le -2$

6. $x > 9$

7. $x < -8$

8. $x > \dfrac{1}{2}$

SOLUTIONS

1. $x \geq 5$ $[5, \infty)$
2. $x < 1$ $(-\infty, 1)$
3. $x \leq 4$ $(-\infty, 4]$
4. $x \geq -10$ $[-10, \infty)$
5. $x \leq -2$ $(\infty, -2]$
6. $x > 9$ $(9, \infty)$
7. $x < -8$ $(-\infty, -8)$
8. $x > \dfrac{1}{2}$ $\left(\dfrac{1}{2}, \infty\right)$

Applied Problems

We use linear inequalities to solve applied problems in much the same way that we used linear equations to solve such problems. There are two important differences, though. Multiplying and dividing both sides of an inequality by a negative number requires that the sign reverse. We must also decide which inequality sign to use: $<$, $>$, $\leq$, and $\geq$. Tables 9-2 and 9-3 should help.

TABLE 9-2

A < B	A > B
A is less than B	A is greater than B
A is smaller than B	A is larger than B
B is greater than A	B is less than A
B is larger than A	A is more than B

TABLE 9-3

A ≤ B	B ≤ A
A is less than or equal to B	B is less than or equal to A
A is not more than B	B is not more than A
B is at least A	A is at least B
B is A or more	A is B or more
A is no greater than B	B is no greater than A
B is no less than A	A is no less than B

Some word problems give two alternatives and ask for what interval of the variable is one alternative more attractive than the other. If the alternative is between two costs, for example, in order for the cost of A to be more attractive than the cost of B, we solve "Cost of A < Cost of B." If the cost of A to be no more than the cost of B (also the cost of A to be at least as attractive as the cost of B), we solve "Cost of A ≤ Cost of B." If the alternative is between two incomes of some kind, for the income of A to be more attractive than the income of B, solve "Income of A > Income of B." If the income of A is to be at least as attractive as the income of B (also the income of A to be no less attractive than the income of B), solve "Income of A ≥ Income of B."

Some of the following examples and practice problems are business problems, so we now review a few business formulas. Revenue is normally the price per unit times the number of units sold. For instance, if an item sells for $3.25 each and x represents the number of units sold, the revenue is represented by $3.25x$ (dollars). The total cost generally includes overhead costs (sometimes called *fixed costs*) and production costs (sometimes called *variable costs*). The overhead cost is a fixed number (no variable). The production cost is usually computed as the cost per unit times the number of units sold. The total cost is usually the overhead costs plus the production costs. Profit is revenue minus cost. If a problem asks how many units must be sold to make a profit, solve "Revenue > Total Cost."

EXAMPLE

A manufacturing plant, which produces compact disks, has monthly overhead costs of $6,000. Each disk costs 18 cents to produce and sells for 30 cents. How many disks must be sold in order for the plant to make a profit?

Let x = number of CDs produced and sold monthly

In order for the plant to make a profit, revenue must be more than cost, so we need to solve the inequality Revenue > Cost, where Cost = $6000 + 0.18x$ and Revenue = $0.30x$.

$$\text{Revenue} > \text{Cost}$$
$$0.30x > 6000 + 0.18x$$
$$-0.18x > \qquad -0.18x$$
$$0.12x > 6000$$
$$x > \frac{6000}{0.12}$$
$$x > 50{,}000$$

The plant should produce and sell more than 50,000 CDs per month in order to make a profit.

Mary inherited $16,000 and will deposit it into two accounts, one paying $6\frac{3}{4}\%$ interest and the other paying $5\frac{1}{2}\%$ interest. What is the most she can deposit into the $5\frac{1}{2}\%$ account so that her interest at the end of a year will be at least $960?

Let x = amount deposited in the $5\frac{1}{2}\%$ account

$$0.055x = \text{interest earned at } 5\tfrac{1}{2}\%$$
$$16,000 - x = \text{amount deposited in the } 6\tfrac{3}{4}\% \text{ account}$$
$$0.0675(16,000 - x) = \text{interest earned at } 6\tfrac{3}{4}\%$$

We want the values of x for which the total interest is $960 or more, so we solve Interest earned at $5\frac{1}{2}\%$ + Interest earned at $6\frac{3}{4}\% \geq 960$.

$$0.055x + 0.0675(16,000 - x) \geq 960$$
$$0.055x + 1080 - 0.0675x \geq 960$$
$$-0.0125x + 1080 \geq 960$$
$$\underline{\quad -1080 \quad -1080 \quad}$$
$$-0.0125x \geq -120$$
$$\frac{-0.0125}{-0.0125}x \leq \frac{-120}{-0.0125}$$
$$x \leq 9600$$

Mary can invest no more than $9600 in the $5\frac{1}{2}\%$ account in order to receive at least $960 interest at the end of the year.

An excavating company can rent a piece of equipment for $45,000 per year. The company could purchase the equipment for monthly costs of $2500 plus $20 for each hour it is used. How many hours per year must the equipment be used to justify purchasing it rather than renting it?

Let x = number of hours per year the equipment is used

The monthly purchase costs amount to $12(2500) = 30,000$ dollars annually. The annual purchase cost $= 30,000 + 20x$.

$$\text{Purchase cost} < \text{Rent cost}$$
$$30,000 + 20x < 45,000$$
$$-30,000 \qquad\qquad -30,000$$
$$20x < 15,000$$
$$x \le \frac{15,000}{20}$$
$$x < 750$$

The equipment should be used less than 750 hours annually to justify purchasing it rather than renting it.

An amusement park sells an unlimited season pass for \$240. A daily ticket sells for \$36. How many times would a customer need to use the ticket in order for the season ticket to cost less than purchasing daily tickets?

Let x = number of daily tickets purchased per season

$$36x = \text{daily ticket cost}$$

$$\text{Season ticket cost} < \text{daily ticket cost}$$
$$240 < 36x$$
$$\frac{240}{36} < x$$
$$6\tfrac{2}{3} < x$$

A customer would need to use the ticket more than $6\tfrac{2}{3}$ times (or 7 or more times) in order for the season ticket to cost less than purchasing daily tickets.

Bank A offers a $6\tfrac{1}{2}\%$ certificate of deposit and Bank B offers a $5\tfrac{3}{4}\%$ certificate of deposit but will give a \$25 bonus at the end of the year. What is the least amount a customer would need to deposit at Bank A to make Bank A's offer no less attractive than Bank B's offer?

Let x = amount to deposit

If x dollars is deposited at Bank A, the interest at the end of the year would be $0.065x$. If x dollars is deposited at Bank B, the interest at the end of the year would be 0.0575x. The total income from Bank B would be 25 + 0.0575x.

$$\text{Income from Bank A} \geq \text{Income from Bank B}$$

$$0.0650x \geq 25 + 0.0575x$$

$$-0.0575x \geq \qquad -0.0575x$$

$$0.0075x \geq 25$$

$$x \geq \frac{25}{0.0075}$$

$$x \geq 3333.33$$

A customer would need to deposit at least $3333.33 in Bank A to earn no less than would be earned at Bank B.

PRACTICE

1. A scholarship administrator is using a $500,000 endowment to purchase two bonds. A corporate bond pays 8% interest per year and a safer treasury bond pays $5\frac{1}{4}$% interest per year. If he needs at least $30,000 annual interest payments, what is the least he can spend on the corporate bond?

2. Kelly sells corn dogs at a state fair. Booth rental and equipment rental total $200 per day. Each corn dog costs 35 cents to make and sells for $2. How many corn dogs should she sell in order to have a daily profit of at least $460?

3. The owner of a snow cone machine pays $200 per month to rent his equipment and $400 per month for a stall in a flea market. Each snow cone costs 25 cents to make and sells for $1.50. How many snow cones does he need to sell in order to make a profit?

4. A tee-shirt stand can sell a certain sports tee shirt for $18. Each shirt costs $8 in materials and labor. Monthly fixed costs are $1500. How many tee shirts must be sold to guarantee a monthly profit of at least $3500?

5. A car rental company rents a certain car for $40 per day with unlimited mileage or $24 per day plus 80 cents per mile. What is the most a customer can drive the car per day for the $24 option to cost no more than the unlimited mileage option?

6. A company providing cell phone service has two plans. The monthly charge on one plan is $75 with unlimited talk and Web access. The monthly charge for the other plan is $15 plus 20 cents per minute of talk and web access. How much time per month would a customer need to use the phone for talk and web access so that the unlimited plan is less expensive than the other plan?

7. Sharon can purchase a pair of ice skates for $60. It costs her $3 to rent a pair each time she goes to the rink. How many times would she need to use the skates to make purchasing them more attractive than buying them?

8. The James family has $210 budgeted each month for electricity. They have a monthly base charge of $28 plus 7 cents per kilowatt-hour. How many kilowatt-hours can they use each month to stay within their budget?

9. A warehouse store charges an annual fee of $40 to shop there. A shopper without paying this fee can still shop there if he pays a 5% buyer's premium on his purchases. How much would a shopper need to spend at the store to make paying the annual $40 fee cost no more than paying the 5% buyer's premium?

10. A sales clerk at an electronics store is given the option for her salary to be changed from a straight annual salary of $25,000 to an annual base salary of $15,000 plus an 8% commission on sales. What would her annual sales level need to be in order for this option to be at least as attractive as the straight salary option?

SOLUTIONS

1. Let x = amount invested in the corporate bond

   $500,000 - x$ = amount invested in the treasury bond

   $0.08x$ = annual interest from the corporate bond

   $0.0525(500,000 - x)$ = annual interest from the treasury bond

   Corporate bond interest + Treasury bond interest $\geq 30,000$

$$0.08x + 0.0525(500,000 - x) \geq 30,000$$
$$0.08x + 26,250 - 0.0525x \geq 30,000$$
$$\underline{-26,250 \qquad\qquad -26,250}$$
$$0.0275x \geq 3750$$
$$x \geq \frac{3750}{0.0275}$$
$$x \geq 136,363.64$$

The administrator should invest at least $136,363.64 in the corporate bond in order to receive at least $30,000 per year in interest payments.

2. Let x = number of corn dogs sold per day

   $2x$ = revenue

   $200 + 0.35x$ = overhead costs + production costs = total cost

   $2x - (200 + 0.35x)$ = profit

   profit ≥ 460

   $2x - (200 + 0.35x) \geq 460$

   $2x - 200 - 0.35x \geq 460$

   $1.65x - 200 \geq 460$

   $\underline{+200 \quad +200}$

   $1.65x \geq 660$

   $x \geq \dfrac{660}{1.65}$

   $x \geq 400$

Kelly needs to sell at least 400 corn dogs in order for her daily profit to be at least $460.

3. Let x = number of snow cones sold per month

   $1.50x$ = revenue

   $600 + .25x$ = overhead costs + production costs = total cost

   Revenue > Cost

   $1.50x > 600 + 0.25x$

   $\underline{-0.25x \qquad\quad -0.25x}$

   $1.25x > 600$

   $x > \dfrac{600}{1.25}$

   $x > 480$

The owner should sell more than 480 snow cones per month to make a profit.

4. Let x = number of tee shirts sold per month

   $18x$ = revenue

   $1500 + 8x$ = overhead costs + production costs = total cost

   $18x - (1500 + 8x)$ = profit

   $$\text{profit} \geq 3500$$
   $$18x - (1500 + 8x) \geq 3500$$
   $$18x - 1500 - 8x \geq 3500$$
   $$10x - 1500 \geq 3500$$
   $$\underline{+1500 \quad +1500}$$
   $$10x \geq 5000$$
   $$x \geq \frac{5000}{10}$$
   $$x \geq 500$$

   At least 500 tee shirts would need to be sold each month to make a monthly profit of at least $3500.

5. Let x = number of average daily miles

   The $24 option costs $24 + 0.80x$ per day (on average).

   $$24 + 0.80x \leq 40$$
   $$\underline{-24 \qquad\qquad -24}$$
   $$0.80x \leq 16$$
   $$x \leq \frac{16}{0.80} = 20$$

   The most a customer could drive is an average of 20 miles per day in order for the $24 plan to cost no more than the $40 plan.

6. Let x represent the number of minutes of service used in a month. The monthly cost for the unlimited plan is 75, and the monthly cost for the other plan is $15 + 0.20x$. The unlimited plan is less expensive than the other when $15 + 0.20x > 75$.

   $$15 + 0.20x > 75$$
   $$0.20x > 60$$
   $$x > \frac{60}{0.20}$$
   $$x > 300$$

   A customer would need to use more than 300 minutes or 5 hours of the phone and web service per month in order to make the unlimited plan less expensive.

7. Let x = number of times Sharon uses her skates

    The cost to rent skates is $3x$.

    $60 < 3x$

    $\dfrac{60}{3} < x$

    $20 < x$ (or $x > 20$)

    Sharon would need to use her skates more than 20 times to justify purchasing them instead of renting them.

8. Let x = number of kilowatt-hours used per month

    $28 + 0.07x$ = monthly bill

    $28 + 0.07x \le 210$

    $\underline{-28 \qquad\qquad -28}$

    $ 0.07x \le 182$

    $x \le \dfrac{182}{0.07}$

    $x \le 2600$

    The James family can use no more than 2600 kilowatt-hours per month in order to keep their electricity costs within their budget.

9. Let x = amount spent at the store annually

    $0.05x$ = extra 5% purchase charge per year

    $40 \le 0.05x$

    $\dfrac{40}{0.05} \le x$

    $800 \le x \left(\text{or } x \ge 800 \right)$

    A shopper would need to spend at least \$800 per year to justify the \$40 annual fee.

10. Let x = annual sales level

    $0.08x$ = annual commission

    $15,000 + 0.08x$ = straight annual salary plus commission

    $15,000 + 0.08x \ge 25,000$

    $\underline{-15,000 \qquad\qquad -15,000}$

    $ 0.08x \ge 10,000$

    $x \ge \dfrac{10,000}{0.08}$

    $x \ge 125,000$

The sales clerk would need an annual sales level of $125,000 or more in order for the salary plus commission option to be at least as attractive as the straight salary option.

Bounded Intervals

A bounded interval has both a left endpoint and a right endpoint. These intervals are represented by double inequalities. The double inequality $a < x < b$ means all real numbers between a and b. In fact, $a < x < b$ stands for *two* separate inequalities. The notation $a < x < b$ stands for "$a < x$ and $x < b$."

An inequality such as $10 < x < 5$ is *never* true because no number x is both *larger* than 10 and *smaller* than 5. In other words, an inequality in the form "larger number $< x <$ smaller number" is meaningless.

The inequality $a < x < b$ is represented on the number line by shading the region between a and b. The inequality is represented with interval notation by (a, b). Figure 9-20 gives the number line region and interval notation for each type of double inequality.

Inequality	Region on the Number Line	Interval
$a < x < b$	$\overset{a \qquad\qquad b}{\circ\text{———}\circ}$ All real numbers between a and b but not including a and b	(a, b)
$a \leq x \leq b$	$\overset{a \qquad\qquad b}{\bullet\text{———}\bullet}$ All real numbers between a and b, including a and b	$[a, b]$
$a < x \leq b$	$\overset{a \qquad\qquad b}{\circ\text{———}\bullet}$ All real numbers between a and b, including b but not a	$(a, b]$
$a \leq x < b$	$\overset{a \qquad\qquad b}{\bullet\text{———}\circ}$ All real numbers between a and b, including a but not b	$[a, b)$

FIGURE 9-20

EXAMPLE

$$3 < x \le 7 \quad (3, 7]$$
$$-4 \le x \le -1 \quad [-4, -1]$$
$$-8 < x < 8 \quad (-8, 8)$$
$$0 \le x < \frac{1}{2} \quad [0, \frac{1}{2})$$
$$-6 < x < 0 \quad (-6, 0)$$

PRACTICE

Give the interval notation for the double inequality.

1. $6 < x < 8$
2. $-4 \le x < 5$
3. $-2 \le x < 2$
4. $0 \le x \le 10$
5. $9 < x \le 11$
6. $\frac{1}{4} \le x \le \frac{1}{2}$
7. $904 < x < 1100$

SOLUTIONS

1. $6 < x < 8$ $(6, 8)$
2. $-4 \le x < 5$ $[-4, 5)$
3. $-2 \le x < 2$ $[-2, 2)$
4. $0 \le x \le 10$ $[0, 10]$
5. $9 < x \le 11$ $(9, 11]$
6. $\frac{1}{4} \le x \le \frac{1}{2}$ $\left[\frac{1}{4}, \frac{1}{2}\right]$
7. $904 < x < 1100$ $(904, 1100)$

Solving Double Inequalities

We solve double inequalities the same way we solve other inequalities except that there are three "sides" to the inequality instead of two.

EXAMPLE

Solve the double inequality and give the solution in interval notation.

$$4 \le 2x \le 12$$

Because we want to solve for *x*, we divide all three quantities by 2.

$$4 \leq 2x \leq 12$$

$$\frac{4}{2} \leq \frac{2}{2}x \leq \frac{12}{2}$$

$$2 \leq x \leq 6 \quad [2, 6]$$

Solve the double inequality and give the solution in interval notation.

$$6 \leq 4x - 2 \leq 10$$

We begin by adding 2 to each quantity and then dividing each quantity by 4.

$$6 \leq 4x - 2 \leq 10$$

$$+2 \qquad +2 \quad +2$$

$$8 \leq 4x \leq 12$$

$$\frac{8}{4} \leq \frac{4}{4}x \leq \frac{12}{4}$$

$$2 \leq x \leq 3 \quad [2, 3]$$

Solve the double inequality and give the solution in interval notation.

$$-6 < -2x + 3 < 1$$

$$-6 < -2x + 3 < 1$$

$$-3 \qquad -3 \ -3$$

$$-9 < -2x < -2$$

$$\frac{-9}{-2} > \frac{-2}{-2}x > \frac{-2}{-2}$$

$$\frac{9}{2} > x > 1 \text{ or } 1 < x < \frac{9}{2} \quad \left(1, \frac{9}{2}\right)$$

Solve the double inequality and give the solution in interval notation.

$$7 \leq 3x + 7 < 4$$

$$7 \leq 3x + 7 < 4$$

$$-7 \qquad -7 \ -7$$

$$0 \leq 3x < -3$$

$$\frac{0}{3} \leq \frac{3}{3}x < \frac{-3}{3}$$

$$0 \leq x < -1 \quad [0, -1)$$

Solve the double inequality and give the solution in interval notation.

$16 < 4(2x - 1) \le 20$

$$16 < 4(2x - 1) \le 20$$

$$16 < 8x - 4 \le 20$$

$$+4 \qquad +4 \quad +4$$

$$20 < 8x \le 24$$

$$\frac{20}{8} < \frac{8}{8}x \le \frac{24}{8}$$

$$\frac{5}{2} < x \le 3 \qquad \left(\frac{5}{2}, 3\right]$$

Solve the double inequality and give the solution in interval notation.

$-2 < \dfrac{-1}{2}(4x - 5) < 2$

$$-2 < \frac{-1}{2}(4x - 5) < 2$$

$$-2 < -2x + \frac{5}{2} < 2$$

$$-\frac{5}{2} \qquad -\frac{5}{2} \quad -\frac{5}{2}$$

$$-\frac{9}{2} < -2x < -\frac{1}{2}$$

$$\frac{-1}{2} \cdot \frac{-9}{2} > \frac{-1}{2}(-2x) > \frac{-1}{2} \cdot \frac{-1}{2}$$

$$\frac{9}{4} > x > \frac{1}{4} \text{ or } \frac{1}{4} < x < \frac{9}{4} \qquad \left(\frac{1}{4}, \frac{9}{4}\right)$$

Solve the double inequality and give the solution in interval notation.

$$2 < \frac{5x - 1}{4} < 6$$

We begin by "clearing the fraction," so we multiply each quantity by 4.

$$2 < \frac{5x-1}{4} < 6$$

$$4(2) < \frac{4}{1} \cdot \frac{5x-1}{4} < 4(6)$$

$$8 < 5x - 1 < 24$$

$$+1 \qquad +1 \quad +1$$

$$9 < 5x < 25$$

$$\frac{9}{5} < \frac{5}{5}x < \frac{25}{5}$$

$$\frac{9}{5} < x < 5 \qquad \left(\frac{9}{5}, 5\right)$$

 PRACTICE

Solve the double inequality and give your solution in interval notation.

1. $14 < 2x < 20$

2. $5 \leq 3x - 1 \leq 8$

3. $-2 \leq 3x - 4 \leq 5$

4. $-4 < -2x + 6 < 4$

5. $0.12 \leq 4x - 1 \leq 1.8$

6. $4 < 3(-2x + 1) \leq 7$

7. $-1 < -6x + 11 < 1$

8. $\frac{7}{8} < \frac{1}{4}x \leq 2$

9. $8 \leq 4.5x - 1 \leq 11$

10. $-6 \leq \frac{2}{3}x + 4 < 0$

11. $-1 < \frac{2x-5}{3} < 1$

 SOLUTIONS

1. $14 < 2x < 20$

$$\frac{14}{2} < \frac{2}{2}x < \frac{20}{2}$$

$$7 < x < 10 \qquad (7,10)$$

2. $5 \le 3x - 1 \le 8$

   $+1$ $+1$ $+1$

   $6 \le 3x \le 9$

   $\dfrac{6}{3} \le \dfrac{3}{3}x \le \dfrac{9}{3}$

   $2 \le x \le 3$ $[2, 3]$

3. $-2 \le 3x - 4 \le 5$

   $+4$ $+4$ $+4$

   $2 \le 3x \le 9$

   $\dfrac{2}{3} \le \dfrac{3}{3}x \le \dfrac{9}{3}$

   $\dfrac{2}{3} \le x \le 3$ $\left[\dfrac{2}{3}, 3\right]$

4. $-4 < -2x + 6 < 4$

   -6 -6 -6

   $-10 < -2x < -2$

   $\dfrac{-10}{-2} > \dfrac{-2}{-2}x > \dfrac{-2}{-2}$

   $5 > x > 1$ or $1 < x < 5$ $(1, 5)$

5. $0.12 \le 4x - 1 \le 1.8$

   $+1.00$ $+1$ $+1.0$

   $1.12 \le 4x \le 2.8$

   $\dfrac{1.12}{4} \le \dfrac{4}{4}x \le \dfrac{2.8}{4}$

   $0.28 \le x \le 0.7$ $[0.28, 0.7]$

6. $4 < 3(-2x + 1) \le 7$

   $4 < -6x + 3 \le 7$

   -3 -3 -3

   $1 < -6x \le 4$

   $\dfrac{1}{-6} > \dfrac{-6}{-6}x \ge \dfrac{4}{-6}$

   $-\dfrac{1}{6} > x \ge -\dfrac{2}{3}$ or $-\dfrac{2}{3} \le x < -\dfrac{1}{6}$ $\left[-\dfrac{2}{3}, -\dfrac{1}{6}\right)$

7.　$-1 < -6x + 11 < 1$

$\quad -11 \qquad\quad -11 \ -11$

$-12 < -6x < -10$

$\dfrac{-12}{-6} > \dfrac{-6}{-6}x > \dfrac{-10}{-6}$

$\quad 2 > x > \dfrac{5}{3}$　or　$\dfrac{5}{3} < x < 2$　$\left(\dfrac{5}{3}, 2\right)$

8.　$\dfrac{7}{8} < \dfrac{1}{4}x \le 2$

$\dfrac{4}{1}\cdot\dfrac{7}{8} < \dfrac{4}{1}\cdot\dfrac{1}{4}x \le 4(2)$

$\quad\dfrac{7}{2} < x \le 8$　$\left(\dfrac{7}{2}, 8\right]$

9.　$8 \le 4.5x - 1 \le 11$

$\ +1 \qquad\quad +1 \ +1$

$\quad 9 \le 4.5x \le 12$

$\dfrac{9}{4.5} \le \dfrac{4.5}{4.5}x \le \dfrac{12}{4.5}$

$\quad 2 \le x \le \dfrac{8}{3}$　$\left[2, \dfrac{8}{3}\right]$

10.　$-6 \le \dfrac{2}{3}x + 4 < 0$

$\qquad -4 \qquad\ -4 \ -4$

$-10 \le \dfrac{2}{3}x < -4$

$\dfrac{3}{2}\cdot\dfrac{-10}{1} \le \dfrac{3}{2}\cdot\dfrac{2}{3}x < \dfrac{3}{2}\cdot\dfrac{-4}{1}$

$\quad -15 \le x < -6$　$[-15, -6)$

11.　$-1 < \dfrac{2x - 5}{3} < 1$

$(3)(-1) < \dfrac{3}{1}\cdot\dfrac{2x - 5}{3} < 3(1)$

$\quad -3 < 2x - 5 < 3$

$\quad +5 \qquad +5 \ +5$

$\qquad 2 < 2x < 8$

$\quad\dfrac{2}{2} < \dfrac{2}{2}x < \dfrac{8}{2}$

$\qquad 1 < x < 4$　$(1, 4)$

Applications of Double Inequalities

Before we work with applied problems, we need to learn how to solve inequalities that involve two variables. We are given a range for one variable and we want to find the corresponding range for the other variable, like the temperature example in the introduction to this chapter. In general, we make a substitution for the variable that is in the inequality.

 EXAMPLE

Give the corresponding interval for x.

$$y = 3x - 2$$

If $7 \leq y \leq 10$, what is the corresponding interval for x? Because $y = 3x - 2$, replace "y" with "$3x - 2$." "$7 \leq y \leq 10$" becomes "$7 \leq 3x - 2 \leq 10$".

$$7 \leq 3x - 2 \leq 10$$
$$+2 \qquad +2 \ +2$$
$$9 \leq 3x \leq 12$$
$$\frac{9}{3} \leq \frac{3}{3}x \leq \frac{12}{3}$$
$$3 \leq x \leq 4$$

Give the corresponding interval for x.

$$y = 4x + 1; \; -3 < y < 3$$

If $-3 < y < 3$, the corresponding interval for x can be found by solving $-3 < 4x + 1 < 3$.

$$-3 < 4x + 1 < 3$$
$$-1 \qquad -1 \ -1$$
$$-4 < 4x < 2$$
$$\frac{-4}{4} < \frac{4}{4}x < \frac{2}{4}$$
$$-1 < x < \frac{1}{2}$$

Give the corresponding interval for x.

$$y = 3 - x; \; 0 \leq y < 4$$

If $0 \leq y < 4$, the corresponding interval for x can be found by solving $0 \leq 3 - x < 4$.

$$0 \leq 3 - x < 4$$
$$-3 \quad -3 \qquad -3$$
$$-3 \leq -x < 1$$
$$-(-3) \geq -(-x) > -1$$
$$3 \geq x > -1 \quad \text{or} \quad -1 < x \leq 3$$

PRACTICE

Give the corresponding interval for x.

1. $y = x - 4$ $-5 < y < 5$

2. $y = 4x - 3$ $0 < y \leq 2$

3. $y = 7 - 2x$ $4 \leq y \leq 10$

4. $y = 8x + 1$ $-5 \leq y \leq -1$

5. $y = \dfrac{2}{3}x - 8$ $4 < y < 6$

6. $y = \dfrac{x - 4}{2}$ $-1 \leq y < 3$

SOLUTIONS

1. $y = x - 4$ $-5 < y < 5$

$$-5 < x - 4 < 5$$
$$+4 \qquad +4 \ +4$$
$$-1 < x < 9$$

2. $y = 4x - 3$ $0 < y \leq 2$

$$0 < 4x - 3 \leq 2$$
$$+3 \qquad +3 \ +3$$
$$3 < 4x \leq 5$$
$$\frac{3}{4} < \frac{4}{4}x \leq \frac{5}{4}$$
$$\frac{3}{4} < x \leq \frac{5}{4}$$

3. $y = 7 - 2x$ $4 \leq y \leq 10$

$$4 \leq 7 - 2x \leq 10$$
$$ -7 -7 -7$$
$$-3 \leq -2x \leq 3$$
$$\frac{-3}{-2} \geq \frac{-2}{-2}x \geq \frac{3}{-2}$$
$$\frac{3}{2} \geq x \geq \frac{-3}{2} \quad \text{or} \quad \frac{-3}{2} \leq x \leq \frac{3}{2}$$

4. $y = 8x + 1$ $-5 \leq y \leq -1$

$$-5 \leq 8x + 1 \leq -1$$
$$-1 -1 -1$$
$$-6 \leq 8x \leq -2$$
$$\frac{-6}{8} \leq \frac{8}{8}x \leq \frac{-2}{8}$$
$$\frac{-3}{4} \leq x \leq \frac{-1}{4}$$

5. $y = \dfrac{2}{3}x - 8$ $4 < y < 6$

$$4 < \frac{2}{3}x - 8 < 6$$
$$+8 \phantom{< \frac{2}{3}x} +8 \phantom{<} +8$$
$$12 < \frac{2}{3}x < 14$$
$$\frac{3}{2} \cdot \frac{12}{1} < \frac{3}{2} \cdot \frac{2}{3}x < \frac{3}{2} \cdot \frac{14}{1}$$
$$18 < x < 21$$

6. $y = \dfrac{x - 4}{2}$ $-1 \leq y < 3$

$$-1 \leq \frac{x - 4}{2} < 3$$
$$2(-1) \leq \frac{2}{1} \cdot \frac{x - 4}{2} < 2(3)$$
$$-2 \leq x - 4 < 6$$
$$+4 +4 \phantom{<} +4$$
$$2 \leq x \leq 10$$

We use double inequalities to solve word problems where the solution is a limited range of values. For example, if a student is paid $15 per hour and wants to earn between $300 and $400 per week, we solve $300 \le \text{Pay} \le 400$, or $300 \le 15x \le 400$, if x represents the number of hours worked in a week.

EXAMPLE

A high school student earns $8 per hour in her summer job. She hopes to earn between $120 and $200 per week. What range of hours will she need to work so that her pay is in this range?

Let x represent the number of hours worked per week. Represent her weekly pay by $p = 8x$. The student wants $120 \le p \le 200$, so we want to solve the inequality $120 \le 8x \le 200$.

$$120 \le 8x \le 200$$
$$\frac{120}{8} \le \frac{8}{8}x \le \frac{200}{8}$$
$$15 \le x \le 25$$

The student would need to work between 15 and 25 hours per week for her pay to range from $120 to $200 per week.

A manufacturing plant produces pencils. It has monthly overhead costs of $60,000. Each gross (144) of pencils costs $3.60 to manufacturer. The company wants to keep total costs between $96,000 and $150,000 per month. How many gross of pencils should the plant produce to keep its costs in this range?

Let x represent the number of gross of pencils manufactured monthly. Production cost is represented by $3.60x$. Represent the total cost by $c = 60,000 + 3.60x$.

The manufacturer wants $96,000 \le c \le 150,000$. This gives us the inequality $96,000 \le 60,000 + 3.60x \le 150,000$.

$$96,000 \le 60,000 + 3.60x \le 150,000$$
$$-60,000 \quad -60,000 \qquad\qquad -60,000$$
$$36,000 \le 3.60x \le 90,000$$
$$\frac{36,000}{3.60} \le \frac{3.60}{3.60}x \le \frac{90,000}{3.60}$$
$$10,000 \le x \le 25,000$$

The manufacturing plant should produce between 10,000 and 25,000 gross per month to keep its monthly costs between $96,000 and $150,000.

PRACTICE

1. According to Hooke's law, the force, F (in pounds), required to stretch a certain spring x inches beyond its natural length is $F = 4.2x$. If $7 \le F \le 14$, what is the corresponding range for x?

2. The relationship between the Fahrenheit and Celsius temperature scales is given by $F = \frac{9}{5}C + 32$. If $5 \le F \le 23$, what is the corresponding range for C?

3. A saleswoman's salary is a combination of an annual base salary of $15,000 plus a 10% commission on sales. What level of sales does she need to maintain in order that her annual salary range from $40,000 to $60,000?

4. The Smith's electric bills consist of a base charge of $20 plus 12 cents per kilowatt-hour. If the Smiths want to keep their electric bill in the $80 to $110 range, what range of kilowatt-hours do they need to maintain?

5. A particular collect call costs $2.10 plus 50 cents per minute. (The company bills in two-second intervals.) How many minutes would a call need to last to keep a charge between $4.50 and $6.00?

SOLUTIONS

1. $7 \le F \le 14$ and $F = 4.2x$.

$$7 \le 4.2x \le 14$$

$$\frac{7}{4.2} \le \frac{4.2}{4.2}x \le \frac{14}{4.2}$$

$$\frac{5}{3} \le x \le \frac{10}{3}$$

If the force is to be kept between 7 and 14 pounds, the spring will stretch between $\frac{5}{3}$ and $\frac{10}{3}$ inches beyond its natural length.

2. $5 \leq F \leq 23$ and $F = \dfrac{9}{5}C + 32.$

$$5 \leq \dfrac{9}{5}C + 32 \leq 23$$

$$-32 \qquad -32 \ -32$$

$$-27 \leq \dfrac{9}{5}C \leq -9$$

$$\dfrac{5}{9} \cdot \dfrac{-27}{1} \leq \dfrac{5}{9} \cdot \dfrac{9}{5}C \leq \dfrac{5}{9} \cdot \dfrac{-9}{1}$$

$$-15 \leq C \leq -5$$

3. Let x represent the saleswoman's annual sales. Let $s = 15,000 + 0.10x$ represent her annual salary. She wants $40,000 \leq 15,000 + 0.10x \leq 60,000.$

$$40,000 \leq 15,000 + 0.10x \leq 60,000$$

$$-15,000 \ -15,000 \qquad -15,000$$

$$25,000 \leq 0.10x \leq 45,000$$

$$\dfrac{25,000}{0.10} \leq \dfrac{0.10x}{0.10} \leq \dfrac{45,000}{0.10}$$

$$250,000 \leq x \leq 450,000$$

She needs to have her annual sales range from $250,000 to $450,000 in order to maintain her annual salary between $40,000 and $60,000.

4. Let x represent the number of kilowatt-hours the Smiths use per month. Then $c = 20 + 0.12x$ represents their monthly electric bill.

$$80 \leq 20 + 0.12x \leq 110$$

$$-20 \ -20 \qquad -20$$

$$60 \leq 0.12x \leq 90$$

$$\dfrac{60}{0.12} \leq \dfrac{0.12x}{0.12} \leq \dfrac{90}{0.12}$$

$$500 \leq x \leq 750$$

The Smiths would need to keep their monthly usage between 500 and 750 kilowatt-hours to keep their monthly bills in the $80 to $110 range.

5. Let x represent the number of minutes the call lasts. Let $c = 2.10 + 0.50x$ represent the total cost of the call.

$$4.50 \le 2.10 + 0.50x \le 6.00$$

$$-2.10 \quad -2.10 \qquad \quad -2.10$$

$$2.40 \le 0.50x \le 3.90$$

$$\frac{2.40}{0.50} \le \frac{0.50x}{0.50} \le \frac{3.90}{0.50}$$

$$4.8 \le x \le 7.8$$

4.8 minutes is 4 minutes 48 seconds and 7.8 minutes is 7 minutes 48 seconds because 0.80 minutes is 0.80(60) seconds = 48 seconds.

A call would need to last between 4 minutes 48 seconds and 7 minutes 48 seconds in order to cost between $4.50 and $6.00.

Summary

In this chapter, we learned how to:

- *Represent an inequality as a region on the number line and with interval notation.* Figure 9-19 shows the relationship between an unbounded interval, its region on the number line, and its representation as an interval.

Inequality	Number Line Region	Interval Notation
$x < a$	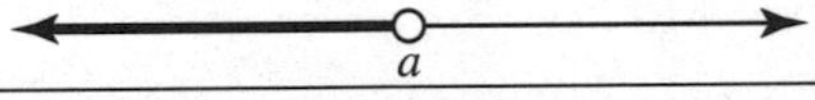 a	$(-\infty, a)$
$x \le a$	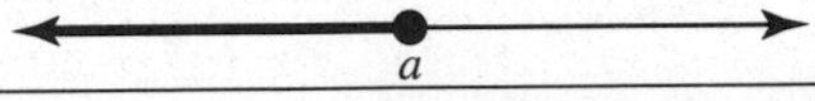 a	$(-\infty, a]$
$x > a$	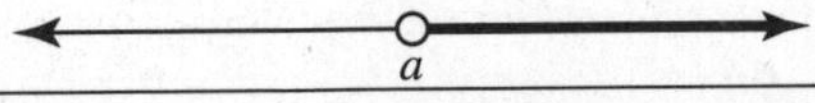 a	(a, ∞)
$x \ge a$	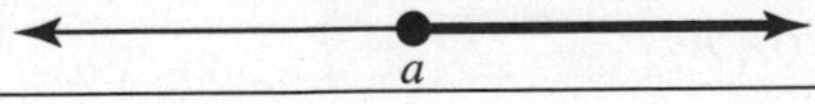 a	$[a, \infty)$

FIGURE 9-19

- *Solve linear inequalities.* We solve a linear inequality in much the same way we solve a linear equation except that we must reverse the inequality symbol if we multiply or divide each side of the inequality by a negative number.

- *Work with double inequalities.* A bounded interval is represented by a double inequality. Figure 9-20 shows the relationship between a bounded interval, its region on the number line, and its representation as an interval.

Inequality	Region on the Number Line	Interval
$a < x < b$	All real numbers between a and b but not including a and b	(a, b)
$a \leq x \leq b$	All real numbers between a and b, including a and b	$[a, b]$
$a < x \leq b$	All real numbers between a and b, including b but not a	$(a, b]$
$a \leq x < b$	All real numbers between a and b, including a but not b	$[a, b)$

FIGURE 9-20

- *Solve double inequalities.* We solve double inequalities the same way we solve single inequalities, except that the inequality has three quantities instead of two "sides." The quantity between the inequality symbols contains the variable. As with other inequalities, we use addition/subtraction and multiplication/division to isolate the variable.

- *Solve applications with inequalities.* The applications in this section are the inequality version of problems we solved in Chapter 8. We must decide which inequality symbol is appropriate. Tables 9-2 and 9-3 summarize the phrases that symbolize the various inequalities.

TABLE 9-2

A < B	A > B
A is less than B	A is greater than B
A is smaller than B	A is larger than B
B is greater than A	B is less than A
B is larger than A	A is more than B

TABLE 9-3

A ≤ B	B ≤ A
A is less than or equal to B	B is less than or equal to A
A is not more than B	B is not more than A
B is at least A	A is at least B
B is A or more	A is B or more
A is no greater than B	B is no greater than A
B is no less than A	A is no less than B

QUIZ

1. The solution to $5x + 6 > 8 - 7x$ is

   A. $x < \dfrac{1}{6}$

   B. $x > -1$

   C. $x < -1$

   D. $x > \dfrac{1}{6}$

2. The inequality $x > -3$ is represented by

   A.   ⟵————————●————————⟶
   -3

   B.   ⟵————————●————————⟶
   -3

   C.   ⟵————————○————————⟶
   -3

   D.   ⟵————————○————————⟶
   -3

   FIGURE 9-21

3. The solution to $6 - 2x \leq 3 + 4\left(\dfrac{1}{2}x + 1\right)$ is

   A. $x \geq -\dfrac{1}{4}$

   B. $x \leq -\dfrac{1}{4}$

   C. $x \leq \dfrac{1}{2}$

   D. $x \geq \dfrac{1}{2}$

4. The inequality $x < 8$ is represented by

   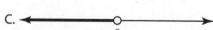

   A.   ⟵————————●————————⟶
   8

   B.   ⟵————————●————————⟶
   8

   C.   ⟵————————○————————⟶
   8

   D.   ⟵————————○————————⟶
   8

   FIGURE 9-22

5. The interval notation for $x \leq 4$ is

   A. $[4, \infty)$

   B. $(4, \infty)$

   C. $(-\infty, 4)$

   D. $(-\infty, 4]$

6. The chair of a scholarship committee needs to decide how to invest a $200,000 gift to his college. This year, he expects a bond mutual fund to pay 5% and a stock mutual fund to pay $7\frac{1}{2}\%$. The committee must fund scholarships totaling $13,000 with this gift. How much can be invested in the bond fund and still have enough to fund the scholarships?

   A. At least $120,000

   B. At most $120,000

   C. At most $80,000

   D. At least $80,000

7. The interval notation for the inequality $-1 \le x < 7$ is

   A. $(-1, 7]$

   B. $(-1, 7)$

   C. $[-1, 7)$

   D. $[-1, 7]$

8. The inequality $-4 \le x \le 4$ is represented by

A. 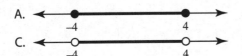   B.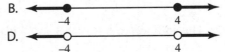

C.   D.

FIGURE 9-23

9. The interval notation for the inequality $3 < x < -2$ is

   A. $(3, -2)$

   B. $(-2, 3)$

   C. $3 < x < -2$ is not an interval of numbers.

10. The solution to $6 < 5x + 1 \le 16$ is

   A. $[1, 3)$

   B. $(1, 3]$

   C. $(1, 3)$

   D. $[1, 3]$

11. The interval notation for $x > 100$ is

   A. $(-\infty, 100)$

   B. $(100, -\infty)$

   C. $(100, \infty)$

   D. $(\infty, 100)$

12. The Martinez family's monthly budget for electricity is $80 to $140. Their electricity provider charges a $20 customer charge plus $0.12 per kilowatt-hour (kWh) of electricity. How much electricity can they use in a month and remain in their budget?

    A. Between 500 and 1000 kWh

    B. Between 500 and 1200 kWh

    C. Between 600 and 1200 kWh

    D. Between 600 and 1300 kWh

13. A father plans to rent a car to travel to another state to attend his son's college graduation ceremony. One rental car agency charges $32 per day with unlimited mileage. Another agency charges $23 per day plus 6 cents per mile. For what average daily mileage is the unlimited mileage plan more attractive?

    A. More than 525 miles

    B. At most 525 miles

    C. At most 150 miles

    D. More than 150 miles

14. The solution to $1 \leq 7 - 2x \leq 15$ is

    A. $-4 \leq x \leq 3$

    B. $-3 \leq x \leq 4$

    C. $4 \leq x \leq 3$

    D. $4 \leq x \leq -3$

15. The solution to $5 < \dfrac{3x + 2}{4} < 8$ is

    A. $-\dfrac{1}{3} < x < \dfrac{2}{3}$

    B. $6 < x < 10$

    C. $-\dfrac{1}{3} < x < 0$

    D. $\dfrac{10}{3} < x < \dfrac{16}{3}$

chapter 10

Quadratic Equations

Linear equations and quadratic equations occur so much in algebra (and calculus) that solving them eventually becomes second nature. We learned a strategy in Chapter 7 that helps us to solve any linear equation. Here, we learn a few strategies for solving quadratic equations. One of those strategies involves the factoring methods that we learned in Chapter 6, and another involves an important formula in mathematics—the quadratic formula. The strategies that we develop here extend to other mathematics courses to solve larger families of equations.

CHAPTER OBJECTIVES

In this chapter, you will

- Solve a quadratic equation by factoring
- Solve a quadratic equation by extracting roots
- Solve a quadratic equation using the quadratic formula
- Solve equations that lead to quadratic equations

Solving Quadratic Equations by Factoring

A quadratic equation is one that we can write in the form $ax^2 + bx + c = 0$ where a, b, and c are numbers and a is not zero (b and/or c might be zero). For instance, we can write the equation $3x^2 + 7x = 4$ in the above form, so it is a quadratic equation.

$$3x^2 + 7x = 4$$
$$ -4 \quad -4$$
$$3x^2 + 7x - 4 = 0$$

We use one of two main approaches to solve quadratic equations. One approach is based on the fact that if the product of two numbers is zero, then at least one of the numbers must be zero. In other words, $wz = 0$ implies $w = 0$ or $z = 0$ (or both $w = 0$ and $z = 0$). To use this fact on a quadratic equation we first make sure that one side of the equation is zero. Once this is done, we factor the nonzero side. We then set each factor equal to zero and solve one or two linear equations.

 EXAMPLES

Solve the equation by factoring.

$$x^2 + 2x - 3 = 0$$

One side of the equation is already 0, so we can begin by factoring the nonzero side. $x^2 + 2x - 3$ factors as $(x + 3)(x - 1)$, so $x^2 + 2x - 3 = 0$ becomes $(x + 3)(x - 1) = 0$.
We now set each factor equal to zero and solve for x.

$$x + 3 = 0 \qquad x - 1 = 0$$
$$-3 \quad -3 \qquad +1 \quad +1$$
$$x = -3 \qquad x = 1$$

We can check our solutions by substituting them into the original equation.

$$x^2 + 2x - 3 = 0$$
$$x = -3: \ (-3)^2 + 2(-3) - 3 = 9 - 6 - 3 = 0\checkmark$$
$$x = 1: \ 1^2 + 2(1) - 3 = 1 + 2 - 3 = 0\checkmark$$

Solve the equation by factoring.

$x^2 + 5x + 6 = 0$

$x^2 + 5x + 6 = 0$ becomes $(x + 2)(x + 3) = 0$

$x + 2 = 0 \qquad x + 3 = 0$

$-2 \ -2 \qquad -3 \ -3$

$x = -2 \qquad x = -3$

Solve the equation by factoring.

$$x^2 + 7x = 8$$

Before the factoring method can work, we must write the equation so that 0 is on one side, so we begin by subtracting 8 from each side.

$$x^2 + 7x = 8$$

$$-8 \ -8$$

$$x^2 + 7x - 8 = 0$$

$x^2 + 7x - 8 = 0$ becomes $(x + 8)(x - 1) = 0$

$x + 8 = 0 \qquad x - 1 = 0$

$-8 \ -8 \qquad +1 \ +1$

$x = -8 \qquad x = 1$

Solve the equation by factoring.

$$x^2 - 16 = 0$$

You might notice that the nonzero side of this equation is the difference of two squares, so $x^2 - 16$ becomes $(x - 4)(x + 4)$. The equation $x^2 - 16 = 0$ becomes $(x - 4)(x + 4) = 0$. Setting each of these factors equal to 0 gives us:

$x - 4 = 0 \qquad x + 4 = 0$

$+4 \ +4 \qquad -4 \ -4$

$x = 4 \qquad x = -4$

Solve the equation by factoring.

$3x^2 - 9x - 30 = 0$

$3x^2 - 9x - 30 = 0$ becomes $3(x^2 - 3x - 10) = 0$ which becomes $3(x - 5)(x + 2) = 0$.

$$
\begin{array}{ll}
x - 5 = 0 & x + 2 = 0 \\
+5 \ +5 & -2 \ -2 \\
x = 5 & x = -2
\end{array}
$$

We do not set the factor 3 equal to zero because "3 = 0" does not lead to a solution.

PRACTICE

Solve the equation by factoring.

1. $x^2 - x - 12 = 0$

2. $x^2 + 7x + 12 = 0$

3. $x^2 + 8x = -15$

4. $x^2 - 10x = -21$

5. $3x^2 - x - 2 = 0$

6. $4x^2 - 8x = 5$

7. $x^2 - 25 = 0$

8. $9x^2 - 16 = 0$

9. $x^2 = 100$

10. $x^2 + 6x + 9 = 0$

11. $x^2 = 0$

12. $5x^2 = 0$

13. $x^2 - \dfrac{1}{9} = 0$

 SOLUTIONS

1. $x^2 - x - 12 = 0$

    $(x - 4)(x + 3) = 0$

$$
\begin{array}{ll}
x - 4 = 0 & x + 3 = 0 \\
+4 \ +4 & -3 \ -3 \\
x = 4 & x = -3
\end{array}
$$

2. $x^2 + 7x + 12 = 0$

$(x+3)(x+4) = 0$

$x+3 = 0 \qquad x+4 = 0$

$-3 \; -3 \qquad -4 \; -4$

$x = -3 \qquad x = -4$

3. $x^2 + 8x = -15$

$+15 \quad +15$

$x^2 + 8x + 15 = 0$

$(x+3)(x+5) = 0$

$x+3 = 0 \qquad x+5 = 0$

$-3 \; -3 \qquad -5 \; -5$

$x = -3 \qquad x = -5$

4. $x^2 - 10x = -21$

$+21 \quad +21$

$x^2 - 10x + 21 = 0$

$(x-3)(x-7) = 0$

$x-3 = 0 \qquad x-7 = 0$

$+3 \; +3 \qquad +7 \; +7$

$x = 3 \qquad x = 7$

5. $3x^2 - x - 2 = 0$

$(3x+2)(x-1) = 0$

$3x+2 = 0 \qquad x-1 = 0$

$-2 \; -2 \qquad +1 \; +1$

$3x = -2 \qquad x = 1$

$x = \dfrac{-2}{3}$

6. $4x^2 - 8x = 5$

$-5 \quad -5$

$4x^2 - 8x - 5 = 0$

$(2x+1)(2x-5) = 0$

$2x+1 = 0 \qquad 2x-5 = 0$

$-1 \; -1 \qquad +5 \; +5$

$2x = -1 \qquad 2x = 5$

$x = \dfrac{-1}{2} \qquad x = \dfrac{5}{2}$

7. $$x^2 - 25 = 0$$
    $$(x - 5)(x + 5) = 0$$

$x - 5 = 0$	$x + 5 = 0$
+5 +5	−5 −5
$x = 5$	$x = -5$

8. $$9x^2 - 16 = 0$$
    $$(3x - 4)(3x + 4) = 0$$

$3x - 4 = 0$	$3x + 4 = 0$
+4 +4	−4 −4
$3x = 4$	$3x = -4$
$x = \dfrac{4}{3}$	$x = \dfrac{-4}{3}$

9. $$x^2 = 100$$
    $$-100 \quad -100$$
    $$x^2 - 100 = 0$$
    $$(x - 10)(x + 10) = 0$$

$x - 10 = 0$	$x + 10 = 0$
+10 +10	−10 −10
$x = 10$	$x = -10$

10. $$x^2 + 6x + 9 = 0$$
    $$(x + 3)(x + 3) = 0$$
    $$x + 3 = 0$$
    $$-3 \quad -3$$
    $$x = -3$$

11. $$x^2 = 0$$
    $$(x)(x) = 0$$
    $$x = 0$$

12. $$5x^2 = 0$$
    $$5(x)(x) = 0$$
    $$x = 0$$

13.
$$x^2 - \frac{1}{9} = 0$$

$$\left(x - \frac{1}{3}\right)\left(x + \frac{1}{3}\right) = 0$$

$$x - \frac{1}{3} = 0 \qquad x + \frac{1}{3} = 0$$

$$+\frac{1}{3} \quad +\frac{1}{3} \qquad -\frac{1}{3} \quad -\frac{1}{3}$$

$$x = \frac{1}{3} \qquad x = -\frac{1}{3}$$

Not all quadratic expressions are as easy to factor as the previous examples and problems. Sometimes we have to multiply or divide both sides of the equation by some number in order to factor the nonzero side of the equation. Because zero multiplied or divided by any nonzero number is still zero, only one side of the equation changes. Keep in mind that not all quadratic expressions can be factored using rational numbers (fractions) or even real numbers. Later, we will learn another method of solving every quadratic equation which bypasses the factoring method.

 EXAMPLES

Solve the equation by factoring.

$$-x^2 + 4x - 3 = 0$$

The equation $-x^2 + 4x - 3 = 0$ is awkward to factor because of the negative sign in front of x^2. Multiplying both sides of the equation by -1 makes the factoring a little easier.

$$-1(-x^2 + 4x - 3) = -1(0)$$

$$x^2 - 4x + 3 = 0$$

$$(x - 3)(x - 1) = 0$$

$$x - 3 = 0 \qquad x - 1 = 0$$

$$+3 \ +3 \qquad +1 \ +1$$

$$x = 3 \qquad x = 1$$

Decimals and fractions in a quadratic equation can be eliminated in the same way; multiplying both sides of the equation either by a power of

10 (for decimals) or the LCD (fractions) can make the nonzero side of the equation easier to factor.

$$0.1x^2 - 1.5x + 5.6 = 0$$

Multiplying both sides of the equation by 10 clears the decimal.

$$10(0.1x^2 - 1.5x + 5.6) = 10(0)$$

$$10(0.1x^2) - 10(1.5x) + 10(5.6) = 0$$

$$x^2 - 15x + 56 = 0$$

$$(x - 8)(x - 7) = 0$$

$$x - 8 = 0 \qquad x - 7 = 0$$

$$+8 \ +8 \qquad\quad +7 + 7$$

$$x = 8 \qquad\quad x = 7$$

Solve the equation by factoring.

$$\frac{3}{4}x^2 + \frac{1}{2}x - \frac{1}{4} = 0$$

We clear the fraction by multiplying both sides of the equation by 4 (the LCD).

$$4\left(\frac{3}{4}x^2 + \frac{1}{2}x - \frac{1}{4}\right) = 4(0)$$

$$4\left(\frac{3}{4}x^2\right) + 4\left(\frac{1}{2}x\right) - 4\left(\frac{1}{4}\right) = 0$$

$$3x^2 + 2x - 1 = 0$$

$$(3x - 1)(x + 1) = 0$$

$$3x - 1 = 0 \qquad x + 1 = 0$$

$$+1 \ +1 \qquad\quad -1 \ -1$$

$$3x = 1 \qquad\quad x = -1$$

$$x = \frac{1}{3}$$

Solve the equation by factoring.

$$\frac{1}{2}x^2 - 3x + 4 = 0$$

$$\frac{1}{2}x^2 - 3x + 4 = 0$$

$$2\left(\frac{1}{2}x^2 - 3x + 4\right) = 2(0)$$

$$x^2 - 6x + 8 = 0$$

$$(x - 4)(x - 2) = 0$$

$$\begin{array}{cc} x - 4 = 0 & x - 2 = 0 \\ +4 \ +4 & +2 \ +2 \\ x = 4 & x = 2 \end{array}$$

Solve the equation by factoring.

$$-2x^2 - 18x - 28 = 0$$

$$-2x^2 - 18x - 28 = 0$$

$$-(-2x^2 - 18x - 28) = -0$$

$$2x^2 + 18x + 28 = 0$$

$$\frac{1}{2}(2x^2 + 18x + 28) = \frac{1}{2}(0)$$

$$x^2 + 9x + 14 = 0$$

$$(x + 7)(x + 2) = 0$$

$$\begin{array}{cc} x + 7 = 0 & x + 2 = 0 \\ -7 \ -7 & -2 \ -2 \\ x = -7 & x = -2 \end{array}$$

If we had multiplied both sides of the original equation by $-\frac{1}{2}$, we would have eliminated one of the steps.

PRACTICE _____

Solve the equation by factoring.

1. $-x^2 - x + 30 = 0$

2. $-9x^2 + 25 = 0$

3. $0.01x^2 + 0.14x + 0.13 = 0$

4. $-0.1x^2 + 1.1x - 2.8 = 0$

5. $\dfrac{1}{5}x^2 + \dfrac{1}{5}x - 6 = 0$

6. $\dfrac{1}{6}x^2 - \dfrac{2}{3}x - \dfrac{16}{3} = 0$

7. $x^2 - \dfrac{1}{2}x - 3 = 0$

8. $-\dfrac{3}{2}x^2 + \dfrac{1}{2}x + 1 = 0$

9. $6x^2 + 18x - 24 = 0$

10. $-10x^2 - 34x - 12 = 0$

 SOLUTIONS

1. $-x^2 - x + 30 = 0$

   $-(-x^2 - x + 30) = -0$

   $x^2 + x - 30 = 0$

   $(x + 6)(x - 5) = 0$

$$x + 6 = 0 \qquad x - 5 = 0$$
$$\underline{-6 \ -6} \qquad \underline{+5 \ +5}$$
$$x = -6 \qquad x = 5$$

2. $-9x^2 + 25 = 0$

   $-(-9x^2 + 25) = -0$

   $9x^2 - 25 = 0$

   $(3x - 5)(3x + 5) = 0$

$$3x - 5 = 0 \qquad 3x + 5 = 0$$
$$\underline{+5 \ +5} \qquad \underline{-5 \ -5}$$
$$3x = 5 \qquad 3x = -5$$
$$x = \dfrac{5}{3} \qquad x = \dfrac{-5}{3}$$

3.

$$0.01x^2 + 0.14x + 0.13 = 0$$

$$100(0.01x^2 + 0.14x + 0.13) = 100(0)$$

$$x^2 + 14x + 13 = 0$$

$$(x + 13)(x + 1) = 0$$

$$x + 13 = 0 \qquad x + 1 = 0$$

$$\underline{-13 \; -13} \qquad \underline{-1 \; -1}$$

$$x = -13 \qquad x = -1$$

4.

$$-0.1x^2 + 1.1x - 2.8 = 0$$

$$-10(-0.1x^2 + 1.1x - 2.8) = -10(0)$$

$$x^2 - 11x + 28 = 0$$

$$(x - 7)(x - 4) = 0$$

$$x - 7 = 0 \qquad x - 4 = 0$$

$$\underline{+7 \; +7} \qquad \underline{+4 \; +4}$$

$$x = 7 \qquad x = 4$$

5.

$$\frac{1}{5}x^2 + \frac{1}{5}x - 6 = 0$$

$$5\left(\frac{1}{5}x^2 + \frac{1}{5}x - 6\right) = 5(0)$$

$$x^2 + x - 30 = 0$$

$$(x + 6)(x - 5) = 0$$

$$x + 6 = 0 \qquad x - 5 = 0$$

$$\underline{-6 \; -6} \qquad \underline{+5 \; +5}$$

$$x = -6 \qquad x = 5$$

6.

$$\frac{1}{6}x^2 - \frac{2}{3}x - \frac{16}{3} = 0$$

$$6\left(\frac{1}{6}x^2 - \frac{2}{3}x - \frac{16}{3}\right) = 6(0)$$

$$x^2 - 4x - 32 = 0$$

$$(x - 8)(x + 4) = 0$$

$$x - 8 = 0 \qquad x + 4 = 0$$

$$\underline{+8 \; +8} \qquad \underline{-4 \; -4}$$

$$x = 8 \qquad x = -4$$

7. $x^2 - \dfrac{1}{2}x - 3 = 0$

$2\left(x^2 - \dfrac{1}{2}x - 3\right) = 2(0)$

$2x^2 - x - 6 = 0$

$(2x + 3)(x - 2) = 0$

$\begin{array}{ll} 2x + 3 = 0 & x - 2 = 0 \\ -3\ -3 & +2\ +2 \\ 2x = -3 & x = 2 \end{array}$

$x = \dfrac{-3}{2}$

8. $-\dfrac{3}{2}x^2 + \dfrac{1}{2}x + 1 = 0$

$-2\left(-\dfrac{3}{2}x^2 + \dfrac{1}{2}x + 1\right) = -2(0)$

$3x^2 - x - 2 = 0$

$(3x + 2)(x - 1) = 0$

$\begin{array}{ll} 3x + 2 = 0 & x - 1 = 0 \\ -2\ -2 & +1\ +1 \\ 3x = -2 & x = 1 \end{array}$

$\dfrac{3}{3}x = \dfrac{-2}{3}$

$x = \dfrac{-2}{3}$

9. $6x^2 + 18x - 24 = 0$

$\dfrac{1}{6}(6x^2 + 18x - 24) = \dfrac{1}{6}(0)$

$x^2 + 3x - 4 = 0$

$(x + 4)(x - 1) = 0$

$\begin{array}{ll} x + 4 = 0 & x - 1 = 0 \\ -4\ -4 & +1\ +1 \\ x = -4 & x = 1 \end{array}$

10.
$$-10x^2 - 34x - 12 = 0$$

$$\frac{-1}{2}(-10x^2 - 34x - 12) = \frac{-1}{2}(0)$$

$$5x^2 + 17x + 6 = 0$$

$$(5x + 2)(x + 3) = 0$$

$$5x + 2 = 0 \qquad x + 3 = 0$$

$$-2\ -2 \qquad -3\ -3$$

$$5x = -2 \qquad x = -3$$

$$x = \frac{-2}{5}$$

Extracting Roots

The second main approach to solve quadratic equations comes from the fact that $x^2 = k$ implies $x = \pm\sqrt{k}$. For instance, if $x^2 = 9$, then $x = 3$ or -3 because $3^2 = 9$ and $(-3)^2 = 9$. This method works best if the equation can be put in the form $ax^2 - c = 0$, where c is not negative. We begin by solving for x^2 and then taking the square root of each side of the equation. This method is sometimes called *extracting roots*, or *the square root method*.

 EXAMPLES

Solve the quadratic equation by extracting roots.

$$x^2 = 16$$

The equation is already written with x^2 isolated on one side, so we can begin by taking the square root of each side.

$$x = \pm\sqrt{16}$$

$$x = \pm 4$$

Solve the quadratic equation by extracting roots.

$$3x^2 = 27$$

Before we can take the square root of each side, we isolate x^2, so we begin by dividing each side of the equation by 3.

$$3x^2 = 27$$

$$\frac{3x^2}{3} = \frac{27}{3}$$

$$x^2 = 9$$

$$x = \pm 3$$

Solve the quadratic equation by extracting roots.

$$25 - x^2 = 0$$
$$25 = x^2$$
$$\pm 5 = x$$

Solve the quadratic equation by extracting roots.

$$4x^2 = 49$$
$$x^2 = \frac{49}{4}$$
$$x = \pm\sqrt{\frac{49}{4}}$$
$$x = \pm\frac{7}{2}$$

In the next example, we need to simplify the solution. Recall that $\sqrt{a}$ is not simplified if the number a has a perfect square as a factor.

$$3x^2 = 36$$
$$x^2 = 12$$
$$x = \pm\sqrt{12}$$
$$x = \pm\sqrt{4 \cdot 3} = \pm(\sqrt{4})(\sqrt{3}) = \pm 2\sqrt{3}$$

 PRACTICE

Solve the quadratic equation by extracting roots.

1. $x^2 - 81 = 0$
2. $64 - x^2 = 0$
3. $4x^2 = 100$
4. $2x^2 = 3$
5. $-6x^2 = -80$

 SOLUTIONS

1. $x^2 - 81 = 0$
$$x^2 = 81$$
$$x = \pm 9$$

2. $64 - x^2 = 0$

$$64 = x^2$$

$$\pm 8 = x$$

3. $4x^2 = 100$

$$x^2 = \frac{100}{4}$$

$$x^2 = 25$$

$$x = \pm 5$$

4. $2x^2 = 3$

$$x^2 = \frac{3}{2}$$

$$x = \pm \sqrt{\frac{3}{2}}$$

$$x = \pm \frac{\sqrt{3}}{\sqrt{2}} \cdot \frac{\sqrt{2}}{\sqrt{2}}$$

$$x = \pm \frac{\sqrt{6}}{2}$$

5. $-6x^2 = -80$

$$x^2 = \frac{-80}{-6}$$

$$x^2 = \frac{40}{3}$$

$$x = \pm \sqrt{\frac{40}{3}}$$

$$x = \pm \frac{\sqrt{40}}{\sqrt{3}} \cdot \frac{\sqrt{3}}{\sqrt{3}} = \pm \frac{\sqrt{120}}{3}$$

$$x = \pm \frac{\sqrt{4 \cdot 30}}{3} = \pm \frac{2\sqrt{30}}{3}$$

Solving Quadratic Equations with the Quadratic Formula

The first two methods we used for solving quadratic equations only work for equations that are factorable or where one side is written as a perfect square. The last method that we will learn works to solve *any* quadratic equation. This method for solving quadratic equations comes from the fact that $x^2 = k$ implies $x = \sqrt{k}, -\sqrt{k}$. Using this fact and a technique called *completing the square*, the equation $ax^2 + bx + c = 0$ can be written in the form $(x \pm \text{number})^2 = $ another number. Taking the square root of each side of this equation and then solving for x gives us the solutions

$$x = \frac{-b + \sqrt{b^2 - 4ac}}{2a}, \frac{-b - \sqrt{b^2 - 4ac}}{2a}$$

These solutions are abbreviated as

$$x = \frac{-b \pm \sqrt{b^2 - 4ac}}{2a}$$

This formula is called the *quadratic formula*. It is very important in algebra and is worth memorizing. You might wonder why we bother factoring quadratic expressions to solve quadratic equations when the quadratic formula solves any quadratic equation. There are two reasons. One, factoring is an important skill in algebra and calculus, used to solve a variety of problems. Two, the factoring method is often easier and faster to use than computing the quadratic formula. We normally use the quadratic formula to solve quadratic equations where the factoring is difficult. For now, we will learn how to evaluate the quadratic formula and simplify the numbers that we get from it. We then use it to solve quadratic equations.

Before the formula can be used, the quadratic formula must be in the form $ax^2 + bx + c = 0$. Once a, b, and c are identified, applying the quadratic formula is simply a matter of performing arithmetic and simplifying the solutions. We begin by identifying a, b, and c.

$2x^2 - x - 7 = 0$	$x = 2, b = -1, c = -7$
$10x^2 - 4 = 0$ is equivalent to $10x^2 + 0x - 4 = 0$	$a = 10, b = 0, c = -4$
$3x^2 + x = 0$ is equivalent to $3x^2 + x + 0 = 0$	$a = 3, b = 1, c = 0$
$4x^2 = 0$ is equivalent to $4x^2 + 0x + 0 = 0$	$a = 4, b = 0, c = 0$
$x^2 + 3x = 4$ is equivalent to $x^2 + 3x - 4 = 0$	$a = 1, b = 3, c = -4$
$-8x^2 = -64$ is equivalent to $8x^2 + 0x - 64 = 0$	$a = 8, b = 0, c = -64$

PRACTICE

Identify a, b, and c for $ax^2 + bx + c = 0$.

1. $2x^2 + 9x + 3 = 0$
2. $-3x^2 + x + 5 = 0$
3. $x^2 - x - 6 = 0$
4. $x^2 - 9 = 0$
5. $2x^2 = 32$
6. $x^2 + x = 0$
7. $x^2 - x = 0$
8. $9x^2 = 10x$
9. $8x^2 + 20x = 9$
10. $4x - 5 - 3x^2 = 0$

SOLUTIONS

1. $2x^2 + 9x + 3 = 0$ $a = 2$ $b = 9$ $c = 3$
2. $-3x^2 + x + 5 = 0$ $a = -3$ $b = 1$ $c = 5$
3. $x^2 - x - 6 = 0$ $a = 1$ $b = -1$ $c = -6$
4. $x^2 - 9 = 0$ $a = 1$ $b = 0$ $c = -9$
5. $2x^2 = 32$ Rewritten: $2x^2 - 32 = 0$
    $a = 2$ $b = 0$ $c = -32$
6. $x^2 + x = 0$ $a = 1$ $b = 1$ $c = 0$
7. $x^2 - x = 0$ $a = 1$ $b = -1$ $c = 0$
8. $9x^2 = 10x$ Rewritten: $9x^2 - 10x = 0$
    $a = 9$ $b = -10$ $c = 0$
9. $8x^2 + 20x = 9$ Rewritten: $8x^2 + 20x - 9 = 0$
    $a = 8$ $b = 20$ $c = -9$
10. $4x - 5 - 3x^2 = 0$ Rewritten: $-3x^2 + 4x - 5 = 0$
    $a = -3$ $b = 4$ $c = -5$

Before we use the quadratic formula to solve quadratic equations, we will work with the arithmetic necessary to simplify the equation before using the formula and with simplifying the expression after we have plugged numbers from the equation into the formula.

The quadratic formula can be messy to compute when any of a, b, or c is a fraction or decimal. We can get around this difficulty by multiplying both sides of the equation by the LCD or some power of 10. This would leave us with a, b, and c as integers (whole numbers or their negatives). To see how this step can save time, we will plug values for a, b, and c without clearing the fractions and then we will clear the fractions, allowing us to plug in simpler values.

 EXAMPLE

Evaluate the quadratic formula.

$$\frac{1}{2}x^2 - \frac{1}{2}x - 1 = 0$$

We use $a = \frac{1}{2}$, $b = -\frac{1}{2}$, $c = -1$ in the quadratic formula.

$$x = \frac{-b \pm \sqrt{b^2 - 4ac}}{2a}$$

$$x = \frac{-\left(-\frac{1}{2}\right) \pm \sqrt{\left(-\frac{1}{2}\right)^2 - 4\left(\frac{1}{2}\right)(-1)}}{2\left(\frac{1}{2}\right)}$$

Simplifying this expression would require several steps. Let us see how simple the formula looks after we have cleared the fractions in the original equation. We now eliminate the fractions by multiplying both sides of the equation by 2.

$$2\left(\frac{1}{2}x^2 - \frac{1}{2}x - 1\right) = 2(0)$$

$$x^2 - x - 2 = 0$$

With the fractions cleared, we can use the quadratic formula with more convenient values: $a = 1$, $b = -1$ and $c = -2$

$$x = \frac{-(-1) \pm \sqrt{(-1)^2 - 4(1)(-2)}}{2(1)}$$

Simplifying this expression is much easier than the expression containing all of the fractions.

Sometimes we must simplify the solutions we find with the quadratic formula. We begin by simplifying the square root if the number under the radical has a perfect square as a factor. For example, $\sqrt{12}$ is not simplified because 12 has 4, a perfect square, as a factor. We simplify $\sqrt{12}$ with the radical properties $\sqrt{ab} = \sqrt{a} \cdot \sqrt{b}$ and $\sqrt{a^2} = a$ (provided a and b are not negative). Thus we obtain $\sqrt{12} = \sqrt{4 \cdot 3} = \sqrt{4} \cdot \sqrt{3} = 2\sqrt{3}$.

Once we simplify the square root, we then see if the numerator and denominator have any common factors. If so, we factor the numerator and divide out the common factor.

**EXAMPLES**

Simplify the fraction.

$$\frac{8 \pm \sqrt{24}}{2}$$

We begin by simplifying the square root: $\sqrt{24} = \sqrt{4 \cdot 6} = 2\sqrt{6}$

$$\frac{8 \pm \sqrt{24}}{2} = \frac{8 \pm 2\sqrt{6}}{2}$$

The denominator is divisible by 2 and each term in the numerator is divisible by 2, so we factor a 2 from each term in the numerator. We then divide 2 from the numerator and denominator.

$$\frac{8 \pm 2\sqrt{6}}{2} = \frac{2(4 \pm \sqrt{6})}{2} = 4 \pm \sqrt{6}$$

Simplify the fractions.

$$\frac{-3 \pm \sqrt{18}}{6} = \frac{-3 \pm \sqrt{9 \cdot 2}}{6} = \frac{-3 \pm 3\sqrt{2}}{6} = \frac{3(-1 \pm \sqrt{2})}{6} = \frac{-1 \pm \sqrt{2}}{2}$$

$$\frac{15 \pm \sqrt{50}}{10} = \frac{15 \pm \sqrt{25 \cdot 2}}{10} = \frac{15 \pm 5\sqrt{2}}{10} = \frac{5(3 \pm \sqrt{2})}{10} = \frac{3 \pm \sqrt{2}}{2}$$

PRACTICE

Simplify the fraction.

1. $\dfrac{6 \pm \sqrt{12}}{2} =$

2. $\dfrac{12 \pm \sqrt{27}}{6} =$

3. $\dfrac{2 \pm \sqrt{48}}{4} =$

4. $\dfrac{20 \pm \sqrt{300}}{10} =$

5. $\dfrac{-6 \pm \sqrt{20}}{-2} =$

SOLUTIONS

1. $\dfrac{6 \pm \sqrt{12}}{2} = \dfrac{6 \pm \sqrt{4 \cdot 3}}{2} = \dfrac{6 \pm 2\sqrt{3}}{2} = \dfrac{2(3 \pm \sqrt{3})}{2} = 3 \pm \sqrt{3}$

2. $\dfrac{12 \pm \sqrt{27}}{6} = \dfrac{12 \pm \sqrt{9 \cdot 3}}{6} = \dfrac{12 \pm 3\sqrt{3}}{6} = \dfrac{3(4 \pm \sqrt{3})}{6} = \dfrac{4 \pm \sqrt{3}}{2}$

3. $\dfrac{2 \pm \sqrt{48}}{4} = \dfrac{2 \pm \sqrt{16 \cdot 3}}{4} = \dfrac{2 \pm 4\sqrt{3}}{4} = \dfrac{2(1 \pm 2\sqrt{3})}{4} = \dfrac{1 \pm 2\sqrt{3}}{2}$

4. $\dfrac{20 \pm \sqrt{300}}{10} = \dfrac{20 \pm \sqrt{100 \cdot 3}}{10} = \dfrac{20 \pm 10\sqrt{3}}{10} = \dfrac{10(2 \pm \sqrt{3})}{10} = 2 \pm \sqrt{3}$

5. $\dfrac{-6 \pm \sqrt{20}}{-2} = \dfrac{-6 \pm \sqrt{4 \cdot 5}}{-2} = \dfrac{-6 \pm 2\sqrt{5}}{-2} = \dfrac{-2(3 \pm \sqrt{5})}{-2} = 3 \pm \sqrt{5}$

Note that the negative of $\pm\sqrt{5}$ is still $\pm\sqrt{5}$.

Now that we can identify a, b, and c in the quadratic formula and can simplify the solutions, we are ready to solve quadratic equations using the formula.

 # Still Struggling

When using the quadratic formula, be careful to put *all* of $-b \pm \sqrt{b^2 - 4ac}$ over $2a$.

EXAMPLES

Solve the quadratic equation with the quadratic formula.

$$2x^2 + 3x + 1 = 0$$

We begin by observing a, b, and c: $a = 2$ $b = 3$ $c = 1$.

$$x = \frac{-3 \pm \sqrt{(3)^2 - 4(2)(1)}}{2(2)} = \frac{-3 \pm \sqrt{9-8}}{4}$$

$$= \frac{-3 \pm \sqrt{1}}{4} = \frac{-3+1}{4}, \frac{-3-1}{4} = \frac{-2}{4}, \frac{-4}{4} = -\frac{1}{2}, -1$$

Solve the quadratic equation with the quadratic formula.

$$x^2 - x - 1 = 0 \quad a = 1 \quad b = -1 \quad c = -1$$

$$x = \frac{-(-1) \pm \sqrt{(-1)^2 - 4(1)(-1)}}{2(1)}$$

$$= \frac{1 \pm \sqrt{1-(-4)}}{2} = \frac{1 \pm \sqrt{1+4}}{2} = \frac{1 \pm \sqrt{5}}{2} = \frac{1+\sqrt{5}}{2}, \frac{1-\sqrt{5}}{2}$$

Solve the quadratic equation with the quadratic formula.

$$x^2 - 18 = 0 \quad a = 1 \quad b = 0 \quad c = -18$$

$$x = \frac{-0 \pm \sqrt{0^2 - 4(1)(-18)}}{2(1)} = \frac{\pm\sqrt{0-(-72)}}{2}$$

$$= \frac{\pm\sqrt{72}}{2} = \frac{\pm\sqrt{36 \cdot 2}}{2} = \frac{\pm 6\sqrt{2}}{2} = \pm 3\sqrt{2} = 3\sqrt{2}, -3\sqrt{2}$$

Solve the quadratic equation with the quadratic formula.

$$2x^2 + 6x = 5$$

Rewrite as $2x^2 + 6x - 5 = 0$ $a = 2$ $b = 6$ $c = -5$

$$x = \frac{-6 \pm \sqrt{6^2 - 4(2)(-5)}}{2(2)} = \frac{-6 \pm \sqrt{36 - (-40)}}{4} = \frac{-6 \pm \sqrt{76}}{4} = \frac{-6 \pm \sqrt{4 \cdot 19}}{4}$$

$$= \frac{-6 \pm 2\sqrt{19}}{4} = \frac{2(-3 \pm \sqrt{19})}{4} = \frac{-3 \pm \sqrt{19}}{2} = \frac{-3+\sqrt{19}}{2}, \frac{-3-\sqrt{19}}{2}$$

Solve the quadratic equation with the quadratic formula.

$$\frac{1}{3}x^2 + x - 2 = 0$$

We begin by multiplying both sides of the equation by 3 to eliminate the fraction.

$$3\left(\frac{1}{3}x^2 + x - 2\right) = 3(0)$$

$x^2 + 3x - 6 = 0$ $a = 1$ $b = 3$ $c = -6$

$$x = \frac{-3 \pm \sqrt{3^2 - 4(1)(-6)}}{2(1)} = \frac{-3 \pm \sqrt{9 - (-24)}}{2} = \frac{-3 \pm \sqrt{33}}{2} = \frac{-3 + \sqrt{33}}{2}, \frac{-3 - \sqrt{33}}{2}$$

Solve the quadratic equation with the quadratic formula.

$$0.1x^2 - 0.8x + 0.21 = 0$$

We can eliminate the decimal numbers by multiplying both sides of the equation by 100.

$100(0.1x^2 - 0.8x + 0.21) = 100(0)$

$10x^2 - 80x + 21 = 0$ $a = 10$ $b = -80$ $c = 21$

$$x = \frac{-(-80) \pm \sqrt{(-80)^2 - 4(10)(21)}}{2(10)} = \frac{80 \pm \sqrt{6400 - 840}}{20} = \frac{80 \pm \sqrt{5560}}{20}$$

$$= \frac{80 \pm \sqrt{4 \cdot 1390}}{20} = \frac{80 \pm 2\sqrt{1390}}{20} = \frac{2(40 \pm \sqrt{1390})}{20} = \frac{40 \pm \sqrt{1390}}{10}$$

$$= \frac{40 + \sqrt{1390}}{10}, \frac{40 - \sqrt{1390}}{10}$$

PRACTICE

Solve the quadratic equation with the quadratic formula.

1. $x^2 - 5x + 3 = 0$

2. $4x^2 + x - 6 = 0$

3. $7x^2 + 3x = 2$

4. $2x^2 = 9$

5. $-3x^2 + 4x + 1 = 0$

6. $9x^2 - x = 10$

7. $10x^2 - 5x = 0$

8. $0.1x^2 - 0.11x - 1 = 0$

9. $\dfrac{1}{3}x^2 + \dfrac{1}{6}x - \dfrac{1}{8} = 0$

10. $80x^2 - 16x - 32 = 0$

11. $18x^2 + 39x + 20 = 0$

12. $x^2 + 10x + 25 = 0$

 SOLUTIONS

1. $x^2 - 5x + 3 = 0$

$$x = \frac{-(-5) \pm \sqrt{(-5)^2 - 4(1)(3)}}{2(1)} = \frac{5 \pm \sqrt{25 - 12}}{2} = \frac{5 \pm \sqrt{13}}{2}$$

2. $4x^2 + x - 6 = 0$

$$x = \frac{-1 \pm \sqrt{1^2 - 4(4)(-6)}}{2(4)} = \frac{-1 \pm \sqrt{1 - (-96)}}{8} = \frac{-1 \pm \sqrt{97}}{8}$$

3. $7x^2 + 3x = 2$ **(Equivalent to $7x^2 + 3x - 2 = 0$)**

$$x = \frac{-3 \pm \sqrt{3^2 - 4(7)(-2)}}{2(7)} = \frac{-3 \pm \sqrt{9 - (-56)}}{14} = \frac{-3 \pm \sqrt{65}}{14}$$

4. $2x^2 = 9$ **(Equivalent to $2x^2 - 9 = 0$)**

$$x = \frac{-0 \pm \sqrt{0^2 - 4(2)(-9)}}{2(2)} = \frac{0 \pm \sqrt{-(-72)}}{4} = \frac{\pm\sqrt{72}}{4} = \frac{\pm\sqrt{36 \cdot 2}}{4}$$

$$= \frac{\pm 6\sqrt{2}}{4} = \frac{\pm 3\sqrt{2}}{2}$$

5. $-3x^2 + 4x + 1 = 0$

$$x = \frac{-4 \pm \sqrt{4^2 - 4(-3)(1)}}{2(-3)} = \frac{-4 \pm \sqrt{16 - (-12)}}{-6} = \frac{4 \pm \sqrt{28}}{-6}$$

$$= \frac{-4 \pm \sqrt{4 \cdot 7}}{-6} = \frac{-4 \pm 2\sqrt{7}}{-6} = \frac{-2(2 \pm \sqrt{7})}{-6} = \frac{2 \pm \sqrt{7}}{3}$$

or $-1(-3x^2 + 4x + 1) = -1(0)$

$$3x^2 - 4x - 1 = 0$$

$$x = \frac{-(-4) \pm \sqrt{(-4)^2 - 4(3)(-1)}}{2(3)} = \frac{4 \pm \sqrt{16 - (-12)}}{6}$$

$$= \frac{4 \pm \sqrt{28}}{6} = \frac{4 \pm \sqrt{4 \cdot 7}}{6} = \frac{4 \pm 2\sqrt{7}}{6} = \frac{2(2 \pm \sqrt{7})}{6} = \frac{2 \pm \sqrt{7}}{3}$$

6. $9x^2 - x = 10$ (Equivalent to $9x^2 - x - 10 = 0$)

$$x = \frac{-(-1) \pm \sqrt{(-1)^2 - 4(9)(-10)}}{2(9)} = \frac{1 \pm \sqrt{1 - (-360)}}{18}$$

$$= \frac{1 \pm \sqrt{361}}{18} = \frac{1 \pm 19}{18} = \frac{1 + 19}{18}, \frac{1 - 19}{18} = \frac{20}{18}, \frac{-18}{18} = \frac{10}{9}, -1$$

7. $10x^2 - 5x = 0$

$$x = \frac{-(-5) \pm \sqrt{(-5)^2 - 4(10)(0)}}{2(10)} = \frac{5 \pm \sqrt{25 - 0}}{20} = \frac{5 \pm \sqrt{25}}{20}$$

$$= \frac{5 \pm 5}{20} = \frac{10}{20}, \frac{0}{20} = \frac{1}{2}, 0$$

8. $0.1x^2 - 0.11x - 1 = 0$

$100(0.1x^2 - 0.11x - 1) = 100(0)$

$10x^2 - 11x - 100 = 0$

$$x = \frac{-(-11) \pm \sqrt{(-11)^2 - 4(10)(-100)}}{2(10)} = \frac{11 \pm \sqrt{121 - (-4000)}}{20}$$

$$= \frac{11 \pm \sqrt{4121}}{20}$$

9. $\dfrac{1}{3}x^2 + \dfrac{1}{6}x - \dfrac{1}{8} = 0$

$$24\left(\frac{1}{3}x^2 + \frac{1}{6}x - \frac{1}{8}\right) = 24(0)$$

$8x^2 + 4x - 3 = 0$

$$x = \frac{-4 \pm \sqrt{4^2 - 4(8)(-3)}}{2(8)} = \frac{-4 \pm \sqrt{16 - (-96)}}{16} = \frac{-4 \pm \sqrt{112}}{16}$$

$$= \frac{-4 \pm \sqrt{16 \cdot 7}}{16} = \frac{-4 \pm 4\sqrt{7}}{16} = \frac{4(-1 \pm \sqrt{7})}{16} = \frac{-1 \pm \sqrt{7}}{4}$$

10. $80x^2 - 16x - 32 = 0$

$$\frac{1}{16}(80x^2 - 16x - 32) = \frac{1}{16}(0)$$

$$5x^2 - x - 2 = 0$$

$$x = \frac{-(-1) \pm \sqrt{(-1)^2 - 4(5)(-2)}}{2(5)} = \frac{1 \pm \sqrt{1 - (-40)}}{10} = \frac{1 \pm \sqrt{41}}{10}$$

11. $18x^2 + 39x + 20 = 0$

$$x = \frac{-39 \pm \sqrt{39^2 - 4(18)(20)}}{2(18)} = \frac{-39 \pm \sqrt{1521 - 1440}}{36}$$

$$= \frac{-39 \pm \sqrt{81}}{36} = \frac{-39 \pm 9}{36} = \frac{-39 + 9}{36}, \frac{-39 - 9}{36} = \frac{-30}{36}, \frac{-48}{36}$$

$$= \frac{-5}{6}, \frac{-4}{3}$$

12. $x^2 + 10x + 25 = 0$

$$x = \frac{-10 \pm \sqrt{10^2 - 4(1)(25)}}{2(1)} = \frac{-10 \pm \sqrt{100 - 100}}{2} = \frac{-10 \pm 0}{2}$$

$$= \frac{-10}{2} = -5$$

Rational Equations That Lead to Quadratic Equations

Some rational equations (an equation with one or more fractions as terms) become quadratic equations once we multiply each term by the least common denominator (LCD). Remember, we *must* be sure that any solutions do not lead to a zero in a denominator in the original equation.

Generally, we use one of two main approaches to clear the denominator(s) in a rational equation. If the equation is in the form of "fraction = fraction," cross-multiply. If the equation is not in this form, we multiply each side of the equation by LCD. Finding the LCD often means factoring each denominator completely. We learned in Chapter 7 to multiply both sides of an equation by the LCD and then to distribute the LCD. In this chapter, we will simply multiply each term by the LCD, eliminating one step. Next, we collect all of the terms on one side of the equation, leaving a 0 on the other side. In the examples and practice problems below, the solutions that lead to a zero in a denominator will be stated.

EXAMPLES

Solve the equation.

$$\frac{x-1}{x^2-2x-8}=-\frac{2}{5}$$

This is in the form "fraction = fraction" so we cross-multiply.

$$5(x-1)=-2(x^2-2x-8)$$
$$5x-5=-2x^2+4x+16$$
$$+2x^2-4x-16 \quad +2x^2-4x-16$$
$$2x^2+x-21=0$$
$$(2x+7)(x-3)=0$$

$$2x+7=0 \qquad\qquad\qquad x-3=0$$
$$-7\ -7 \qquad\qquad\qquad\quad +3\ +3$$
$$2x=-7 \qquad\qquad\qquad\quad x=3$$
$$x=\frac{-7}{2}$$

Solve the equation.

$$\frac{4}{x^2-2x}+\frac{x+1}{x}=\frac{-4}{x-2}$$

We begin by factoring each denominator so that we can find the LCD. After identifying the LCD, we will multiply each of the three terms by the LCD.

$$\frac{4}{x(x-2)}+\frac{x+1}{x}=\frac{-4}{x-2} \qquad LCD=x(x-2)$$

$$\frac{x(x-2)}{1}\cdot\frac{4}{x(x-2)}+\frac{x(x-2)}{1}\cdot\frac{x+1}{x}=\frac{x(x-2)}{1}\cdot\frac{-4}{x-2}$$

$$4+(x-2)(x+1)=-4x$$
$$4+x^2-x-2=-4x$$
$$x^2-x+2=-4x$$
$$+4x \qquad\qquad +4x$$
$$x^2+3x+2=0$$
$$(x+2)(x+1)=0$$

$$x+2=0 \qquad\qquad x+1=0$$
$$-2\ -2 \qquad\qquad\quad -1\ -1$$
$$x=-2 \qquad\qquad\quad x=-1$$

Solve the equation.

$$\frac{3}{x-4} = \frac{-6x+24}{x^2-2x-8}$$

We begin with cross-multiplication.

$3(x^2-2x-8) = (x-4)(-6x+24)$ This is the cross-multiplication step.

$3x^2-6x-24 = -6x^2+48x-96$ Use the FOIL method on the right.

$+6x^2-48x+96+6x^2-48x+96$ Rewrite with 0 on one side.

$9x^2-54x+72 = 0$

$$\frac{1}{9}(9x^2-54x+72) = \frac{1}{9}(0)$$ Each term is divisible by 9.

$$x^2-6x+8 = 0$$

$$(x-4)(x-2) = 0$$

$$x-4 = 0 \qquad x-2 = 0$$

$$x = 4 \qquad x = 2$$

We cannot let x be 4 because $x = 4$ leads to a zero in the denominator of $\frac{3}{x-4}$.

The only solution, then, is $x = 2$.

PRACTICE

Solve the equation. (Because all of these problems factor, factoring is used in the solutions. If factoring takes too long on some of these, you may use the quadratic formula.)

1. $\dfrac{3x}{x-4} = \dfrac{-3x}{2}$

2. $\dfrac{x-1}{2x+3} = \dfrac{6}{x-2}$

3. $\dfrac{x+2}{x-3} + \dfrac{2x+1}{x^2-9} = \dfrac{12x+3}{x+3}$

4. $\dfrac{2x-1}{x+1} - \dfrac{3x}{x-2} = \dfrac{-8x-7}{x^2-x-2}$

5. $\dfrac{4x+1}{x-1} + \dfrac{x-5}{1-x} = \dfrac{24}{x}$

6. $\dfrac{2x}{x+1} + \dfrac{3}{x} = \dfrac{2}{x^2+x}$

7. $\dfrac{2x}{2x+1} - \dfrac{3}{x} = \dfrac{-3x-4}{3x}$

8. $\dfrac{2}{x-5} + \dfrac{1}{3x} = \dfrac{-8}{3x+15}$

9. $\dfrac{1}{x-4} + \dfrac{2}{x+1} - \dfrac{3}{x+3} = \dfrac{x-3}{x^2-3x-4}$

10. $\dfrac{4}{x-1} + \dfrac{3}{x+1} = \dfrac{1}{2x} - \dfrac{6}{x^2-1}$

 **SOLUTIONS**

1.
$$\frac{3x}{x-4} = \frac{-3x}{2}$$
$$3x(2) = (x-4)(-3x)$$
$$6x = -3x^2 + 12x$$
$$+3x^2 - 12x \quad +3x^2 - 12x$$
$$3x^2 - 6x = 0$$
$$3x(x-2) = 0$$
$$3x = 0 \qquad x-2 = 0$$
$$x = \frac{0}{3} \qquad +2 \quad +2$$
$$x = 0 \qquad\quad x = 2$$

2.
$$\frac{x-1}{2x+3} = \frac{6}{x-2}$$

$$(x-1)(x-2) = 6(2x+3)$$

$$x^2 - 3x + 2 = 12x + 18$$

$$-12x - 18 \quad -12x - 18$$

$$x^2 - 15x - 16 = 0$$

$$(x-16)(x+1) = 0$$

$$x - 16 = 0 \qquad x + 1 = 0$$

$$+16 \ +16 \qquad -1 \ -1$$

$$x = 16 \qquad x = -1$$

3. $$\frac{x+2}{x-3} + \frac{2x+1}{x^2-9} = \frac{12x+3}{x+3}$$

Denominator factored: $\dfrac{x+2}{x-3} + \dfrac{2x+1}{(x-3)(x+3)} = \dfrac{12x+3}{x+3}$

$$LCD = (x-3)(x+3)$$

$$(x-3)(x+3) \cdot \frac{x+2}{x-3} + (x-3)(x+3) \cdot \frac{2x+1}{(x-3)(x+3)}$$

$$= (x-3)(x+3) \cdot \frac{12x+3}{x+3}$$

$$(x+3)(x+2) + 2x + 1 = (x-3)(12x+3)$$

$$x^2 + 5x + 6 + 2x + 1 = 12x^2 - 33x - 9$$

$$x^2 + 7x + 7 = 12x^2 - 33x - 9$$

$$-x^2 \ -7x \ -7 \qquad -x^2 \ -7x \ -7$$

$$0 = 11x^2 - 40x - 16$$

$$0 = (11x+4)(x-4)$$

$$11x + 4 = 0 \qquad x - 4 = 0$$

$$-4 \ -4 \qquad +4 \ +4$$

$$11x = -4 \qquad x = 4$$

$$x = \frac{-4}{11}$$

4. $\dfrac{2x-1}{x+1} - \dfrac{3x}{x-2} = \dfrac{-8x-7}{x^2-x-2}$

Denominator factored: $\dfrac{2x-1}{x+1} - \dfrac{3x}{x-2} = \dfrac{-8x-7}{(x-2)(x+1)}$

$\text{LCD} = (x+1)(x-2)$

$(x+1)(x-2) \cdot \dfrac{2x-1}{x+1} - (x+1)(x-2) \cdot \dfrac{3x}{x-2}$

$\qquad\qquad = (x+1)(x-2) \cdot \dfrac{-8x-7}{(x-2)(x+1)}$

$(x-2)(2x-1) - 3x(x+1) = -8x-7$

$2x^2 - 5x + 2 - 3x^2 - 3x = -8x - 7$

$-x^2 - 8x + 2 = -8x - 7$

$\underline{+x^2 + 8x - 2 \quad\ +x^2 + 8x - 2}$

$0 = x^2 - 9$

$0 = (x-3)(x+3)$

$x - 3 = 0 \qquad x + 3 = 0$

$\underline{+3 \ +3} \qquad \underline{-3 \ -3}$

$x = 3 \qquad\quad x = -3$

5. $\dfrac{4x+1}{x-1} + \dfrac{x-5}{1-x} = \dfrac{24}{x}$

Using the fact that $1 - x = -(1 - x)$ allows us to write the second denominator the same as the first.

$$\dfrac{4x+1}{x-1} + \dfrac{x-5}{-(x-1)} = \dfrac{24}{x}$$

$$\dfrac{4x+1}{x-1} + \dfrac{-(x-5)}{x-1} = \dfrac{24}{x} \quad \text{LCD} = x(x-1)$$

$$x(x-1) \cdot \dfrac{4x+1}{x-1} + x(x-1) \cdot \dfrac{-(x-5)}{x-1} = x(x-1) \cdot \dfrac{24}{x}$$

$$x(4x-1) + (-x)(x-5) = 24(x-1)$$

$$x(4x+1) - x(x-5) = 24(x-1)$$

$$4x^2 + x - x^2 + 5x = 24x - 24$$

$$3x^2 + 6x = 24x - 24$$

$$-24x + 24 \quad -24x + 24$$

$$3x^2 - 18x + 24 = 0$$

$$\frac{1}{3}(3x^2 - 18x + 24) = \frac{1}{3}(0)$$

$$x^2 - 6x + 8 = 0$$

$$(x - 4)(x - 2) = 0$$

$$x - 4 = 0 \quad x - 2 = 0$$

$$+4 \; +4 \quad +2 \; +2$$

$$x = 4 \quad x = 2$$

6. $\dfrac{2x}{x+1} + \dfrac{3}{x} = \dfrac{2}{x^2 + x}$ Denominator factored: $\dfrac{2x}{x+1} + \dfrac{3}{x} = \dfrac{2}{x(x+1)}$

LCD $= x(x+1)$

$$x(x+1) \cdot \frac{2x}{x+1} + x(x+1) \cdot \frac{3}{x} = x(x+1) \cdot \frac{2}{x(x+1)}$$

$$2x^2 + 3(x+1) = 2$$

$$2x^2 + 3x + 3 = 2$$

$$-2 \; -2$$

$$2x^2 + 3x + 1 = 0$$

$$(2x+1)(x+1) = 0$$

$$2x + 1 = 0 \quad x + 1 = 0$$

$$-1 \; -1 \qquad -1 \; -1$$

$2x = -1 \qquad x = -1$ But $x = -1$ leads to a zero

in a denominator, so

$x = \dfrac{-1}{2}$ $x = -1$ is *not* a solution.

7. $\dfrac{2x}{2x+1} - \dfrac{3}{x} = \dfrac{-3x-4}{3x}$ LCD $= 3x(2x+1)$

$$3x(2x+1) \cdot \dfrac{2x}{2x+1} - 3x(2x+1) \cdot \dfrac{3}{x} = 3x(2x+1) \cdot \dfrac{-3x-4}{3x}$$

$$3x(2x) - 3(2x+1)3 = (2x+1)(-3x-4)$$

$$6x^2 - 9(2x+1) = -6x^2 - 11x - 4$$

$$6x^2 - 18x - 9 = -6x^2 - 11x - 4$$

$$\underline{+6x^2 + 11x + 4 \quad +6x^2 + 11x + 4}$$

$$12x^2 - 7x - 5 = 0$$

$$(12x+5)(x-1) = 0$$

$$12x + 5 = 0 \qquad x - 1 = 0$$

$$\underline{-5 \;\; -5} \qquad \underline{+1 \;\; +1}$$

$$12x = -5 \qquad x = 1$$

$$x = \dfrac{-5}{12}$$

8. $\dfrac{2}{x-5} + \dfrac{1}{3x} = \dfrac{-8}{3x+15}$

Denominator factored: $\dfrac{2}{x-5} + \dfrac{1}{3x} = \dfrac{-8}{3(x+5)}$

LCD $= 3x(x-5)(x+5)$

$$3x(x-5)(x+5) \cdot \dfrac{2}{x-5} + 3x(x-5)(x+5) \cdot \dfrac{1}{3x}$$

$$= 3x(x-5)(x+5) \cdot \dfrac{-8}{3(x+5)}$$

$$3x(x+5)2 + (x-5)(x+5) = x(x-5)(-8)$$

$$6x(x+5) + (x-5)(x+5) = -8x(x-5)$$

$$6x^2 + 30x + x^2 - 25 = -8x^2 + 40x$$

$$7x^2 + 30x - 25 = -8x^2 + 40x$$

$$\underline{+8x^2 - 40x \qquad +8x^2 - 40x}$$

$$15x^2 - 10x - 25 = 0$$

$$\dfrac{1}{5}(15x^2 - 10x - 25) = \dfrac{1}{5}(0)$$

$$3x^2 - 2x - 5 = 0$$

$$(3x-5)(x+1) = 0$$

$$3x - 5 = 0 \qquad x + 1 = 0$$

$$\underline{+5 \;\; +5} \qquad \underline{-1 \;\; -1}$$

$$3x = 5 \qquad x = -1$$

$$x = \dfrac{5}{3}$$

9. $\dfrac{1}{x-4}+\dfrac{2}{x+1}-\dfrac{3}{x+3}=\dfrac{x-3}{x^2-3x-4}$

Denominator factored: $\dfrac{1}{x-4}+\dfrac{2}{x+1}-\dfrac{3}{x+3}=\dfrac{x-3}{(x-4)(x+1)}$

$\text{LCD}=(x-4)(x+1)(x+3)$

$(x-4)(x+1)(x+3)\cdot\dfrac{1}{x-4}+(x-4)(x+1)(x+3)\cdot\dfrac{2}{x+1}$

$-(x-4)(x+1)(x+3)\cdot\dfrac{3}{x+3}$

$=(x-4)(x+1)(x+3)\cdot\dfrac{x-3}{(x-4)(x+1)}$

$(x+1)(x+3)+2[(x-4)(x+3)]-3[(x-4)(x+1)]=(x+3)(x-3)$

$x^2+4x+3+2(x^2-x-12)-3(x^2-3x-4)=x^2-9$

$x^2+4x+3+2x^2-2x-24-3x^2+9x+12=x^2-9$

$11x-9=\ x^2-9$

$\underline{-11x+9\ \ -11x+9}$

$0=x^2-11x$

$0=x(x-11)$

$x=0\qquad x-11=0$

$\underline{+11\ +11}$

$x=11$

10. $\dfrac{4}{x-1}+\dfrac{3}{x+1}=\dfrac{1}{2x}-\dfrac{6}{x^2-1}$

Denominator factored: $\dfrac{4}{x-1}+\dfrac{3}{x+1}=\dfrac{1}{2x}-\dfrac{6}{(x-1)(x+1)}$

$\text{LCD}=2x(x-1)(x+1)$

$2x(x-1)(x+1)\cdot\dfrac{4}{x-1}+2x(x-1)(x+1)\dfrac{3}{x+1}$

$=2x(x-1)(x+1)\cdot\dfrac{1}{2x}-2x(x-1)(x+1)\dfrac{6}{(x-1)(x+1)}$

$2x(x+1)(4)+2x(x-1)(3)=(x-1)(x+1)-2x(6)$

$8x(x+1)+6x(x-1)=x^2-1-12x$

$8x^2+8x+6x^2-6x=x^2-12x-1$

$14x^2+2x=x^2-12x-1$

$\underline{-x^2+12x+1\ \ -x^2+12x+1}$

$$13x^2 + 14x + 1 = 0$$
$$(13x + 1)(x + 1) = 0$$

$$13x + 1 = 0 \qquad x + 1 = 0$$
$$\underline{-1 \quad -1} \qquad \underline{-1 \quad -1}$$
$$13x = -1 \qquad x = -1 \qquad$$ But $x = -1$ leads to a zero
in a denominator, so
$$x = \frac{-1}{13} \qquad\qquad\qquad x = -1$$ is *not* a solution.

Summary

In this chapter, we learned how to:

- *Write quadratic equations in the form* $ax^2 + bx + c = 0$.
- *Solve quadratic equations by factoring.* Once the equation is written in the above form, we factor the nonzero side, set each factor equal to 0, and then solve for x.
- *Solve a quadratic equation with the quadratic formula.* The quadratic formula $x = \dfrac{-b \pm \sqrt{b^2 - 4ac}}{2a}$ solves *every* quadratic equation. Once we identify a, b, and c, we put these values in the formula and then perform the arithmetic.
- *Simplify an equation to make factoring and using the quadratic formula easier.* If an equation has decimal numbers in it, we multiply each side of the equation by a power of 10 large enough to eliminate any decimal point. If the coefficient of x is not 1 (that is, $a \neq 1$) we might be able to divide each side of the equation by this number to make the factoring method or the quadratic formula easier. If the quadratic formula has fractions in it, we can multiply both sides of the equation by the LCD to clear the fraction(s), making the quadratic formula easier to use.
- *Solve rational equations that lead to quadratic equations.* Sometimes when a rational equation is in the form "fraction = fraction," we can cross-multiply and be left with a quadratic equation. If the rational equation is not in this form, then we multiply each term by the LCD and then solve the quadratic equation. With either of these two methods, we could have an extraneous solution, so we must make certain that the solution(s) do not cause a 0 in a denominator in the original equation.

QUIZ

1. If $(x - 8)(x + 9) = 0$, then

   A. $x = 8, 9$

   B. $x = -8, -9$

   C. $x = 8, -9$

   D. $x = -8, 9$

2. To apply the quadratic formula $3x^2 + 4x = 8$,

   A. $a = 3, b = 4, c = 8$

   B. $a = 3, b = -4, c = 8$

   C. $a = 3, b = 4, c = -8$

   D. $a = 3, b = -4, c = -8$

3. Simplify $\dfrac{6 \pm \sqrt{40}}{4}$.

   A. $\dfrac{3 \pm \sqrt{10}}{2}$

   B. $\dfrac{3 \pm \sqrt{40}}{2}$

   C. $3 \pm \sqrt{10}$

   D. $-3 \pm \sqrt{10}$

4. If $x^2 - x - 9 = 0$, then

   A. $x = \dfrac{-1 \pm \sqrt{37}}{2}$

   B. $x = \dfrac{1 \pm \sqrt{37}}{2}$

   C. $x = \dfrac{-1 \pm \sqrt{35}}{2}$

   D. $x = \dfrac{1 \pm \sqrt{35}}{2}$

5. If $x^2 + 9x + 18 = 0$, then

   A. $x = 3, 6$

   B. $x = -3, -6$

   C. $x = 2, 9$

   D. $x = -2, -9$

6. If $3x^2 - 2x - 2 = 0$, then

   A. $x = \dfrac{-1 \pm 2\sqrt{7}}{3}$

   B. $x = \dfrac{1 \pm 2\sqrt{7}}{3}$

   C. $x = 2 \pm \dfrac{\sqrt{7}}{3}$

   D. $x = \dfrac{1 \pm \sqrt{7}}{3}$

7. If $9x^2 - 1 = 0$, then

   A. $x = \dfrac{1}{3}$

   B. $x = \pm \dfrac{1}{3}$

   C. $x = -\dfrac{1}{3}$

   D. $x = 3$

8. If $\dfrac{2}{x - 6} = \dfrac{2x}{2x - 15}$, then

   A. $x = 3, 5$

   B. $x = 4 \pm \sqrt{62}$

   C. $x = 5, 6$

   D. There is no solution.

9. If $\dfrac{x+1}{x+3} + \dfrac{28}{x^2-9} = \dfrac{6}{x-3}$, then

    A. $x = 2, 6$

    B. $x = 7, 1$

    C. $x = 2 \pm \sqrt{29}$

    D. There is no solution.

10. If $\dfrac{4}{x-3} + \dfrac{x}{2x+1} = 1$, then

    A. $x = 5 \pm 8\sqrt{2}$

    B. $x = \pm\sqrt{3}$

    C. $x = 5 \pm 4\sqrt{2}$

    D. There is no solution.

Quadratic Applications

Now that we can solve a larger family of equations, we can solve a larger family of applied problems. In addition to the problem types that we worked in Chapter 8, we will solve distance problems involving cars/runners/etc. that are traveling in paths at right angles to each other; problems involving the height of a falling object; and revenue problems. We will learn a formula that predicts the height of a falling object t seconds after it is released. The revenue problem involves a price that is changing. We know something about how the change in a price causes a change in sales and are asked what price to charge to bring in some specified revenue. As we did in Chapter 8, we begin with number sense problems.

CHAPTER OBJECTIVES

In this chapter, you will

- Solve number sense problems with quadratic equations
- Solve revenue problems with quadratic equations
- Solve falling object problems
- Solve distance, work, and geometry problems with quadratic equations

Number Sense Problems

Most of the problems in this chapter aren't much different from the word problems in Chapter 8. The only difference is that we use quadratic equations to solve them. Because most quadratic equations have two solutions, some of the applied problems also have two solutions, too. Many, though, have one solution, and we have to decide which one of them is the correct solution.

 EXAMPLES

The product of two consecutive positive numbers is 240. Find the numbers.

We let x represent the first number. Because the numbers are consecutive, the next number is one more than the first: $x + 1$ represents the next number. The product of these two numbers is $x(x + 1)$, which equals 240.

$$x(x+1) = 240$$
$$x^2 + x = 240$$
$$x^2 + x - 240 = 0$$
$$(x-15)(x+16) = 0$$
$$x - 15 = 0 \quad (x+16 = 0 \text{ leads to a negative solution})$$
$$+15 \ +15$$
$$x = 15$$

The consecutive positive numbers are 15 and $15 + 1 = 16$.

The product of two consecutive even numbers is 528. What are the numbers?

We let x represent the first number. Consecutive even numbers (and consecutive odd numbers) differ by two, so we let $x + 2$ represent the second number. Their product is $x(x + 2)$.

$$x(x+2) = 528$$
$$x^2 + 2x = 528$$
$$x^2 + 2x - 528 = 0$$
$$(x-22)(x+24) = 0$$

$$x - 22 = 0 \qquad x + 24 = 0$$
$$+22 \ +22 \qquad -24 \ -24$$
$$x = 22 \qquad x = -24$$

The two solutions are 22 and 24, and –24 and –22.

Two positive numbers differ by 5. Their product is 104. Find the two numbers.

We let x represent the first number. If x differs from the other number by 5, then the other number could either be $x + 5$ or $x - 5$; it does not matter which representation we use. To see why it does not matter, we will work this problem with both representations.

Let $x + 5$ represent the other number	Let $x - 5$ represent the other number
$x(x + 5) = 104$	$x(x - 5) = 104$
$x^2 + 5x = 104$	$x^2 - 5x = 104$
$x^2 + 5x - 104 = 0$ $(x + 13)(x - 8) = 0$ $x - 8 = 0$ $(x + 13 = 0$ leads $+8 \;\; +8$ to a negative $x = 8$ solution)	$x^2 - 5x - 104 = 0$ $(x - 13)(x + 8) = 0$ $x - 13 = 0$ $(x + 8 = 0$ leads to a $+13 \;\; +13$ negative solution) $x = 13$
The numbers are 8 and $8 + 5 = 13$.	The numbers are 13 and $13 - 5 = 8$.

PRACTICE

1. The product of two consecutive odd numbers is 399. Find the numbers.

2. The product of two consecutive numbers is 380. Find the numbers.

3. The product of two consecutive numbers is 650. Find the numbers.

4. The product of two consecutive even numbers is 288. Find the numbers.

5. Two numbers differ by 7. Their product is 228. Find the numbers.

SOLUTIONS

1. Let $x =$ first number and $x + 2 =$ second number.

$$x(x + 2) = 399$$
$$x^2 + 2x = 399$$
$$x^2 + 2x - 399 = 0$$
$$(x - 19)(x + 21) = 0$$

$x - 19 = 0$ $x + 21 = 0$
$x = 19$ $x = -21$
$x + 2 = 21$ $x + 2 = -19$

There are two solutions: 19 and 21, and −21 and −19.

2. Let x = first number and $x + 1$ = second number.

$$x(x + 1) = 380$$
$$x^2 + x = 380$$
$$x^2 + x - 380 = 0$$
$$(x + 20)(x - 19) = 0$$

$$x + 20 = 0 \qquad x - 19 = 0$$
$$x = -20 \qquad x = 19$$
$$x + 1 = -19 \qquad x + 1 = 20$$

There are two solutions: 19 and 20, and −19 and −20.

3. Let x = first number and $x + 1$ = second number.

$$x(x + 1) = 650$$
$$x^2 + x = 650$$
$$x^2 + x - 650 = 0$$
$$(x - 25)(x + 26) = 0$$

$$x - 25 = 0 \qquad x + 26 = 0$$
$$x = 25 \qquad x = -26$$
$$x + 1 = 26 \qquad x + 1 = -25$$

There are two solutions: 25 and 26, and −25 and −26.

4. Let x = first number and $x + 2$ = second number.

$$x(x + 2) = 288$$
$$x^2 + 2x = 288$$
$$x^2 + 2x - 288 = 0$$
$$(x - 16)(x + 18) = 0$$

$$x - 16 = 0 \qquad x + 18 = 0$$
$$x = 16 \qquad x = -18$$
$$x + 2 = 18 \qquad x + 2 = -16$$

There are two solutions: 16 and 18, and −16 and −18.

5. Let x = first number and $x + 7$ = second number.

$$x(x+7) = 228$$
$$x^2 + 7x = 228$$
$$x^2 + 7x - 228 = 0$$
$$(x-12)(x+19) = 0$$

$x - 12 = 0$	$x + 19 = 0$
$x = 12$	$x = -19$
$x + 7 = 19$	$x + 7 = -12$

There are two solutions: 12 and 19, −12 and −19.

Revenue Problems

A common business application of quadratic equations occurs when raising a price results in lower sales or lowering a price results in higher sales. The obvious question is what price brings in the most revenue. This problem is addressed in pre-calculus and calculus. The problem addressed here is finding a price that would bring some specific revenue.

These revenue problems involve raising (or lowering) a price by a certain number of increments and sales decreasing (or increasing) by a certain amount for each incremental change in the price. For instance, suppose for each increase of $10 in the price, two customers are lost. The price and sales level both depend on the *number* of $10 increases. If the price is increased by $10(1) = $10, two customers are lost. If the price is increased by $2(10) = $20, 2(2) = 4 customers will be lost, and if the price is increased by $3(10) = $30, 2(3) = 6 customers will be lost. In general, if the price increases by $10x$, then $2x$ customers will be lost. The variable represents the *number* of incremental increases (or decreases) of the price.

For the following problems, we use the revenue formula, $R = PQ$, where R represents the revenue, P represents the price, Q represents the number sold, and x represents the number of incremental increases or decreases in the price. If the price is increased, then P is the current price plus the x times the increment. If the price is decreased, then P is the current price minus x times the increment. If sales decrease, then Q is the current sales level minus x times the incremental loss. If sales increase, then Q is the current sales level plus x times the incremental change. For now, we focus on how to represent the revenue in terms of x.

EXAMPLES

Represent the revenue formula in terms of x, the number incremental changes in the price.

A department store sells 20 music players per week at $80 each. The manager believes that for each decrease of $5 in the price, six more players will be sold.

Let x represent the number of $5 decreases in the price. Then the price will decrease by $5x$: $P = 80 - 5x$.

The sales level will increase by six for each $5 decrease in the price—the sales level will increase by $6x$: $Q = 20 + 6x$.

Thus, $R = PQ$ becomes $R = (80 - 5x)(20 + 6x)$.

A rental company manages an office complex with 16 offices. Each office can be rented if the monthly rent is $1000. For each $200 increase in the rent, one tenant will be lost.

Let x represent the number of $200 increases in the price.

$$P = 1000 + 200x \quad Q = 16 - 1x \quad R = (1000 + 200x)(16 - x)$$

A grocery store sells 300 pounds of bananas each day when they are priced at 45 cents per pound. The produce manager observes that for each 5-cent decrease in the price per pound of bananas, an additional 50 pounds are sold. Let x represent the number of 5 cent decreases in the price.

$$P = 45 - 5x \quad Q = 300 + 50x \quad R = (45 - 5x)(300 + 50x)$$

(The revenue will be in cents instead of dollars.)

A music storeowner sells 60 newly released CDs per day when the price is $12 per CD. For each $1.50 decrease in the price, the store will sell an additional 16 CDs each week. Let x represent the number of $1.50 decreases in the price.

$$P = 12.00 - 1.50x \quad Q = 60 + 16x \quad R = (12.00 - 1.50x)(60 + 16x)$$

PRACTICE

Let x represent the number of increases/decreases in the price.

1. The owner of an apartment complex knows he can rent all 50 apartments when the monthly rent is $400. He thinks that for each $25 increase in the rent, he will lose two tenants.

   $P = $ _____

   $Q = $ _____ $R = $ _____

2. A grocery store sells 4000 gallons of milk per week when the price is $2.80 per gallon. Customer research indicates that for each $0.10 decrease in the price, 200 more gallons of milk will be sold.

   $P =$ _____

   $Q =$ _____ $R =$ _____

3. A movie theater's concession stand sells an average of 500 buckets of popcorn each weekend when the price is $4 per bucket. The manager knows from experience that for every $0.05 decrease in the price, 20 more buckets of popcorn will be sold each weekend.

   $P =$ _____

   $Q =$ _____ $R =$ _____

4. An automobile repair shop performs 40 oil changes per day when the price is $30. Industry research indicates that the shop will lose 5 customers for each $2 increase in the price.

   $P =$ _____

   $Q =$ _____ $R =$ _____

5. A fast food restaurant sells an average of 250 orders of onion rings each week when the price is $1.50 per order. The manager believes that for each $0.05 decrease in the price, 10 more orders will be sold.

   $P =$ _____

   $Q =$ _____ $R =$ _____

6. A shoe store sells a certain athletic shoe for $40 per pair. The store averages sales of 80 pairs each week. The store owner's past experience leads him to believe that for each $2 increase in the price of the shoe, one less pair would be sold each week.

   $P =$ _____

   $Q =$ _____ $R =$ _____

✔ SOLUTIONS _____

1. $P = 400 + 25x$ $Q = 50 - 2x$

   $R = (400 + 25x)(50 - 2x)$

2. $P = 2.80 - 0.10x$ $Q = 4000 + 200x$

   $R = (2.80 - 0.10x)(4000 + 200x)$

3. $P = 4 - 0.05x$ $Q = 500 + 20x$

   $R = (4 - 0.05x)(500 + 20x)$

4. $P = 30 + 2x$ $\qquad\qquad\qquad$ $Q = 40 - 5x$
   $R = (30 + 2x)(40 - 5x)$

5. $P = 1.50 - 0.05x$ $\qquad\qquad$ $Q = 250 + 10x$
   $R = (1.50 - 0.05x)(250 + 10x)$

6. $P = 40 + 2x$ $\qquad\qquad\qquad$ $Q = 80 - 1x$
   $R = (40 + 2x)(80 - x)$

For the problems in this section, we will be given some specific revenue and are asked to find the price that brings in this revenue. Because we found the revenue equations earlier, all we need to do here is to solve the quadratic equation. Some of these problems have *two* solutions.

EXAMPLES

A department store sells 20 music players per week at $80 each. The manager believes that for each decrease of $5 in the price, six more players will be sold.

Let x represent the number of $5 decreases in the price.

$$P = 80 - 5x \quad Q = 20 + 6x \quad R = (80 - 5x)(20 + 6x).$$

What price should the manager charge if the revenue needs to be $2240?

$R = (80 - 5x)(20 + 6x)$ becomes $2240 = (80 - 5x)(20 + 6x)$

$$2240 = (80 - 5x)(20 + 6x)$$

$$2240 = 1600 + 380x - 30x^2$$

$$30x^2 - 380x + 640 = 0$$

$$\frac{1}{10}(30x^2 - 380x + 640) = \frac{1}{10}(0)$$

$$3x^2 - 38x + 64 = 0$$

$$(3x - 32)(x - 2) = 0$$

$$3x - 32 = 0 \qquad x - 2 = 0$$

$$3x = 32 \qquad\quad x = 2$$

$$x = \frac{32}{3}$$

If $x = \dfrac{32}{3}$, the price for each player is $P = 80 - 5\left(\dfrac{32}{3}\right) = \$26.67$.

If $x = 2$, the price for each player is $P = 80 - 5(2) = \$70$.

A rental company manages an office complex with 16 offices. Each office can be rented if the monthly rent is $1000. For each $200 increase in the rent, one tenant will be lost.

Let x represent the number of $200 increases in the price.

$$P = 1000 + 200x \quad Q = 16 - 1x \quad R = (1000 + 200x)(16 - x)$$

What should the monthly rent be if the rental company needs $20,800 each month in revenue?

$$R = (1000 + 200x)(16 - x)$$
$$20,800 = (1000 + 200x)(16 - x)$$
$$20,800 = 16,000 + 2200x - 200x^2$$
$$200x^2 - 2200x + 4800 = 0$$
$$\frac{1}{200}(200x^2 - 2200x + 4800) = \frac{1}{200}(0)$$
$$x^2 - 11x + 24 = 0$$
$$(x - 3)(x - 8) = 0$$
$$x - 3 = 0 \qquad x - 8 = 0$$
$$x = 3 \qquad x = 8$$

If $x = 3$, the monthly rent will be $1000 + 200(3) = \$1600$. If $x = 8$, the monthly rent will be $1000 + 200(8) = \$2600$.

A grocery store sells 300 pounds of bananas each day when they are priced at 45 cents per pound. The produce manager observes that for each 5-cent decrease in the price per pound of bananas, an additional 50 pounds are sold.

Let x represent the number of 5 cent decreases in the price.

$$P = 45 - 5x \quad Q = 300 + 50x \quad R = (45 - 5x)(300 + 50x)$$

What should the price of bananas be for weekly sales to be $140? How many bananas (in pounds) will be sold at this price (these prices)? (The revenue will be in terms of cents, so $140 becomes 14,000 cents.)

$$R = (45 - 5x)(300 + 50x)$$
$$14,000 = (45 - 5x)(300 + 50x)$$
$$14,000 = 13,500 + 750x - 250x^2$$
$$250x^2 - 750x + 500 = 0$$

$$\frac{1}{250}(250x^2 - 750x + 500) = \frac{1}{250}(0)$$

$$x^2 - 3x + 2 = 0$$

$$(x - 2)(x - 1) = 0$$

$$x - 2 = 0 \qquad x - 1 = 0$$

$$x = 2 \qquad\qquad x = 1$$

If $x = 2$, the price per pound will be $45 - 5(2) = 35$ cents. The number of pounds sold each week will be $300 + 50(2) = 400$. If $x = 1$, the price per pound will be $45 - 5(1) = 40$ cents and the number of pounds sold each week will be $300 + 50(1) = 350$.

A music storeowner sells 60 newly released CDs per day when the cost is $12 per CD. For each $1.50 decrease in the price, the store will sell an additional 16 CDs per week.

Let x represent the number of $1.50 decreases in the price.

$$P = 12.00 - 1.50x \qquad Q = 60 + 16x \qquad R = (12.00 - 1.50x)(60 + 16x)$$

What should the price be if the storeowner needs revenue of $810 per week for the sale of these CDs? How many will be sold at this price (these prices)?

$$R = (12.00 - 1.50x)(60 + 16x)$$

$$810 = (12.00 - 1.50x)(60 + 16x)$$

$$810 = 720 + 102x - 24x^2$$

$$24x^2 - 102x + 90 = 0$$

$$\frac{1}{6}(24x^2 - 102x + 90) = \frac{1}{6}(0)$$

$$4x^2 - 17x + 15 = 0$$

$$(4x - 5)(x - 3) = 0$$

$$4x - 5 = 0 \qquad x - 3 = 0$$

$$4x = 5 \qquad\qquad x = 3$$

$$x = \frac{5}{4} = 1.25$$

When $x = 1.25$, the price should be $12 - 1.50(1.25) = \$10.13$ and the number sold would be $60 + 16(1.25) = 80$. If $x = 3$, the price should be $12 - 1.50(3) = \$7.50$ and the number sold would be $60 + 16(3) = 108$.

PRACTICE

1. The owner of an apartment complex knows he can rent all 50 apartments when the monthly rent is $400. He thinks that for each $25 increase in the rent, he will lose two tenants. What should the rent be for the revenue to be $20,400?

2. A grocery store sells 4000 gallons of milk per week when the price is $2.80 per gallon. Customer research indicates that for each $0.10 decrease in the price, 200 more gallons of milk will be sold. What does the price need to be so that weekly milk sales reach $11,475?

3. A movie theater's concession stand sells an average of 500 buckets of popcorn each weekend when the price is $4 per bucket. The manager knows from experience that for every $0.05 decrease in the price, 20 more buckets of popcorn will be sold each weekend. What should the price be so that $2450 worth of popcorn is sold? How many buckets will be sold at this price (these prices)?

4. An automobile repair shop performs 40 oil changes per day when the price is $30. Industry research indicates that the shop will lose 5 customers for each $2 increase in the price. What would the shop have to charge in order for the daily revenue from oil changes to be $1120? How many oil changes will the shop perform each day?

5. A fast food restaurant sells an average of 250 orders of onion rings each week when the price is $1.50 per order. The manager believes that for each $0.05 decrease in the price, 10 more orders are sold. If the manager wants $378 weekly revenue from onion ring sales, what should she charge for onion rings?

6. A shoe store sells a certain athletic shoe for $40 per pair. The store averages sales of 80 pairs each week. The store owner's past experience leads him to believe that for each $2 increase in the price of the shoe, one less pair would be sold each week. What price would result in $3648 weekly sales?

SOLUTIONS

1. $P = 400 + 25x$ $Q = 50 - 2x$ $R = (400 + 25x)(50 - 2x)$

   $20{,}400 = (400 + 25x)(50 - 2x)$

   $20{,}400 = 20{,}000 + 450x - 50x^2$

   $50x^2 - 450x + 400 = 0$

$$\frac{1}{50}(50x^2 - 450x + 400) = \frac{1}{50}(0)$$

$$x^2 - 9x + 8 = 0$$

$$(x - 8)(x - 1) = 0$$

$$x - 8 = 0 \qquad x - 1 = 0$$

$$x = 8 \qquad x = 1$$

If $x = 1$, the rent should be $400 + 25(1) = \$425$. If $x = 8$, the rent should be $400 + 25(8) = \$600$.

2. $P = 2.80 - 0.10x \qquad Q = 4000 + 200x$

$R = (2.80 - 0.10x)(4000 + 200x)$

$11,475 = (2.80 - 0.10x)(4000 + 200x)$

$11,475 = 11,200 + 160x - 20x^2$

$20x^2 - 160x + 275 = 0$

$\frac{1}{5}(20x^2 - 160x + 275) = \frac{1}{5}(0)$

$4x^2 - 32x + 55 = 0$

$(2x - 5)(2x - 11) = 0$

$$2x - 5 = 0 \qquad 2x - 11 = 0$$

$$2x = 5 \qquad 2x = 11$$

$$x = \frac{5}{2} \qquad x = \frac{11}{2}$$

$$x = 2.5 \qquad x = 5.5$$

If $x = 2.50$, the price should be $2.80 - 0.10(2.5) = \$2.55$. If $x = 5.5$, the price should be $2.80 - 0.10(5.50) = \$2.25$.

3. $P = 4 - 0.05x \qquad Q = 500 + 20x \qquad R = (4 - 0.05x)(500 + 20x)$

$2450 = (4 - 0.05x)(500 + 20x)$

$2450 = 2000 + 55x - x^2$

$x^2 - 55x + 450 = 0$

$(x - 45)(x - 10) = 0$

$$x - 45 = 0 \qquad x - 10 = 0$$

$$x = 45 \qquad x = 10$$

If $x = 45$, the price should be $4 - 0.05(45) = \$1.75$ and $500 + 20(45) = 1400$ buckets would be sold. If $x = 10$, the price should be $4 - 0.05(10) = \$3.50$ and $500 + 20(10) = 700$ buckets would be sold.

4. $P = 30 + 2x$ $Q = 40 - 5x$ $R = (30 + 2x)(40 - 5x)$

$1120 = (30 + 2x)(40 - 5x)$

$1120 = 1200 - 70x - 10x^2$

$10x^2 + 70x - 80 = 0$

$\dfrac{1}{10}(10x^2 + 70x - 80) = \dfrac{1}{10}(0)$

$x^2 + 7x - 8 = 0$

$(x - 1)(x + 8) = 0$

$x - 1 = 0$ $x + 8 = 0$

    $x = 1$ $x = -8$ ($x = -8$ is not a solution)

The price should be $30 + 2(1) = \$32$. There would be $40 - 5(1) = 35$ oil changes performed each day.

5. $P = 1.50 - 0.05x$ $Q = 250 + 10x$

$R = (1.50 - 0.05x)(250 + 10x)$

$378 = (1.50 - 0.05x)(250 + 10x)$

$378 = 375 + 2.5x - 0.5x^2$

$0.5x^2 - 2.5x + 3 = 0$

$2(0.5x^2 - 2.5x + 3) = 2(0)$

$x^2 - 5x + 6 = 0$

$(x - 2)(x - 3) = 0$

$x - 2 = 0$ $x - 3 = 0$

    $x = 2$ $x = 3$

If $x = 2$, the price should be $1.50 - 0.05(2) = \$1.40$. If $x = 3$, the price should be $1.50 - 0.05(3) = \$1.35$.

6. $P = 40 + 2x$ $Q = 80 - 1x$ $R = (40 + 2x)(80 - x)$

$3648 = (40 + 2x)(80 - x)$

$3648 = 3200 + 120x - 2x^2$

$2x^2 - 120x + 448 = 0$

$\dfrac{1}{2}(2x^2 - 120x + 448) = \dfrac{1}{2}(0)$

$x^2 - 60x + 224 = 0$

$$x = \frac{-(-60) \pm \sqrt{(-60)^2 - 4(1)(224)}}{2(1)} = \frac{60 \pm \sqrt{3600 - 896}}{2}$$

$$= \frac{60 \pm \sqrt{2704}}{2} = \frac{60 \pm 52}{2} = 4, 56$$

($x = 56$ is not likely to be a solution—the price would be \$152!) If the price of the shoes are $40 + 2(4) = \$48$ per pair, the revenue will be \$3648.

More Work Problems

You might recall from Chapter 8 that we can solve *work* problems by filling in the table below with information given in the problem. In the work formula $Q = rt$ (Q = quantity, r = rate, and t = time), Q is usually "1." We then solve the equation: Worker 1's Rate + Worker 2's Rate = Together Rate.

The information given in the problem is usually the time one or both workers need to complete the job. We want the *rates* not the *times*. We can solve for r in $Q = rt$ to get the rates.

$$Q = rt$$

$$\frac{Q}{t} = r$$

Because Q is usually "1," $\frac{1}{t} = r$.

The equation to solve is usually:

$$\frac{1}{\text{Worker 1's time}} + \frac{1}{\text{Worker 2's time}} = \frac{1}{\text{Together time}}$$

Worker	Quantity	Rate	Time
Worker 1	1	$\dfrac{1}{\text{Worker 1's time}}$	Worker 1's time
Worker 2	1	$\dfrac{1}{\text{Worker 2's time}}$	Worker 2's time
Together	1	$\dfrac{1}{\text{Together time}}$	Together time

EXAMPLE

Together John and Michael can paint a wall in 18 minutes. Alone John needs 15 minutes more to paint the wall than Michael needs. How much time does each John and Michael need to paint the wall by himself?

Let t represent the number of minutes Michael needs to paint the wall. Then $t + 15$ represents the number of minutes John needs to paint the wall.

Worker	Quantity	Rate	Time
Michael	1	$\dfrac{1}{t}$	t
John	1	$\dfrac{1}{t + 15}$	$t + 15$
Together	1	$\dfrac{1}{18}$	18

The equation to solve is $\dfrac{1}{t} + \dfrac{1}{t + 15} = \dfrac{1}{18}$. The LCD is $18t(t + 15)$.

$$\frac{1}{t} + \frac{1}{t+15} = \frac{1}{18}$$

$$18t(t+15) \cdot \frac{1}{t} + 18t(t+15) \cdot \frac{1}{t+15} = 18t(t+15) \cdot \frac{1}{18}$$

$$18(t+15) + 18t = t(t+15)$$

$$18t + 270 + 18t = t^2 + 15t$$

$$36t + 270 = t^2 + 15t$$

$$0 = t^2 - 21t - 270$$

$$0 = (t - 30)(t + 9)$$

$t - 30 = 0 \qquad t + 9 = 0 \qquad$ (This does not lead to solution.)

$t = 30$

Michael needs 30 minutes to paint the wall by himself and John needs $30 + 15 = 45$ minutes.

PRACTICE

1. Alex and Tina working together can peel a bag of potatoes in 6 minutes. By herself Tina needs 5 minutes more than Alex to peel the potatoes by herself. How long would each need to peel the potatoes if they were to work alone?

2. Together Rachel and Jared can wash a car in 16 minutes. Working alone Rachel needs 24 minutes longer than Jared does to wash the car. How long would it take for each Rachel and Jared to wash the car?

3. Two printing presses working together can print a magazine order in 6 hours. Printing Press I can complete the job alone in 5 fewer hours than Printing Press II. How long would each press need to print the run by itself?

4. Together two pipes can fill a small reservoir in 2 hours. Working alone Pipe I can fill the reservoir in 1 hour 40 minutes less time than Pipe II can. How long would each pipe need to fill the reservoir by itself?

5. John and Gary together can unload a truck in 1 hour 20 minutes. Working alone John needs 36 minutes more to unload the truck than Gary needs. How long would each John and Gary need to unload the truck by himself?

 **SOLUTIONS**

1. Let t represent the number of minutes Alex needs to peel the potatoes. Tina needs $t + 5$ minutes to complete the job alone.

Worker	Quantity	Rate	Time
Alex	1	$\dfrac{1}{t}$	t
Tina	1	$\dfrac{1}{t+5}$	$t + 5$
Together	1	$\dfrac{1}{6}$	6

The equation to solve is $\dfrac{1}{t} + \dfrac{1}{t+5} = \dfrac{1}{6}$. The LCD is $6t(t + 5)$.

$$\frac{1}{t} + \frac{1}{t+5} = \frac{1}{6}$$

$$6t(t+5) \cdot \frac{1}{t} + 6t(t+5) \cdot \frac{1}{t+5} = 6t(t+5) \cdot \frac{1}{6}$$

$$6(t+5)+6t = t(t+5)$$
$$6t+30+6t = t^2+5t$$
$$12t+30 = t^2+5t$$
$$0 = t^2-7t-30$$
$$0 = (t-10)(t+3)$$

$t-10=0 \qquad t+3=0 \qquad$ **(This does not lead to a solution.)**

$\qquad t=10$

Alex can peel the potatoes in 10 minutes and Tina can peel them in $10+5=15$ minutes.

2. Let t represent the number of minutes Jared needs to wash the car by himself. The time Rachel needs to wash the car by herself is $t+24$.

Worker	Quantity	Rate	Time
Jared	1	$\frac{1}{t}$	t
Rachel	1	$\frac{1}{t+24}$	$t+24$
Together	1	$\frac{1}{16}$	16

The equation to solve is $\dfrac{1}{t}+\dfrac{1}{t+24}=\dfrac{1}{16}$. The LCD is $16t(t+24)$.

$$\frac{1}{t}+\frac{1}{t+24}=\frac{1}{16}$$
$$16t(t+24)\cdot\frac{1}{t}+16t(t+24)\cdot\frac{1}{t+24}=16t(t+24)\cdot\frac{1}{16}$$
$$16(t+24)+16t = t(t+24)$$
$$16t+384+16t = t^2+24t$$
$$32t+384 = t^2+24t$$
$$0 = t^2-8t-384$$
$$0 = (t-24)(t+16)$$

$t-24=0 \qquad t+16=0 \qquad$ **(This does not lead to a solution.)**

$\qquad t=24$

Jared needs 24 minutes to wash the car alone and Rachel needs $24+24=$ 48 minutes.

3. Let t represent the number of hours Printing Press II needs to print the run by itself. Because Printing Press I needs 5 fewer hours than Printing Press II, $t - 5$ represents the number of hours Printing Press I needs to complete the run by itself.

Worker	Quantity	Rate	Time
Press I	1	$\dfrac{1}{t-5}$	$t - 5$
Press II	1	$\dfrac{1}{t}$	t
Together	1	$\dfrac{1}{6}$	6

The equation to solve is $\dfrac{1}{t-5} + \dfrac{1}{t} = \dfrac{1}{6}$. The LCD is $6t(t-5)$.

$$\frac{1}{t-5} + \frac{1}{t} = \frac{1}{6}$$

$$6t(t-5) \cdot \frac{1}{t-5} + 6t(t-5) \cdot \frac{1}{t} = 6t(t-5) \cdot \frac{1}{6}$$

$$6t + 6(t-5) = t(t-5)$$

$$6t + 6t - 30 = t^2 - 5t$$

$$12t - 30 = t^2 - 5t$$

$$0 = t^2 - 17t + 30$$

$$0 = (t-15)(t-2)$$

$t - 15 = 0 \qquad t - 2 = 0 \qquad$ (This cannot be a solution because
$\qquad t = 15 \qquad\quad t = 2 \qquad$ $2 - 5$ is negative.)

Printing Press II can print the run alone in 15 hours and Printing Press I needs $15 - 5 = 10$ hours.

4. Let t represent the number of hours Pipe II needs to fill the reservoir alone. Pipe I needs 1 hour 40 minutes less to do the job, so $t - 1\frac{40}{60} = t - 1\frac{2}{3} = t - \frac{5}{3}$ represents the time Pipe I needs to fill the reservoir by itself.

Worker	Quantity	Rate	Time
Pipe I	1	$\dfrac{1}{t-\frac{5}{3}}$	$t-\dfrac{5}{3}$
Pipe II	1	$\dfrac{1}{t}$	t
Together	1	$\dfrac{1}{2}$	2

The equation to solve is $\dfrac{1}{t-\frac{5}{3}}+\dfrac{1}{t}=\dfrac{1}{2}$. The LCD is $2t\left(t-\dfrac{5}{3}\right)$.

$$\frac{1}{t-\frac{5}{3}}+\frac{1}{t}=\frac{1}{2}$$

$$2t\left(t-\frac{5}{3}\right)\cdot\frac{1}{t-\frac{5}{3}}+2t\left(t-\frac{5}{3}\right)\cdot\frac{1}{t}=2t\left(t-\frac{5}{3}\right)\cdot\frac{1}{2}$$

$$2t+2\left(t-\frac{5}{3}\right)=t\left(t-\frac{5}{3}\right)$$

$$2t+2t-\frac{10}{3}=t^2-\frac{5}{3}t$$

$$4t-\frac{10}{3}=t^2-\frac{5}{3}t$$

$$3\left(4t-\frac{10}{3}\right)=3\left(t^2-\frac{5}{3}t\right)$$

$$12t-10=3t^2-5t$$

$$0=3t^2-17t+10$$

$$0=(t-5)(3t-2)$$

$$t-5=0 \qquad 3t-2=0$$

$$t=5 \qquad\quad 3t=2$$

$$t=\frac{2}{3}$$

($t=\frac{2}{3}$ cannot be a solution because $t-\frac{5}{3}$ would be negative.)

Pipe II can fill the reservoir in 5 hours and Pipe I can fill it in

$5-\dfrac{5}{3}=\dfrac{5}{1}\cdot\dfrac{3}{3}-\dfrac{5}{3}=\dfrac{15}{3}-\dfrac{5}{3}=\dfrac{10}{3}=3\frac{1}{3}$ hours or 3 hours 20 minutes.

5. Let t represent the number of hours Gary needs to unload the truck by himself. John needs 36 minutes more than Gary needs to unload the truck by himself, so John needs $\frac{36}{60}$ more hours or $\frac{3}{5}$ more hours. The number of hours John needs to unload the truck by himself is $t + \frac{3}{5}$.

Together they can unload the truck in 1 hour 20 minutes, which is $1\frac{1}{3} = \frac{4}{3}$ hours. This means that the Together rate is $\dfrac{1}{\frac{4}{3}} = 1 \div \frac{4}{3} = 1 \cdot \frac{3}{4} = \frac{3}{4}$.

Worker	Quantity	Rate	Time
John	1	$\dfrac{1}{t + \frac{3}{5}}$	$t + \dfrac{3}{5}$
Gary	1	$\dfrac{1}{t}$	t
Together	1	$\dfrac{3}{4}$	$\dfrac{4}{3}$

The equation to solve is $\dfrac{1}{t + \frac{3}{5}} + \dfrac{1}{t} = \dfrac{3}{4}$. The LCD is $4t\left(t + \frac{3}{5}\right)$.

$$\frac{1}{t + \frac{3}{5}} + \frac{1}{t} = \frac{3}{4}$$

$$4t\left(t + \frac{3}{5}\right) \cdot \frac{1}{t + \frac{3}{5}} + 4t\left(t + \frac{3}{5}\right) \cdot \frac{1}{t} = 4t\left(t + \frac{3}{5}\right) \cdot \frac{3}{4}$$

$$4t + 4\left(t + \frac{3}{5}\right) = 3t\left(t + \frac{3}{5}\right)$$

$$4t + 4t + \frac{12}{5} = 3t^2 + \frac{9}{5}t$$

$$8t + \frac{12}{5} = 3t^2 + \frac{9}{5}t$$

$$5\left(8t + \frac{12}{5}\right) = 5\left(3t^2 + \frac{9}{5}t\right)$$

$$40t + 12 = 15t^2 + 9t$$

$$0 = 15t^2 - 31t - 12$$

$$0 = (5t - 12)(3t + 1)$$

$$5t - 12 = 0 \qquad 3t + 1 = 0 \qquad \text{(This does not lead to a solution.)}$$

$$5t = 12$$

$$t = \frac{12}{5} = 2\tfrac{2}{5}$$

Gary needs $2\tfrac{2}{5}$ hours or 2 hours 24 minutes to unload the truck. John needs 2 hours 24 minutes + 36 minutes = 3 hours to unload the truck.

The Height of a Falling Object

We can compute the height of an object dropped, thrown or fired straight upward with quadratic equations. The general formula is $h = -16t^2 + v_0 t + h_0$, where h is the object's height (in feet), t is time (in seconds), h_0 is the object's initial height (that is, its height at $t = 0$ seconds) and v_0 is the object's initial velocity (that is, its speed at $t = 0$ seconds) in feet per second. If the object is tossed, thrown, or fired upward, v_0 is positive. If the object is thrown downward, v_0 is negative. If the object is dropped, v_0 is zero. The object reaches the ground when $h = 0$. (The effect of air resistance is ignored.)

Typical questions are:

When will the object be _____ feet high?

When will the object reach the ground?

What is the object's height after _____ seconds?

We begin by finding the time it takes for an object to hit the ground after being dropped, making $v_0 = 0$ and $h = 0$.

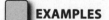 **EXAMPLES**

An object is dropped from a height of 1600 feet. How long will it take for the object to hit the ground?

Because the object is dropped, the initial velocity, v_0, is zero: $v_0 = 0$. The object is dropped from a height of 1600 feet, so $h_0 = 1600$. The formula $h = -16t^2 + v_0 t + h_0$ becomes $h = -16t^2 + 1600$. The object hits the ground

when $h = 0$, so $h_0 = -16t^2 + 1600$ becomes $0 = -16t^2 + 1600$. We solve this for t.

$$0 = -16t^2 + 1600$$
$$16t^2 = 1600$$
$$t^2 = 100$$
$$t = \pm\sqrt{100}$$
$$t = 10 \quad (t = -10 \text{ is not a solution})$$

The object will hit the ground 10 seconds after it is dropped.

A ball is dropped from the top of a four-story building. The building is 48 feet tall. How long will it take for the ball to reach the ground?

Because the object is dropped, the initial velocity, v_0, is zero: $v_0 = 0$. The object is dropped from a height of 48 feet, so $h_0 = 48$. The formula $h = -16t^2 + v_0 t + h_0$ becomes $h = -16t^2 + 48$. The object hits the ground when $h = 0$.

$$h = -16t^2 + 48$$
$$0 = -16t^2 + 48$$
$$16t^2 = 48$$
$$t^2 = 3$$
$$t = \sqrt{3} \quad (t = -\sqrt{3} \text{ is not a solution})$$
$$t \approx 1.73$$

The ball will reach the ground in about 1.73 seconds.

PRACTICE

1. An object is dropped from over a bay. If the ball is dropped 56 feet above the surface of the water, how long will it take for the object to reach the water?

2. An object is dropped from the top of a 240-foot-tall observation tower. How long will it take for the object to reach the ground?

3. A ball is dropped from a sixth floor window at a height 70 feet. When will the ball hit the ground?

4. An object falls from the top of a 100-foot communications tower. How long will it take for the object to hit the ground?

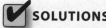 **SOLUTIONS**

For all of these problems, both a negative t and a positive t will be solutions for the quadratic equations. Only the positive t will be a true solution.

1. For the formula $h = -16t^2 + v_0t + h_0$, $h_0 = 56$ and $v_0 = 0$ (because the object is being dropped). The object reaches the ground when $h = 0$.

    $$h = -16t^2 + 56$$
    $$0 = -16t^2 + 56$$
    $$16t^2 = 56$$
    $$t^2 = \frac{56}{16}$$
    $$t^2 = \frac{7}{2}$$
    $$t = \sqrt{\frac{7}{2}}$$
    $$t \approx 1.87$$

    The object will reach the water in about 1.87 seconds.

2. For the formula $h = -16t^2 + v_0t + h_0$, $h_0 = 240$ and $v_0 = 0$ (because the object is being dropped). The object reaches the ground when $h = 0$.

    $$h = -16t^2 + 240$$
    $$0 = -16t^2 + 240$$
    $$16t^2 = 240$$
    $$t^2 = \frac{240}{16}$$
    $$t^2 = 15$$
    $$t = \sqrt{15}$$
    $$t \approx 3.87$$

    The object will reach the ground in about 3.87 seconds.

3. For the formula $h = -16t^2 + v_0t + h_0$, $h_0 = 70$ and $v_0 = 0$ (because the object is being dropped). The object reaches the ground when $h = 0$.

    $$h = -16t^2 + 70$$
    $$0 = -16t^2 + 70$$
    $$16t^2 = 70$$

$$t^2 = \frac{70}{16}$$

$$t^2 = \frac{35}{8}$$

$$t = \sqrt{\frac{35}{8}}$$

$$t \approx 2.09$$

The ball will hit the ground in about 2.09 seconds.

4. For the formula $h = -16t^2 + v_0t + h_0$, $h_0 = 100$ and $v_0 = 0$ (because the object is being dropped). The object reaches the ground when $h = 0$.

$$h = -16t^2 + 100$$

$$0 = -16t^2 + 100$$

$$16t^2 = 100$$

$$t^2 = \frac{100}{16}$$

$$t^2 = \frac{25}{4}$$

$$t = \frac{5}{2}$$

$$t = 2.5$$

The object will hit the ground after 2.5 seconds.

We now work with finding the length of time it takes for an object to reach a certain height after being dropped. We let h represent the height in question.

EXAMPLE

An object is dropped from the roof of a 60-foot building. When will it reach a height of 28 feet?

In the formula $h = -16t^2 + v_0t + h_0$, h_0 is 60 and v_0 is zero (because the object is dropped). The object reaches a height of 28 feet when $h = 28$.

$$h = -16t^2 + 60$$

$$28 = -16t^2 + 60$$

$$16t^2 = 32$$

$$t^2 = \frac{32}{16}$$

$$t^2 = 2$$

$$t = \sqrt{2} \qquad (t = -\sqrt{2} \text{ is not a solution})$$

$$t \approx 1.41$$

The object will reach a height of 28 feet after about 1.41 seconds.

PRACTICE

1. A ball is dropped from a height of 50 feet. How long after it is dropped will it reach a height of 18 feet?

2. A small object falls from a height of 200 feet. How long will it take to reach a height of 88 feet?

3. A small object is dropped from a 10th floor window (at a height of 110 feet). How long will it take for the object to pass the 3rd floor window (at a height of 35 feet)?

4. An object is dropped from 120 feet. How long will it take for the object to have fallen from 100 feet? (Hint: the height the object has reached after it has fallen 100 feet is 120 – 100 = 20 feet.)

SOLUTIONS

Negative values of t will not be solutions.

1. In the formula $h = -16t^2 + v_0 t + h_0$, $h_0 = 50$ and $v_0 = 0$.

    $$h = -16t^2 + 50$$

    We want to find t when $h = 18$.

    $$18 = -16t^2 + 50$$

    $$16t^2 = 32$$

    $$t^2 = \frac{32}{16}$$

    $$t^2 = 2$$

    $$t = \sqrt{2}$$

    $$t \approx 1.41$$

    The ball reaches a height of 18 feet about 1.41 seconds after it is dropped.

2. In the formula $h = -16t^2 + v_0t + h_0$, $h_0 = 200$ and $v_0 = 0$.

$h = -16t^2 + 200$

We want to find t when $h = 88$.

$88 = -16t^2 + 200$

$16t^2 = 112$

$t^2 = \dfrac{112}{16}$

$t^2 = 7$

$t \approx 2.65$

The object will reach a height of 88 feet after about 2.65 seconds.

3. In the formula $h = -16t^2 + v_0t + h_0$, $h_0 = 110$ and $v_0 = 0$.

$h = -16t^2 + 110$

We want to find t when $h = 35$.

$35 = -16t^2 + 110$

$16t^2 = 75$

$t^2 = \dfrac{75}{16}$

$t = \sqrt{\dfrac{75}{16}}$

$t \approx 2.17$

The object will pass the third floor window after about 2.17 seconds.

4. In the formula $h = -16t^2 + v_0t + h_0$, $h_0 = 120$ and $v_0 = 0$.

$h = -16t^2 + 120$

The object has fallen 100 feet when the height is 120 – 100 = 20 feet, so we want to find t when $h = 20$.

$20 = -16t^2 + 120$

$16t^2 = 100$

$t^2 = \dfrac{100}{16}$

$t = \sqrt{\dfrac{100}{16}}$

$t = \dfrac{10}{4}$

$t = 2.5$

The object will have fallen 100 feet 2.5 seconds after it is dropped.

Finally, we solve height problems for objects that are not dropped. We are given their initial velocity and height. We are either asked to find the time that the object will reach a specific height or to find when it will hit the ground. Because v_0 is not 0, either we will solve the quadratic equation by factoring or by using the quadratic formula.

 EXAMPLE

An object is tossed up in the air from the ground at the rate of 40 feet per second. How long will it take for the object to hit the ground?

In the formula $h = -16t^2 + v_0t + h_0$, $v_0 = 40$ and $h_0 = 0$.

$$h = -16t^2 + 40t$$

We want to find t when $h = 0$.

$$0 = -16t^2 + 40t$$
$$0 = t(-16t + 40)$$
$$-16t + 40 = 0 \quad t = 0 \quad \text{(This is when the object is tossed.)}$$
$$40 = 16t$$
$$\frac{40}{16} = t$$
$$\frac{5}{2} = t$$
$$2.5 = t$$

The object will hit the ground after 2.5 seconds.

A projectile is fired upward from the ground at an initial velocity of 60 feet per second. When will the projectile be 44 feet above the ground?

In the formula $h = -16t^2 + v_0t + h_0$, $v_0 = 60$ and $h_0 = 0$.

$$h = -16t^2 + 60t$$

We want to find t when $h = 44$.

$$44 = -16t^2 + 60t$$
$$16t^2 - 60t + 44 = 0$$

$$\frac{1}{4}(16t^2 - 60t + 44) = \frac{1}{4}(0)$$
$$4t^2 - 15t + 11 = 0$$
$$(t - 1)(4t - 11) = 0$$

$$4t - 11 = 0$$

$$t - 1 = 0 \qquad 4t = 11$$
$$t = 1 \qquad t = \frac{11}{4}$$
$$t = 2.75$$

The projectile will be 44 feet off the ground at 1 second (on the way up) and again at 2.75 seconds (on the way down).

PRACTICE

1. An object on the ground is thrown upward at the rate of 25 feet per second. After how much time will the object hit the ground?

2. A projectile is fired upward from the ground at the rate of 150 feet per second. How long will it take the projectile to fall back to the ground?

3. An object is thrown upward from the top of a 50-foot building. Its initial velocity is 20 feet per second. When will the object be 55 feet off the ground?

4. A projectile is fired upward from the top of a 36-foot building. Its initial velocity is 80 feet per second. When will it be 90 feet above the ground?

SOLUTIONS

1. In the formula $h = -16t^2 + v_0 t + h_0$, $v_0 = 25$ and $h_0 = 0$.

$h = -16t^2 + 25t$

We want to find t when $h = 0$.

$0 = -16t^2 + 25t$

$0 = t(-16t^2 + 25)$

$$-16t + 25 = 0 \quad t = 0 \quad \text{(This is when the object is thrown.)}$$

$$-16t = -25$$

$$t = \frac{-25}{-16}$$

$$t = \frac{25}{16}$$

$$t = 1.5625$$

The object will hit the ground after 1.5625 seconds.

2. In the formula $h = -16t^2 + v_0t + h_0$, $v_0 = 150$ and $h_0 = 0$.

$$h = -16t^2 + 150t$$

We want to find t when $h = 0$.

$$0 = -16t^2 + 150t$$

$$0 = t(-16t + 150)$$

$$-16t + 150 = 0 \quad t = 0 \quad \text{(This is when the projectile is fired.)}$$

$$-16t = -150$$

$$t = \frac{-150}{-16}$$

$$t = \frac{75}{8}$$

$$t = 9.375$$

The object will fall back to the ground after 9.375 seconds.

3. In the formula $h = -16t^2 + v_0t + h_0$, $v_0 = 20$ and $h_0 = 50$.

$$h = -16t^2 + 20t + 50$$

We want to find t when $h = 55$.

$$55 = -16t^2 + 20t + 50$$

$$0 = -16t^2 + 20t - 5$$

$$t = \frac{-20 \pm \sqrt{(20)^2 - 4(-16)(-5)}}{2(-16)} = \frac{-20 \pm \sqrt{400 - 320}}{-32}$$

$$= \frac{-20 \pm \sqrt{80}}{-32}$$

$$\approx \frac{-20 \pm 8.94}{-32} \approx 0.35, 0.90$$

The object will reach a height of 55 feet at about 0.35 seconds (on its way up) and again at about 0.90 seconds (on its way down).

4. In the formula $h = -16t^2 + v_0 t + h_0$, $v_0 = 80$ and $h_0 = 36$.

$$h = -16t^2 + 80t + 36$$

We want to find t when $h = 90$.

$$90 = -16t^2 + 80t + 36$$

$$0 = -16t^2 + 80t - 54$$

$$\frac{-1}{2}(0) = \frac{-1}{2}(-16t^2 + 80t - 54)$$

$$0 = 8t^2 - 40t + 27 = 0$$

$$t = \frac{-(-40) \pm \sqrt{(-40)^2 - 4(8)(27)}}{2(8)} = \frac{40 \pm \sqrt{1600 - 864}}{16}$$

$$= \frac{40 \pm \sqrt{736}}{16}$$

$$\approx \frac{40 \pm 27.13}{16} \approx 0.80, 4.20$$

The object will reach a height of 90 feet after about 0.80 seconds (on its way up) and again at about 4.20 seconds (on its way down).

Problems in Geometry

To solve word problems involving geometric shapes, we begin with the formula or formulas referred to in the problem. For example, after reading "The perimeter of a rectangular room ..." we write $P = 2L + 2W$. We then read the problem and substitute the numbers given in the problem for the variables. For example, after reading "The perimeter of the room is 50 feet ..." we replace P with 50. The formula then becomes and $50 = 2L + 2W$. In the statement, "Its width is two-thirds its length," we write $W = \frac{2}{3}L$ and the equation $50 = 2L + 2W$ becomes $50 = 2L + 2\left(\frac{2}{3}L\right)$.

The formulas we need in this section are listed below.

Rectangle Formulas

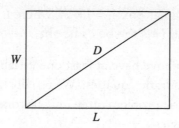

FIGURE 11-1

- Area $A = LW$
- Perimeter (the length around its sides) $P = 2L + 2W$
- Diagonal $D^2 = L^2 + W^2$

Triangle Formulas

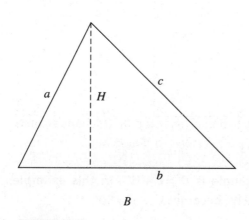

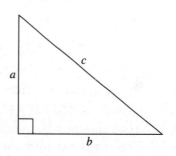

Right triangle

FIGURE 11-2

- Area $A = \dfrac{1}{2}BH$
- Perimeter $P = a + b + c$ (for any triangle)
- Pythagorean theorem $a^2 + b^2 = c^2$ (for right triangles only)

Formulas for Other Shapes

- Volume of a right circular cylinder $V = \pi r^2 h$, where r is the cylinder's radius, h is the cylinder's height.
- Surface area of a sphere (ball) is $SA = 4\pi r^2$, where r is the sphere's radius.

- Area of a circle $A = \pi r^2$, where r is the circle's radius.
- Volume of a rectangular box $V = LWH$, where L is the box's length, W is the box's width, and H is the box's height.

Most of the formulas above have at least one variable that is squared, which will have us solving a quadratic equation. We usually have two solutions to this equation but the geometric problem often only has one solution. We begin with squares and other rectangles.

 EXAMPLES

A square has a diameter of 50 cm. What is the length of each side?

Let x represent the length of each side.

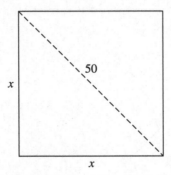

FIGURE 11-3

The diagonal formula for a rectangle is $D^2 = L^2 + W^2$. In this example, $D = 50$, $L = x$, and $W = x$, so $D^2 = L^2 + W^2$ becomes $x^2 + x^2 = 50^2$.

$$x^2 + x^2 = 50^2$$
$$2x^2 = 2500$$
$$x^2 = \frac{2500}{2}$$
$$x^2 = 1250$$
$$x = \sqrt{1250} \quad (x = -\sqrt{1250} \text{ is not a solution})$$
$$x = \sqrt{25^2 \cdot 2}$$
$$x = 25\sqrt{2}$$

Each side is $25\sqrt{2}$ cm long.

A rectangle is 1 inch longer than it is wide. Its diameter is 5 inches. What are its dimensions?

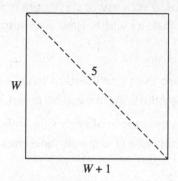

FIGURE 11-4

The diagonal formula for a rectangle is $D^2 = L^2 + W^2$. In this example, $D = 5$ and $L = W + 1$. $D^2 = L^2 + W^2$ becomes $5^2 = (W + 1)^2 + W^2$.

$$5^2 = (W + 1)^2 + W^2$$
$$25 = (W + 1)(W + 1) + W^2$$
$$25 = W^2 + 2W + 1 + W^2$$
$$25 = 2W^2 + 2W + 1$$
$$0 = 2W^2 + 2W - 24$$
$$\frac{1}{2}(0) = \frac{1}{2}(2W^2 + 2W - 24)$$
$$0 = W^2 + W - 12$$
$$0 = (W - 3)(W + 4)$$

$$W - 3 = 0 \qquad W + 4 = 0$$
$$W = 3 \qquad W = -4 \qquad \text{(not a solution)}$$

The width is 3 inches in the length is $3 + 1 = 4$ inches.

PRACTICE

1. The diameter of a square is 60 feet. What is the length of its sides?

2. A rectangle has one side 14 cm longer than the other. Its diameter is 34 cm. What are its dimensions?

3. The length of a rectangle is 7 inches more than its width. The diagonal is 17 inches. What are its dimensions?

4. The width of a rectangle is three-fourths its length. The diagonal is 10 inches. What are its dimensions?

5. The diameter of a rectangular classroom is 34 feet. The room's length is 14 feet longer than its width. How wide and long is the classroom?

SOLUTIONS

When there is more than one solution to an equation and one of them is not valid, only the valid solution will be given.

1. Let x represent the length of each side (in feet). The diagonal is 60 feet, so $D = 60$. The formula $D^2 = L^2 + W^2$ becomes $60^2 = x^2 + x^2$.

$$60^2 = x^2 + x^2$$
$$3600 = 2x^2$$
$$\frac{3600}{2} = x^2$$
$$1800 = x^2$$
$$\sqrt{1800} = x$$
$$\sqrt{30^2 \cdot 2} = x$$
$$30\sqrt{2} = x$$

The length of the square's sides is $30\sqrt{2}$ feet, or approximately 42.4 feet.

2. The length is 14 cm more than the width, so $L = W + 14$. The diameter is 34 cm, so $D = 34$. The formula $D^2 = L^2 + W^2$ becomes $34^2 = (W + 14)^2 + W^2$.

$$34^2 = (W + 14)^2 + W^2$$
$$1156 = (W + 14)(W + 14) + W^2$$
$$1156 = W^2 + 28W + 196 + W^2$$
$$1156 = 2W^2 + 28W + 196$$
$$0 = 2W^2 + 28W - 960$$
$$\frac{1}{2}(0) = \frac{1}{2}(2W^2 + 28W - 960)$$
$$0 = W^2 + 14W - 480$$
$$0 = (W - 16)(W + 30)$$

$W - 16 = 0 \qquad W + 30 = 0 \qquad$ (This does not lead to a solution.)

$\qquad W = 16$

The width is 16 cm and the length is $16 + 14 = 30$ cm.

3. The length is 7 inches more than the width, so $L = W + 7$. The diagonal is 17 inches. The formula $D^2 = L^2 + W^2$ becomes $17^2 = (W + 7)^2 + W^2$.

$$17^2 = (W + 7)^2 + W^2$$

$$289 = (W + 7)(W + 7) + W^2$$

$$289 = W^2 + 14W + 49 + W^2$$

$$289 = 2W^2 + 14W + 49$$

$$0 = 2W^2 + 14W - 240$$

$$\frac{1}{2}(0) = \frac{1}{2}(2W^2 + 14W - 240)$$

$$0 = W^2 + 7W - 120$$

$$0 = (W - 8)(W + 15)$$

$$W - 8 = 0 \qquad W + 15 = 0 \qquad \text{(This does not lead to a solution.)}$$

$$W = 8$$

The rectangle's width is 8 inches and its length is $8 + 7 = 15$ inches.

4. The width is three-fourths its length, so $W = \frac{3}{4}L$. The diagonal is 10 inches, so the formula $D^2 = L^2 + W^2$ becomes $10^2 = L^2 + \left(\frac{3}{4}L\right)^2$.

$$10^2 = L^2 + \left(\frac{3}{4}\right)^2 L^2$$

$$100 = L^2 + \frac{9}{16}L^2$$

$$100 = L^2\left(1 + \frac{9}{16}\right)$$

$$100 = L^2\left(\frac{16}{16} + \frac{9}{16}\right)$$

$$100 = \frac{25}{16}L^2$$

$$\frac{16}{25}(100) = L^2$$

$$64 = L^2$$

$$\sqrt{64} = L$$

$$8 = L$$

The rectangle's length is 8 inches and its width is $8\left(\frac{3}{4}\right) = 6$ inches.

5. The classroom's length is 14 feet more than its width, so $L = W + 14$. The diameter is 34 feet. The formula $D^2 = L^2 + W^2$ becomes $34^2 = (W + 14)^2 + W^2$.

$$34^2 = (W + 14)^2 + W^2$$
$$1156 = (W + 14)(W + 14) + W^2$$
$$1156 = W^2 + 28W + 196 + W^2$$
$$1156 = 2W^2 + 28W + 196$$
$$0 = 2W^2 + 28W - 960$$
$$\frac{1}{2}(0) = \frac{1}{2}(2W^2 + 28W - 960)$$
$$0 = W^2 + 14W - 480$$
$$0 = (W - 16)(W + 30)$$
$$W - 16 = 0 \qquad W + 30 = 0 \qquad \text{(This does not lead to a solution.)}$$
$$W = 16$$

The classroom is 16 feet wide and $16 + 14 = 30$ feet long.

We now work with other geometric shapes.

EXAMPLES

The area of a triangle is 40 in^2. Its height is four-fifths the length of its base. What are its base and height?

The area is 40 and $H = \frac{4}{5}B$ so the formula $A = \frac{1}{2}BH$ becomes $40 = \frac{1}{2}B\left(\frac{4}{5}B\right)$.

$$40 = \frac{1}{2}B\left(\frac{4}{5}B\right)$$
$$40 = \frac{1}{2} \cdot \frac{4}{5}B^2$$
$$40 = \frac{2}{5}B^2$$
$$\frac{5}{2} \cdot 40 = B^2$$
$$100 = B^2$$
$$10 = B$$

The triangle's base is 10 inches long and its height is $\left(\frac{4}{5}\right)(10) = 8$ inches.

The hypotenuse of a right triangle is 34 feet. The sum of the lengths of the two legs is 46 feet. Find the lengths of the legs. (The legs of a right triangle are the sides that form the 90° angle.)

The sum of the lengths of the legs is 46 feet, so if we let a and b represent the lengths of the legs, $a + b = 46$, so $a = 46 - b$. The hypotenuse is 34 feet so if c is the length of the hypotenuse, then the formula $a^2 + b^2 = c^2$ becomes $(46 - b)^2 + b^2 = 34^2$.

$$(46 - b)^2 + b^2 = 34^2$$
$$(46 - b)(46 - b) + b^2 = 1156$$
$$2116 - 92b + b^2 + b^2 = 1156$$
$$2b^2 - 92b + 2116 = 1156$$
$$2b^2 - 92b + 960 = 0$$
$$\frac{1}{2}(2b^2 - 92b + 960) = \frac{1}{2}(0)$$
$$b^2 - 46b + 480 = 0$$
$$(b - 30)(b - 16) = 0$$
$$b - 30 = 0 \qquad b - 16 = 0$$
$$b = 30 \qquad\qquad b = 16$$

One leg is 30 feet long and the other is 46 − 30 = 16 feet long.

A can's height is 4 inches and its volume is 28 in³. What is the can's radius?

The volume formula for a right circular cylinder is $V = \pi r^2 h$. The can's volume is 28 in³ and its height is 4 inches, so $V = \pi r^2 h$ becomes $28 = \pi r^2(4)$.

$$28 = \pi r^2 (4)$$
$$\frac{28}{4\pi} = r^2$$
$$\sqrt{\frac{28}{4\pi}} = r$$
$$1.493 \approx r$$

The can's radius is about 1.493 inches.

The volume of a box is 72 cm³. Its height is 3 cm. Its length is 1.5 times its width. What are the length and width of the box?

The formula for the volume of the box is $V = LWH$. The volume is 72, the height is 3 and the length is 1.5 times the width ($L = 1.5W$) so the formula becomes $72 = (1.5W)W(3)$. We now solve this equation for W.

$$72 = (1.5W)W(3)$$

$$72 = 4.5W^2$$

$$\frac{72}{4.5} = W^2$$

$$16 = W^2$$

$$4 = W$$

The box's width is 4 cm and its length is $(1.5)(4) = 6$ cm.

The surface area of a ball is 314 in². What is the ball's diameter?

The formula for the surface area of a sphere is $SA = 4\pi r^2$. The area is 314, so the formula becomes $314 = 4\pi r^2$.

$$314 = 4\pi r^2$$

$$\frac{314}{4\pi} = r^2$$

$$\sqrt{\frac{314}{4\pi}} = r$$

$$5 \approx r$$

The radius of the ball is approximately 5 inches. The diameter is twice the radius, so the diameter is approximately 10 inches.

The manufacturer of a drinking cup that is 6 inches tall is considering increasing its radius. The cup has straight sides (the top is the same size as the bottom). If the radius is increased by 1 inch, the new volume would be 169.6 in³. What is the cup's current radius?

The formula for the volume of a right circular cylinder is $V = \pi r^2 h$. The cup's height is 6. If the cup's radius is increased, the volume would be 169.6.

Let x represent the cup's current radius. Then the radius of the new cup would be $x + 1$. The volume formula becomes $169.6 = \pi(x+1)^2 6$.

$$169.6 = \pi(x+1)^2 6$$

$$\frac{169.6}{6\pi} = (x+1)^2 \quad \left(\frac{169.6}{6\pi} \approx 9\right)$$

$$9 = (x+1)^2$$

$$9 = (x+1)(x+1)$$

$$9 = x^2 + 2x + 1$$

$$0 = x^2 + 2x - 8$$

$$0 = (x-2)(x+4)$$

$x - 2 = 0 \qquad x + 4 = 0 \qquad$ (This does not lead to a solution.)

$\qquad x = 2$

The cup's current radius is approximately 2 inches.

PRACTICE

1. The area of a triangle is 12 in². The length of its base is two-thirds its height. What are the base and height?

2. The area of a triangle is 20 cm². The height is 3 cm more than its base. What are the base and height?

3. The sum of the base and height of a triangle is 14 inches. The area is 20 in². What are the base and height?

4. The hypotenuse of a right triangle is 85 cm long. One leg is 71 cm longer than the other. What are the lengths of its legs?

5. The manufacturer of a food can wants to increase the capacity of one of its cans. The can is 5 inches tall and its diameter is 6 inches. The manufacturer wants to increase the can's capacity by 50% and wants the can's height to remain 5 inches. How much does the diameter need to increase?

6. A pizza restaurant advertises that its large pizza is 20% larger than the competition's large pizza. The restaurant's large pizza is 16" in diameter. What is the diameter of the competition's large pizza?

✔ SOLUTIONS

1. The area formula for a triangle is $A = \frac{1}{2}BH$. The area is 12. The length of its base is two-thirds its height, so $B = \frac{2}{3}H$. The formula becomes $12 = \frac{1}{2}\left(\frac{2}{3}H\right)H$.

$$12 = \frac{1}{2}\left(\frac{2}{3}H\right)H$$

$$12 = \frac{1}{2}\cdot\frac{2}{3}H^2$$

$$12 = \frac{1}{3}H^2$$

$$3(12) = H^2$$

$$36 = H^2$$

$$6 = H$$

The height of the triangle is 6 inches. Its base is $\left(\frac{2}{3}\right)6 = 4$ inches.

2. The formula for the area of a triangle is $A = \frac{1}{2}BH$. The area is 20. The height is 3 cm more than the base, so $H = B + 3$. The formula becomes $20 = \frac{1}{2}B(B + 3)$.

$$20 = \frac{1}{2}B(B+3)$$

$$2(20) = B(B+3)$$

$$40 = B(B+3)$$

$$40 = B^2 + 3B$$

$$0 = B^2 + 3B - 40$$

$$0 = (B-5)(B+8)$$

$$B - 5 = 0 \qquad B + 8 = 0 \qquad \text{(This does not lead to a solution.)}$$

$$B = 5$$

The triangle's base is 5 cm and its height is $5 + 3 = 8$ cm.

3. The formula for the area of a triangle is $A = \frac{1}{2}BH$. The area is 20. The equation $B + H = 14$ gives us so $H = 14 - B$. The formula becomes $20 = \frac{1}{2}B(14 - B)$.

$$20 = \frac{1}{2}B(14 - B)$$

$$40 = B(14 - B)$$

$$40 = 14B - B^2$$

$$0 = 14B - B^2 - 40$$

$$-(0) = -(14B - B^2 - 40)$$

$$0 = -14B + B^2 + 40$$

$$0 = B^2 - 14B + 40$$

$$0 = (B - 10)(B - 4)$$

$$B - 10 = 0 \qquad B - 4 = 0$$

$$B = 10 \qquad B = 4$$

There are two triangles that satisfy the conditions. If the base is 10 inches, the height is $14 - 10 = 4$ inches. If the base is 4 inches, the height is $14 - 4 = 10$ inches.

4. By the Pythagorean theorem, $a^2 + b^2 = c^2$. The hypotenuse is c, so $c = 85$. One leg is 71 longer than the other so $a = b + 71$ ($b = a + 71$ also works). The Pythagorean theorem becomes $85^2 = (b + 71)^2 + b^2$.

$$85^2 = (b + 71)^2 + b^2$$

$$7225 = (b + 71)(b + 71) + b^2$$

$$7225 = b^2 + 142b + 5041 + b^2$$

$$7225 = 2b^2 + 142b + 5041$$

$$0 = 2b^2 + 142b - 2184$$

$$\frac{1}{2}(0) = \frac{1}{2}(2b^2 + 142b - 2184)$$

$$0 = b^2 + 71b - 1092$$

$$0 = (b - 13)(b + 84)$$

$$b - 13 = 0 \qquad b + 84 = 0 \qquad \text{(This does not lead to a solution.)}$$

$$b = 13$$

The shorter leg is 13 cm and the longer leg is $13 + 71 = 84$ cm.

5. Because the can's diameter is 6, the radius is 3. Let x represent the increase in the radius of the can. The radius of the new can is $3 + x$. The volume of the current can is $V = \pi r^2 h = \pi(3)^2 5 = 45\pi$. To increase the volume by 50% means to add half of 45π to itself; the new volume would be $45\pi + \dfrac{1}{2}45\pi = \dfrac{90\pi}{2} + \dfrac{45\pi}{2} = \dfrac{135\pi}{2}$. The volume formula for the new can becomes $\dfrac{135\pi}{2} = \pi(3+x)^2 5$.

$$\frac{135\pi}{2} = \pi(3+x)^2 5$$

$$\frac{1}{5\pi} \cdot \frac{135\pi}{2} = (3+x)^2$$

$$\frac{27}{2} = (3+x)^2$$

$$\frac{27}{2} = (3+x)(3+x)$$

$$\frac{27}{2} = 9 + 6x + x^2$$

$$2\left(\frac{27}{2}\right) = 2(9 + 6x + x^2)$$

$$27 = 18 + 12x + 2x^2$$

$$27 = 2x^2 + 12x + 18$$

$$0 = 2x^2 + 12x - 9$$

$$x = \frac{-12 \pm \sqrt{12^2 - 4(2)(-9)}}{2(2)} = \frac{-12 \pm \sqrt{144 + 72}}{4}$$

$$= \frac{-12 \pm \sqrt{216}}{4} = \frac{-12 \pm \sqrt{6^2 \cdot 6}}{4} = \frac{-12 \pm 6\sqrt{6}}{4}$$

$$= \frac{2(-6 \pm 3\sqrt{6})}{4} = \frac{-6 \pm 3\sqrt{6}}{2} \approx 0.674$$

(The other solution is negative.)

The manufacturer should increase the can's radius by about 0.674 inches. Because the diameter is twice the radius, the manufacturer should increase the can's diameter by about 2(0.674) = 1.348 inches.

6. A pizza's shape is circular so we need the area formula for a circle which is $A = \pi r^2$. The radius is half the diameter, so the restaurant's large pizza has a radius of 8 inches. The area of the restaurant's large pizza is $\pi(8)^2 = 64\pi \approx 201$. The restaurant's large pizza is 20% larger than the competition's large pizza. Let A represent the area of the competition's large pizza. Then 201 is 20% more than A:

$$201 = A + 0.20A = A(1 + 0.20) = 1.20A$$

$$201 = 1.20A$$

$$\frac{201}{1.20} = A$$

$$167.5 = A$$

$$A = \pi r^2$$

$$167.5 = \pi r^2$$

$$\frac{167.5}{\pi} = r^2$$

$$\sqrt{\frac{167.5}{\pi}} = r$$

$$7.3 \approx r$$

The competition's radius is approximately 7.3 inches, so its diameter is approximately $2(7.3) = 14.6$ inches.

Distance Problems

There are several distance problems that quadratic equations can help us solve. One of these types is the "stream" problem. In this type of problem, a vehicle travels the same distance up and back; in one direction, the "stream's" average speed is added to the vehicle's speed, and in the other, the "stream's" average speed is subtracted from the vehicle's speed. Another type involves two bodies moving away from each other where their paths form a right angle (for instance, one travels north and the other west). Finally, the last type is where a vehicle makes a round trip that takes longer in one direction than in the other. In all of these types, the formula $D = rt$ is key.

"Stream" distance problems usually involve boats (traveling upstream or downstream) and planes (traveling against a headwind or with a tailwind). The boat or plane generally travels in one direction then turns around and travels in the opposite direction. The distance upstream and downstream is usually the same. If r represents the boat's or plane's average speed traveling without the "stream," then "$r +$ stream's speed" represents the boat's or plane's average speed traveling with the stream, and "$r -$ stream's speed" represents the boat's or plane's average speed traveling against the stream. Usually, we are told the total trip time, that is, the time downstream and upstream. If this is the case, then we use the model "Total time = Time downstream + Time upstream," where $\text{Time downstream} = \dfrac{\text{Distance downstream}}{\text{Average speed + stream's speed}}$ and $\text{Time upstream} = \dfrac{\text{Distance upstream}}{\text{Average speed} - \text{stream's speed}}$. You might recognize that these representations come from $D = rt$ and $t = \dfrac{D}{r}$. These facts give us the following equation to solve:

$$\text{Total time} = \frac{\text{Distance downstream}}{\text{Average speed + stream's speed}} + \frac{\text{Distance upstream}}{\text{Average speed} - \text{stream's speed}}$$

EXAMPLE

Miami and Pittsburgh are 1000 miles apart. A plane flew into a 50-mph headwind from Miami to Pittsburgh. On the return flight the 50-mph wind became a tailwind. The plane was in the air a total of $4\frac{1}{2}$ hours for the round trip. What would have been the plane's average speed without the wind?

Let r represent the plane's average speed (in mph) without the wind. The plane's average speed against the wind is $r - 50$ (from Miami to Pittsburgh) and the plane's average speed with the wind is $r + 50$ (from Pittsburgh to Miami). The distance from Miami to Pittsburgh is 1000 miles. With this information we can use $t = \dfrac{D}{r}$ to compute the time in the air in each direction. The time in the air from Miami to Pittsburgh is $\dfrac{1000}{r - 50}$. The time in the air from

Pittsburgh to Miami is $\dfrac{1000}{r+50}$. The time in the air from Miami to Pittsburgh plus the time in the air from Pittsburgh to Miami is $4\frac{1}{2} = 4.5$ hours. The equation to solve is $\dfrac{1000}{r-50} + \dfrac{1000}{r+50} = 4.5$. The LCD is $(r-50)(r+50)$.

$$(r-50)(r+50)\dfrac{1000}{r-50} +$$

$$(r-50)(r+50)\dfrac{1000}{r+50} = (r-50)(r+50)(4.5)$$

$$1000(r+50) + 1000(r-50) = 4.5[(r-50)(r+50)]$$

$$1000r + 50{,}000 + 1000r - 50{,}000 = 4.5(r^2 - 2500)$$

$$2000r = 4.5r^2 - 11{,}250$$

$$0 = 4.5r^2 - 2000r - 11{,}250$$

$$r = \dfrac{-(-2000) \pm \sqrt{(-2000)^2 - 4(4.5)(-11{,}250)}}{2(4.5)}$$

$$= \dfrac{2000 \pm \sqrt{4{,}000{,}000 + 202{,}500}}{9} = \dfrac{2000 \pm \sqrt{4{,}202{,}500}}{9}$$

$$= \dfrac{2000 \pm 2050}{9} = \dfrac{2000 + 2050}{9}, \dfrac{2000 - 2050}{9} = 450$$

$(-\frac{50}{9}$ is not a solution.) The plane's average speed without the wind is 450 mph.

PRACTICE

1. A flight from Dallas to Chicago is 800 miles. A plane flew with a 40-mph tailwind from Dallas to Chicago. On the return trip, the plane flew against the same 40-mph wind. The plane was in the air a total of 5.08 hours for the flight from Dallas to Chicago and the return flight. What would have been the plane's speed without the wind?

2. A flight from Houston to New Orleans faced a 50-mph headwind, which became a 50-mph tailwind on the return flight. The total time in the air was $1\frac{3}{4}$ hours. The distance between Houston and New Orleans is 300 miles. How long was the plane in flight from Houston to New Orleans?

3. A small motorboat traveled 15 miles downstream then turned around and traveled 15 miles back. The total trip took 2 hours. The stream's speed is 4 mph. How fast would the boat have traveled in still water?

4. A plane on a flight from Denver to Indianapolis flew with a 20-mph tailwind. On the return flight, the plane flew into a 20-mph headwind. The distance between Denver and Indianapolis is 1000 miles and the plane was in the air a total of $5\frac{1}{2}$ hours. What would have been the plane's average speed without the wind?

5. A plane flew from Minneapolis to Atlanta, a distance of 900 miles, against a 30 mph-headwind. On the return flight, the 30-mph wind became a tailwind. The plane was in the air for a total of $5\frac{1}{2}$ hours. What would the plane's average speed have been without the wind?

SOLUTIONS

1. Let r represent the plane's average speed (in mph) with no wind. Then the average speed from Dallas to Chicago (with the tailwind) is $r + 40$, and the average speed from Chicago to Dallas is $r - 40$ (against the headwind). The distance between Dallas and Chicago is 800 miles. The time in the air from Dallas to Chicago plus the time in the air from Chicago to Dallas is 5.08 hours. The time in the air from Dallas to Chicago is $\frac{800}{r + 40}$. The time in the air from Chicago to Dallas is $\frac{800}{r - 40}$. The equation to solve is $\frac{800}{r + 40} + \frac{800}{r - 40} = 5.08$. The LCD is $(r + 40)(r - 40)$.

$$(r + 40)(r - 40)\frac{800}{r + 40} +$$

$$(r + 40)(r - 40)\frac{800}{r - 40} = (r + 40)(r - 40)(5.08)$$

$$800(r - 40) + 800(r + 40) = 5.08[(r - 40)(r + 40)]$$

$$800r - 32{,}000 + 800r + 32{,}000 = 5.08(r^2 - 1600)$$

$$1600r = 5.08r^2 - 8128$$

$$0 = 5.08r^2 - 1600r - 8128$$

$$r = \frac{-(-1600) \pm \sqrt{(-1600)^2 - 4(5.08)(-8128)}}{2(5.08)}$$

$$= \frac{1600 \pm \sqrt{2{,}560{,}000 + 165{,}160.96}}{10.16}$$

$$= \frac{1600 \pm \sqrt{2{,}725{,}160.96}}{10.16} \approx \frac{1600 \pm 1650.806155}{10.16}$$

$$\approx 320 \left(\frac{1600 - 1650.806155}{10.16} \text{ is negative} \right)$$

The plane's average speed without the wind would have been about 320 mph.

2. Let r represent the plane's average speed without the wind. The average speed from Houston to New Orleans (against the headwind) is $r - 50$, and the average speed from New Orleans to Houston (with the tailwind) is $r + 50$. The distance between Houston and New Orleans is 300 miles. The time in the air from Houston to New Orleans is $\dfrac{300}{r - 50}$, and the time in the air from New Orleans to Houston is $\dfrac{300}{r + 50}$. The time in the air from Houston to New Orleans plus the time in the air from New Orleans to Houston is $1\frac{3}{4} = \frac{7}{4}$ hours. The equation to solve is $\dfrac{300}{r - 50} + \dfrac{300}{r + 50} = \dfrac{7}{4}$. The LCD is $4(r - 50)(r + 50)$.

$$4(r - 50)(r + 50)\frac{300}{r - 50} +$$

$$4(r - 50)(r + 50)\frac{300}{r + 50} = 4(r - 50)(r + 50)\frac{7}{4}$$

$$1200(r + 50) + 1200(r - 50) = 7[(r - 50)(r + 50)]$$

$$1200r + 60{,}000 + 1200r - 60{,}000 = 7(r^2 - 2500)$$

$$2400r = 7r^2 - 17{,}500$$

$$0 = 7r^2 - 2400r - 17{,}500$$

$$r = \frac{-(-2400) \pm \sqrt{(-2400)^2 - 4(7)(-17{,}500)}}{2(7)}$$

$$= \frac{2400 \pm \sqrt{5{,}760{,}000 + 490{,}000}}{14}$$

$$= \frac{2400 \pm \sqrt{6{,}250{,}000}}{14} = \frac{2400 \pm 2500}{14}$$

$$= 350 \left(\frac{2400 - 2500}{14} \text{ is negative} \right)$$

The average speed of the plane without the wind was 350 mph. We want the time in the air from Houston to New Orleans: $\dfrac{300}{r - 50} = \dfrac{300}{350 - 50} = \dfrac{300}{300} = 1$ hour. The plane was in flight from Houston to New Orleans for 1 hour.

3. Let r represent the boat's speed in still water. The average speed downstream is $r + 4$ and the average speed upstream is $r - 4$. The boat was in the water a total of 2 hours. The distance traveled in each direction is 15 miles. The time the boat traveled downstream is $\dfrac{15}{r + 4}$ hours, and it traveled upstream $\dfrac{15}{r - 4}$ hours. The time the boat traveled upstream plus the time it traveled downstream equals 2 hours. The equation to solve is $\dfrac{15}{r + 4} + \dfrac{15}{r - 4} = 2$. The LCD is $(r + 4)(r - 4)$.

$$(r+4)(r-4)\frac{15}{r+4}+(r+4)(r-4)\frac{15}{r-4}=2(r+4)(r-4)$$
$$15(r-4)+15(r+4)=2[(r+4)(r-4)]$$
$$15r-60+15r+60=2(r^2-16)$$
$$30r=2r^2-32$$
$$0=2r^2-30r-32$$
$$\frac{1}{2}(0)=\frac{1}{2}(2r^2-30r-32)$$
$$0=r^2-15r-16$$
$$0=(r-16)(r+1)$$

$r-16=0 \qquad r+1=0 \qquad$ (This does not lead to a solution.)

$\quad r=16$

The boat's average speed in still water is 16 mph.

4. Let r represent the plane's average speed without the wind. The plane's average speed from Denver to Indianapolis is $r + 20$, and the plane's average speed from Indianapolis to Denver is $r - 20$. The total time in flight is $5\frac{1}{2}$ hours and the distance between Denver and Indianapolis is 1000 miles. The time in the air from Denver to Indianapolis is $\dfrac{1000}{r + 20}$ hours and the time in the air from Indianapolis to Denver is $\dfrac{1000}{r - 20}$ hours. The time in the air from Denver to Indianapolis plus the time in the air from Indianapolis to Denver is 5.5 hours. The equation to solve is $\dfrac{1000}{r + 20} + \dfrac{1000}{r - 20} = 5.5$. The LCD is $(r + 20)(r - 20)$.

$$(r+20)(r-20)\frac{1000}{r+20}+(r+20)(r-20)\frac{1000}{r-20}=(r+20)(r-20)(5.5)$$
$$1000(r-20)+1000(r+20)=5.5[(r-20)(r+20)]$$

$$1000r - 20{,}000 + 1000r + 20{,}000 = 5.5(r^2 - 400)$$

$$2000r = 5.5r^2 - 2200$$

$$0 = 5.5r^2 - 2000r - 2200$$

$$r = \frac{-(-2000) \pm \sqrt{(-2000)^2 - 4(5.5)(-2200)}}{2(5.5)}$$

$$= \frac{2000 \pm \sqrt{4{,}000{,}000 + 48{,}400}}{11}$$

$$= \frac{2000 \pm \sqrt{4{,}048{,}400}}{11} \approx \frac{2000 \pm 2012.063617}{11} \approx 365,$$

$$\left(\frac{2000 - 2012.063617}{11} \text{ is negative} \right)$$

The plane would have averaged about 365 mph without the wind.

5. Let r represent the plane's average speed without the wind. The plane's average speed from Minneapolis to Atlanta (against the headwind) is $r - 30$. The plane's average speed from Atlanta to Minneapolis (with the tailwind) is $r + 30$. The total time in the air is $5\frac{1}{2}$ hours and the distance between Atlanta to Minneapolis is 900 miles. The time in the air from Minneapolis to Atlanta is $\dfrac{900}{r - 30}$ hours, and the time in the air from Atlanta to Minneapolis is $\dfrac{900}{r + 30}$ hours. The time in the air from Minneapolis to Atlanta plus the time in the air from Minneapolis to Atlanta is $5\frac{1}{2} = 5.5$ hours. The equation to solve is $\dfrac{900}{r - 30} + \dfrac{900}{r + 30} = 5.5$. The LCD is $(r - 30)(r + 30)$.

$$(r - 30)(r + 30)\frac{900}{r - 30} + (r - 30)(r + 30)\frac{900}{r + 30} = (r - 30)(r + 30)(5.5)$$

$$900(r + 30) + 900(r - 30) = 5.5[(r - 30)(r + 30)]$$

$$900r + 27{,}000 + 900r - 27{,}000 = 5.5(r^2 - 900)$$

$$1800r = 5.5r^2 - 4950$$

$$0 = 5.5r^2 - 1800r - 4950$$

$$r = \frac{-(-1800) \pm \sqrt{(-1800)^2 - 4(5.5)(-4950)}}{2(5.5)}$$

$$= \frac{1800 \pm \sqrt{3{,}240{,}000 + 108{,}900}}{11} = \frac{1800 \pm \sqrt{3{,}348{,}900}}{11}$$

$$= \frac{1800 \pm 1830}{11} = 330 \left(\frac{1800 - 1830}{11} \text{ is negative} \right)$$

The plane's average speed without the wind was 330 mph.

If we need to find the distance between two bodies traveling at right angles away from each other, we must use the Pythagorean theorem, $a^2 + b^2 = c^2$ in combination with the distance formula, $D = rt$. We use the distance formula to represent the lengths of sides of a right triangle. For example, if a car is traveling 60 mph, then the length that it has traveled is $60t$ miles, where t is in hours. Because the bodies are traveling at right angles to each other, their paths form two sides of a right triangle and the distance between them represents the hypotenuse of the right triangle. This is where the Pythagorean theorem comes in, with a and b representing the distance traveled by each body (in terms of t) and c representing the distance between them.

 EXAMPLE

A car passes under a railway trestle at the same time a train is crossing the trestle. The car is headed south at an average speed of 40 mph. The train is traveling east at an average speed of 30 mph. After how long will the car and train be 10 miles apart?

Let t represent the number of hours after the train and car pass each other. (Because the rate is given in miles per hour, time must be given in hours.) The distance traveled by the car after t hours is $40t$ and that of the train is $30t$.

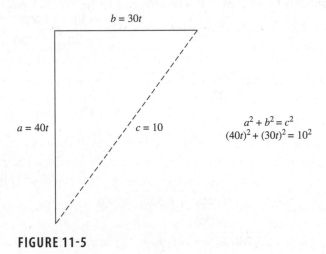

FIGURE 11-5

$$(40t)^2 + (30t)^2 = 10^2$$

$$1600t^2 + 900t^2 = 100$$

$$2500t^2 = 100$$

$$t^2 = \frac{100}{2500}$$

$$t = \sqrt{\frac{100}{2500}}$$

$$t = \frac{10}{50} = \frac{1}{5}$$

After $\frac{1}{5}$ of an hour (or 12 minutes) the car and train will be 10 miles apart.

PRACTICE

1. A car and plane leave an airport at the same time. The car travels eastward at an average speed of 45 mph. The plane travels southward at an average speed 200 mph. When will they be 164 miles apart?

2. Two joggers begin jogging from the same point. One jogs south at the rate of 8 mph and the other jogs east at a rate of 6 mph. When will they be 5 miles apart?

3. A cross-country bicyclist crosses a railroad track just after a train passed. The train is traveling southward at an average speed of 60 mph. The bicyclist is traveling westward at an average speed of 11 mph. When will they be 244 miles apart?

4. A motor scooter and car left a parking lot at the same time. The motor scooter traveled north at 24 mph. The car traveled west at 45 mph. How long did it take for the scooter and car to be 34 miles apart?

5. Two cars pass each other at 4:00 at an overpass. One car is headed north at an average speed of 60 mph and the other is headed east at an average speed of 50 mph. At what time will the cars be 104 miles apart? Give your solution to the nearest minute.

SOLUTIONS

1. Let t represent the number of hours each has traveled. The plane's distance after t hours is 200t and the car's distance is 45t.

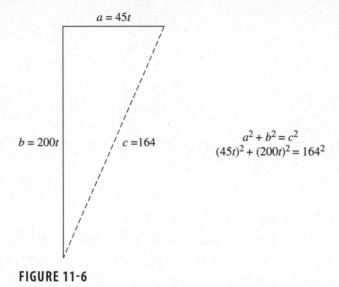

FIGURE 11-6

$$(45t)^2 + (200t)^2 = 164^2$$

$$2025t^2 + 40{,}000t^2 = 26{,}896$$

$$42{,}025t^2 = 26{,}896$$

$$t^2 = \frac{26{,}896}{42{,}025}$$

$$t = \sqrt{\frac{26{,}896}{42{,}025}}$$

$$t = 0.80$$

The car and plane will be 164 miles apart after 0.80 hours or 48 minutes.

2. Let t represent the number of hours after the joggers began jogging. The distance covered by the southbound jogger after t hours is $8t$, and the distance covered by the eastbound jogger is $6t$.

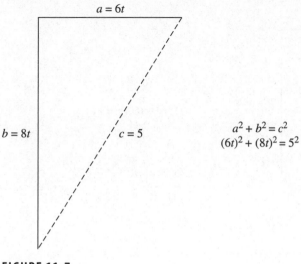

$$a = 6t$$

$$b = 8t \qquad c = 5$$

$$a^2 + b^2 = c^2$$
$$(6t)^2 + (8t)^2 = 5^2$$

FIGURE 11-7

$$(6t)^2 + (8t)^2 = 5^2$$

$$36t^2 + 64t^2 = 25$$

$$100t^2 = 25$$

$$t^2 = \frac{25}{100}$$

$$t^2 = \frac{1}{4}$$

$$t = \sqrt{\frac{1}{4}}$$

$$t = \frac{1}{2}$$

The joggers will be five miles apart after $\frac{1}{2}$ hour or 30 minutes.

3. Let t represent the number of hours after the cyclist crosses the track. The distance traveled by the bicycle after t hours is $11t$ and the distance traveled by the train is $60t$.

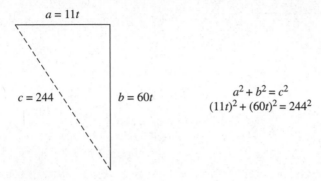

$$a^2 + b^2 = c^2$$
$$(11t)^2 + (60t)^2 = 244^2$$

FIGURE 11-8

$$(11t)^2 + (60t)^2 = 244^2$$
$$121t^2 + 3600t^2 = 59{,}536$$
$$3721t^2 = 59{,}536$$
$$t^2 = \frac{59{,}536}{3721}$$
$$t^2 = 16$$
$$t = \sqrt{16}$$
$$t = 4$$

After 4 hours the cyclist and train will be 244 miles apart.

4. Let t represent the number of hours after the scooter and car left the parking lot. The car's distance after t hours is $45t$. The scooter's distance is $24t$.

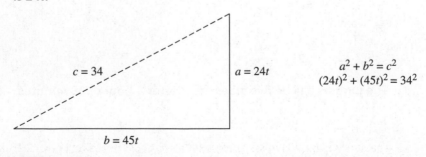

$$a^2 + b^2 = c^2$$
$$(24t)^2 + (45t)^2 = 34^2$$

FIGURE 11-9

$$(24t)^2 + (45t)^2 = 34^2$$
$$576t^2 + 2025t^2 = 1156$$
$$2601t^2 = 1156$$
$$t^2 = \frac{1156}{2601}$$
$$t^2 = \frac{4}{9}$$
$$t = \sqrt{\frac{4}{9}}$$
$$t = \frac{2}{3}$$

The car and scooter will be 34 miles apart after $\frac{2}{3}$ of an hour or 40 minutes.

5. Let t represent the number of hours after the cars passed the overpass. The northbound car's distance after t hours is $60t$ and the eastbound car's distance is $50t$.

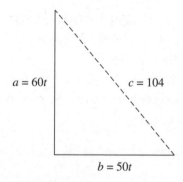

$$a^2 + b^2 = c^2$$
$$(60t)^2 + (50t)^2 = 104^2$$

FIGURE 11-10

$$(60t)^2 + (50t)^2 = 104^2$$
$$3600t^2 + 2500t^2 = 10{,}816$$
$$6100t^2 = 10{,}816$$
$$t^2 = \frac{10{,}816}{6100}$$
$$t^2 = \frac{2704}{1525}$$
$$t = \sqrt{\frac{2704}{1525}}$$
$$t \approx 1.33$$

The cars will be 104 miles apart after about 1.33 hours or 1 hour 20 minutes. The time will be about 5:20.

Round-Trip Problems

In the following problems, people are making a round trip. The average speed in each direction is different, and the total trip time is given. The equation we want to solve is

Time to destination + Time on return trip = Total trip time

To get the time to and from the destination, we again use $t = \dfrac{D}{r}$, from $D = rt$. The equation that we want to solve becomes

$$\frac{\text{Distance}}{\text{Rate to destination}} + \frac{\text{Distance}}{\text{Rate on return trip}} = \text{Total trip time}$$

 EXAMPLE

A jogger jogged seven miles to a park then jogged home. He jogged 1 mph faster to the park than he jogged on the way home. The round trip took 2 hours 34 minutes. How fast did he jog to the park?

Let r represent the jogger's average speed on the way home. He jogged 1 mph faster to the park, so $r + 1$ represents his average speed to the park. The distance to the park is 7 miles, so $D = 7$.

Time to the park + Time home = 2 hours 34 minutes

The time to the park is represented by $t = \dfrac{7}{r+1}$. The time home is represented by $t = \dfrac{7}{r}$. The round trip is 2 hours 34 minutes $= 2\frac{34}{60} = 2\frac{17}{30} = \dfrac{77}{30}$ hours.

The equation to solve becomes $\dfrac{7}{r+1} + \dfrac{7}{r} = \dfrac{77}{30}$.

The LCD is $30r(r+1)$.

$$30r(r+1)\frac{7}{r+1} + 30r(r+1)\frac{7}{r} = 30r(r+1)\frac{77}{30}$$

$$210r + 210(r+1) = 77r(r+1)$$

$$210r + 210r + 210 = 77r^2 + 77r$$

$$420r + 210 = 77r^2 + 77r$$

$$0 = 77r^2 - 343r - 210$$

$$\frac{1}{7}(0) = \frac{1}{7}(77r^2 - 343r - 210)$$

$$0 = 11r^2 - 49r - 30$$

$$0 = (r - 5)(11r + 6)$$

$$r - 5 = 0 \qquad 11r + 6 = 0 \qquad \text{(This does not lead to a solution.)}$$

$$r = 5$$

The jogger's average speed to the park was $5 + 1 = 6$ mph.

PRACTICE

1. A man rode his bike 6 miles to work. The wind reduced his average speed on the way home by 2 mph. The round trip took 1 hour 21 minutes. How fast was he riding on the way to work?

2. On a road trip a saleswoman traveled 120 miles to visit a customer. She averaged 15 mph faster to the customer than on the return trip. She spent a total of 4 hours 40 minutes driving. What was her average speed on the return trip?

3. A couple walked on the beach from their house to a public beach four miles away. They walked 0.2 mph faster on the way home than on the way to the public beach. They walked for a total of 2 hours 35 minutes. How fast did they walk home?

4. A family drove from Detroit to Buffalo, a distance of 215 miles, for the weekend. They averaged 10 mph faster on the return trip. They spent a total of seven hours on the road. What was their average speed on the trip from Detroit to Buffalo? (Give your solution accurate to one decimal place.)

5. Boston and New York are 190 miles apart. A professor drove from his home in Boston to a conference in New York. On the return trip, he faced heavy traffic and averaged 17 mph slower than on his way to New York. He spent a total of 8 hours 5 minutes on the road. How long did his trip from Boston to New York last?

SOLUTIONS

1. Let r represent the man's average speed on the way to work. Then $r - 2$ represents the man's average speed on his way home. The distance each way is 6 miles, so the time he rode to work is $\frac{6}{r}$, and the time he rode

home is $\dfrac{6}{r-2}$. The total time is 1 hour 21 minutes $= 1\frac{21}{60} = 1\frac{7}{20} = \dfrac{27}{20}$ hours. The equation to solve is $\dfrac{6}{r} + \dfrac{6}{r-2} = \dfrac{27}{20}$. The LCD is $20r(r-2)$.

$$20r(r-2)\dfrac{6}{r} + 20r(r-2)\dfrac{6}{r-2} = 20r(r-2)\dfrac{27}{20}$$

$$120(r-2) + 120r = 27r(r-2)$$

$$120r - 240 + 120r = 27r^2 - 54r$$

$$240r - 240 = 27r^2 - 54r$$

$$0 = 27r^2 - 294r + 240$$

$$\dfrac{1}{3}(0) = \dfrac{1}{3}(27r^2 - 294r + 240)$$

$$0 = 9r^2 - 98r + 80$$

$$0 = (r-10)(9r-8)$$

$$r - 10 = 0 \qquad 9r - 8 = 0 \qquad \text{(This does not lead to a solution.)}$$

$$r = 10$$

The man's average speed on his way to work was 10 mph.

2. Let r represent the saleswoman's average speed on her return trip. Her average speed on the way to the customer is $r + 15$. The distance each way is 120 miles. She spent a total of 4 hours 40 minutes $= 4\frac{40}{60} = 4\frac{2}{3} = \dfrac{14}{3}$ hours driving. The time spent driving to the customer is $\dfrac{120}{r+15}$. The time spent driving on the return trip is $\dfrac{120}{r}$. The equation to solve is $\dfrac{120}{r+15} + \dfrac{120}{r} = \dfrac{14}{3}$. The LCD is $3r(r+15)$.

$$3r(r+15) \cdot \dfrac{120}{r+15} + 3r(r+15) \cdot \dfrac{120}{r} = 3r(r+15) \cdot \dfrac{14}{3}$$

$$360r + 360(r+15) = 14r(r+15)$$

$$360r + 360r + 5400 = 14r^2 + 210r$$

$$720r + 5400 = 14r^2 + 210r$$

$$0 = 14r^2 - 510r - 5400$$

$$\dfrac{1}{2}(0) = \dfrac{1}{2}(14r^2 - 510r - 5400)$$

$$0 = 7r^2 - 255r - 2700$$

$$r = \frac{-(-255) \pm \sqrt{(-255)^2 - 4(7)(-2700)}}{2(7)}$$

$$= \frac{255 \pm \sqrt{65,025 + 75,600}}{14} = \frac{255 \pm \sqrt{140,625}}{14} = \frac{255 \pm 375}{14}$$

$$r = 45 \quad \left(r = \frac{255 - 375}{14} \text{ is not a solution} \right)$$

The saleswoman averaged 45 mph on her return trip.

3. Let r represent the couple's average rate on their way home, then $r - 0.2$ represents the couple's average speed to the public beach. The distance to the public beach is 4 miles. They walked for a total of 2 hours 35 minutes $= 2\frac{35}{60} = 2\frac{7}{12} = \frac{31}{12}$ hours. The time spent walking to the public beach is $\dfrac{4}{r - 0.2}$. The time spent walking home is $\dfrac{4}{r}$. The equation to solve is $\dfrac{4}{r - 0.2} + \dfrac{4}{r} = \dfrac{31}{12}$. The LCD is $12r(r - 0.2)$.

$$12r(r - 0.2)\frac{4}{r - 0.2} + 12r(r - 0.2)\frac{4}{r} = 12r(r - 0.2)\frac{31}{12}$$

$$48r + 48(r - 0.2) = 31r(r - 0.2)$$

$$48r + 48r - 9.6 = 31r^2 - 6.2r$$

$$96r - 9.6 = 31r^2 - 6.2r$$

$$0 = 31r^2 - 102.2r + 9.6$$

(Multiplying by 10 to clear the decimals would result in fairly large numbers for the quadratic formula.)

$$r = \frac{-(-102.2) \pm \sqrt{(-102.2)^2 - 4(31)(9.6)}}{2(31)}$$

$$= \frac{102.2 \pm \sqrt{10,444.84 - 1190.4}}{62} = \frac{102.2 \pm \sqrt{9254.44}}{62}$$

$$= \frac{102.2 \pm 96.2}{62} = \frac{16}{5}, \frac{3}{31} \quad \left(\frac{3}{31} \text{ is not a solution.} \right)$$

The couple walked home at the rate of $\dfrac{16}{5} = 3.2$ mph.

4. Let r represent the average speed from Detroit to Buffalo. The average speed from Buffalo to Detroit is $r + 10$. The distance from Detroit to Buffalo is 215 miles and the total time the family spent driving is 7 hours. The time spent driving from Detroit to Buffalo is $\dfrac{215}{r}$. The time

spent driving from Buffalo to Detroit is $\dfrac{215}{r+10}$. The equation to solve is
$\dfrac{215}{r} + \dfrac{215}{r+10} = 7$. The LCD is $r(r+10)$.

$$r(r+10)\frac{215}{r} + r(r+10)\frac{215}{r+10} = r(r+10)7$$

$$215(r+10) + 215r = 7r(r+10)$$

$$215r + 2150 + 215r = 7r^2 + 70r$$

$$430r + 2150 = 7r^2 + 70r$$

$$0 = 7r^2 - 360r - 2150$$

$$r = \frac{-(-360) \pm \sqrt{(-360)^2 - 4(7)(-2150)}}{2(7)}$$

$$= \frac{360 \pm \sqrt{129,600 + 60,200}}{14} = \frac{360 \pm \sqrt{189,800}}{14}$$

$$\approx \frac{360 \pm 435.66}{14} \approx 56.8 \text{ mph} \quad \left(\frac{360 - 435.66}{14} \text{ is not a solution.}\right)$$

The family averaged 56.8 mph from Detroit to Buffalo.

5. Let r represent the average speed on his trip from Boston to New York.
Because his average speed was 17 mph slower on his return trip,
$r - 17$ represents his average speed on his trip from New York to
Boston. The distance between Boston and New York is 190 miles. The
time on the road from Boston to New York is $\dfrac{190}{r}$ and the time on the
road from New York to Boston is $\dfrac{190}{r-17}$. The time on the road from
Boston to New York plus the time on the road from New York to Boston is
8 hours 5 minutes $= 8\frac{5}{60} = 8\frac{1}{12} = \frac{97}{12}$ hours. The equation to solve is
$\dfrac{190}{r} + \dfrac{190}{r-17} = \dfrac{97}{12}$. The LCD is $12r(r-17)$.

$$12r(r-17)\frac{190}{r} + 12r(r-17)\frac{190}{r-17} = 12r(r-17)\frac{97}{12}$$

$$12(190)(r-17) + 12(190)r = 97r(r-17)$$

$$2280(r-17) + 2280r = 97r^2 - 1649r$$

$$2280r - 38,760 + 2280r = 97r^2 - 1649r$$

$$4560r - 38,760 = 97r^2 - 1649r$$

$$0 = 97r^2 - 6209r + 38,760$$

$$r = \frac{-(-6209) \pm \sqrt{(-6209)^2 - 4(97)(38{,}760)}}{2(97)}$$

$$= \frac{6209 \pm \sqrt{38{,}551{,}681 - 15{,}038{,}880}}{194} = \frac{6209 \pm \sqrt{23{,}512{,}801}}{194}$$

$$= \frac{6209 \pm 4849}{194} = \frac{6209 + 4849}{194}, \frac{6209 - 4849}{194} = 7\tfrac{1}{97}, 57$$

The rate cannot be $7\tfrac{1}{97}$ because the total round trip is only 8 hours 5 minutes. The professor's average speed from Boston to New York is 57 mph. We want his time on the road from Boston to New York. His time on the road from Boston to New York is $\dfrac{190}{r} = \dfrac{190}{57} = 3\tfrac{1}{3}$ hours or 3 hours 20 minutes.

Summary

In this chapter, we learned how to:

- *Solve number sense problems that involved products.* We learned how to solve number sense problems for which we are told either the sum/difference between two numbers or that they were consecutive and are then told their product. We solve the product equation, which is reduced to one variable with the information given in the problem. The product equation is a quadratic equation.

- *Solve revenue problems.* When given information about how a price increase/decrease affects sales, we are asked what price will bring in some desired revenue. We let x represent the number of increases/decreases in the price, so the price and quantity sold can be represented by x. We then use the model $R = PQ$, where P represents the price, and Q represents the quantity. The revenue equation becomes a quadratic equation.

- *Solve work, distance, and geometry problems with quadratic equations.* We solved some problem types, such as work, distance, and geometry, in this chapter using the same strategies that we learned in Chapter 8. The only difference is that the models are quadratic equations.

- *Solve special distance problems.* We used the Pythagorean theorem to solve distance problems in which the bodies were moving on paths that form a right angle. For stream problems, we used the following model.

$$\text{Total time} = \frac{\text{Distance downstream}}{\text{Average speed} + \text{stream's speed}} +$$
$$\frac{\text{Distance upstream}}{\text{Average speed} - \text{stream's speed}}$$

- *Solve falling object problems.* If an object is dropped, thrust upward, or thrown downward, its height t seconds after release, is $h = -16t^2 + v_0 t + h_0$, where v_0 is the object's initial velocity (in feet per second); and h_0, its initial height (in feet). We are given enough information to eliminate all variables except t and then we solve the equation for t.

QUIZ

1. The Marshalls own a hotel in a small coastal city. Their daughter, Ava, can clean all the rooms in four for fewer hours than her little brother, Aidan, can clean them. Working together, they can clean all the rooms in 2 hours 40 minutes. Working alone, how long would Ava need to clean all the rooms?

    A. 4 hours
    B. 5 hours
    C. 6 hours
    D. 7 hours

2. A truck, traveling west at 42 mph, goes over an overpass at the same instant a northbound train is passes underneath the overpass. The train is traveling at 40 mph. At these speeds, how long will it take the truck and train to be 29 miles apart?

    A. 20 minutes
    B. 30 minutes
    C. 40 minutes
    D. 50 minutes

3. The area of a rectangle is 90 cm^2. If its length is 13 cm longer than its width, how long is the rectangle?

    A. 15 cm
    B. 16 cm
    C. 17 cm
    D. 18 cm

4. The sum of two numbers is 18 and their product is 77. What is the larger number?

    A. 10
    B. 11
    C. 12
    D. 13

5. An object is dropped from a height of 125 feet. It will hit the ground in about

    A. 2.8 seconds.
    B. 3 seconds.
    C. 3.2 seconds.
    D. The answer cannot be determined.

6. The volume of a rectangular box is 432 in^3. Its length is 8 inches, and its height is $1\frac{1}{2}$ times its width. How tall is the box?

    A. 4 inches
    B. 5 inches
    C. 7 inches
    D. 9 inches

7. Marco drove 210 miles to be a best man at his friend's wedding. The drive there and back took a total of 6 hours 25 minutes. On the way back, he faced traffic and averaged 12 mph slower than the trip to the wedding. What was Marco's average speed on the return trip?

   A. 60 mph
   B. 62 mph
   C. 64 mph
   D. 66 mph

8. A campus sorority sells school shirts at all home basketball games. At $20 per shirt, they average 80 shirts per game. According to a research project conducted by a marketing class, they will sell 10 more shirts for each $1 decrease in the price. What price should they charge so that their average revenue per game is $1960?

   A. $14
   B. $15
   C. $16
   D. $18

9. The hypotenuse of a right triangle is 37 meters. The base is 23 meters longer than its height. What is the height of the triangle?

   A. 12 m
   B. 15 m
   C. 18 m
   D. 20 m

10. An object is tossed upward from the ground at the rate of 48 feet per second. After how long will the object be 10 feet off the ground?

    A. About 0.20 and 2.80 seconds.
    B. About 0.35 and 2.65 seconds.
    C. About 0.23 and 2.77 seconds.
    D. The object will never reach a height of 10 feet.

11. A small motorboat traveled downstream for 8 miles and then turned around and traveled back upstream for 8 miles. The total trip took 3 hours. If the stream's rate is 2 mph, how fast would the motorboat have traveled in still water?

    A. 3 mph
    B. 4 mph
    C. 5 mph
    D. 6 mph

Final Exam

1. Brooklyn's grade in her accounting class is based on four tests and a final exam. The final exam counts twice as much as a regular test. Brooklyn's test grades are 82, 91, 86, and 89. What must she make on the final exam to receive a course grade of 90?

    A. 96   **B.** 98   **C.** 100   **D.** She would need a grade higher than 100.

2. If $5(2x - 3) - 2\left(\dfrac{1}{2}x - 1\right) = 2x + 5$, then

    A. $x = \dfrac{18}{7}$ **B.** $x = \dfrac{22}{7}$ **C.** $x = \dfrac{10}{7}$ **D.** $x = 3$

3. $\dfrac{x - 8}{x^2 - 64} =$

    A. $\dfrac{1}{x - 8}$ **B.** $\dfrac{1}{x + 8}$ **C.** $\dfrac{x}{x - 8}$ **D.** $\dfrac{x - 8}{x^2 - 64}$ cannot be simplified.

4. $(5x^4)^3 =$

    A. $5x^7$ **B.** $5x^{12}$ **C.** $125x^7$ **D.** $125x^{12}$

5. A rectangle's length is 3 m longer than its width. The area is 180 m². How long is the rectangle?

    A. 12 m   **B.** 13 m   **C.** 14 m   **D.** 15 m

6. If $\dfrac{3}{4}x - \dfrac{1}{6} = \dfrac{5}{2}$, then

   A. $x = \dfrac{32}{9}$ B. $x = \dfrac{2}{3}$ C. $x = \dfrac{10}{3}$ D. $x = \dfrac{34}{9}$

7. A cash register contains $5.99 in dimes, nickels, quarters, and pennies. There are 2 more dimes than nickels, 3 more quarters than nickels, and twice as many pennies as nickels. How many dimes are there?

   A. 12   B. 14   C. 16   D. 20

8. $\sqrt[3]{x^4} =$

   A. $x^{4/3}$ B. $x^{3/4}$ C. $-x^{-4/3}$ D. $x^{-3/4}$

9. A small boat traveled 6 miles upstream and later returned. The stream current was 3 mph, and the total travel time was 1 hour, 4 minutes. What would the boat's speed have been in still water?

   A. 9 mph   B. 10 mph   C. 11 mph   D. 12 mph

10. If $3x^2 + 2x - 7 = 0$, then

   A. $x = -2 \pm \dfrac{\sqrt{22}}{3}$ B. $x = \dfrac{-1 \pm 2\sqrt{22}}{3}$ C. $x = \dfrac{-1 \pm \sqrt{22}}{3}$
   D. There are no solutions.

11. Lisa drove 36 miles to visit her sister. For the first part of the drive, she averaged 40 mph and then traffic became lighter, allowing her to average 48 mph. The total trip lasted 50 minutes. How far did she travel on the first part of her trip, the part for which she averaged 40 mph?

   A. 16 miles   B. 18 miles   C. 20 miles   D. 22 miles

12. Which of the following is the solution to $5 - 2x > 3$?
   A. $(-1, \infty)$ B. $(-\infty, -1)$ C. $(-\infty, 1)$ D. $(1, \infty)$

13. $x^2 - \dfrac{9}{25} =$

   A. $\left(x - \dfrac{3}{5}\right)^2$ B. $\left(x - \dfrac{3}{5}\right)\left(x + \dfrac{3}{5}\right)$ C. $(x - 9)\left(x - \dfrac{1}{25}\right)$

   D. $x^2 - \dfrac{9}{25}$ cannot be factored.

14. Janna wants to rent a car to go to a job interview in another state. She has a choice of two plans. At \$36 per day, she has unlimited miles. At \$24 per day, she will have to pay 5 cents per mile. For what average daily mileage is the unlimited mileage plan less expensive?

A. At most 180 miles   B. More than 180 miles   C. At most 240 miles
D. More than 240 miles

15. $27x^6 =$

A. $\dfrac{1}{3}(3x^2)^4$ B. $(3x^3)^2$ C. $3(3x^2)^3$ D. $(3x^2)^3$

16. If $\dfrac{3}{8}(4x+1) = \dfrac{5}{4}$, then

A. $x = \dfrac{3}{4}$ B. $x = \dfrac{7}{12}$ C. $x = -\dfrac{3}{4}$ D. $x = \dfrac{1}{96}$

17. A couple has a \$35 gift certificate at a popular restaurant. The sales tax rate is 8%. They plan to tip 20% (before tax). What is the most they can order so that the gift certificate pays for the entire meal?

A. \$27.34   B. \$26.69   C. \$28.53   D. \$25.87

18. If $x^2 - 5x - 50 = 0$, then

A. $x = 5, -10$ B. $x = -5, 10$ C. $x = 5, 10$ D. $x = -5, -10$

19. The sum of two consecutive positive integers is 25. What is their product?

A. 132   B. 182   C. 154   D. 156

20. $(3x+5)(2x+3) =$

A. $5x^2 + 19x + 8$ B. $6x^2 + 19x + 15$ C. $6x^2 + 15$ D. $9x^2 + 19x + 15$

21. $-2x^4 - 6x^3y + 10x^2y^2 + 16x^3y^2 =$

A. $-2x^2(x^2 + 3xy - 5y^2 - 8xy^2)$ B. $-2x^2(x^2 + 3xy + 5y^2 + 8xy^2)$
C. $-2x^2(x^2 - 3xy - 5y^2 - 8xy^2)$ D. $-2x^2(x^2 - 3xy - 5y^2 + 8xy^2)$

22. An experienced quality assurance employee can inspect a box of bolts in 10 minutes. A trainee needs 15 minutes to inspect the same sized box. If they work together, how long will it take them to inspect the box of bolts?

A. 5 minutes   B. 6 minutes   C. 8 minutes   D. There is not enough information to answer the question.

23. $\dfrac{2x}{x^2 - x - 12} + \dfrac{3}{x^2 - 16} =$

A. $\dfrac{2x^2 - 5x + 12}{(x+3)(x-4)^2}$ B. $\dfrac{2x^2 - 7x + 12}{(x+3)(x-4)^2}$ C. $\dfrac{2x^2 + 11x + 9}{(x+3)(x-4)(x+4)}$

D. $\dfrac{2x^2 + 9x + 11}{(x+3)(x-4)(x+4)}$

24. The division problem $3.28\overline{)10.574}$ can be rewritten as

A. $3280\overline{)10{,}574}$ B. $328\overline{)10{,}574}$ C. $328\overline{)105{,}740}$ D. $3280\overline{)1{,}057{,}400}$

25. Completely factor $2x^3 + 5x^2 - 8x - 20$.
   A. $(x^2 - 4)(2x + 5)$ B. $(x - 2)(x + 2)(2x - 5)$ C. $(x - 2)(x + 2)(2x + 5)$
   D. $2x^3 + 5x^2 - 8x - 20$ cannot be factored.

26. The difference of two positive numbers is 3 and their product is 340. What is the smaller number?
   A. 16   B. 17   C. 18   D. 19

27. $20x^3\sqrt{x} - 15x\sqrt{x} + 5x^2\sqrt{x} =$
   A. $-5x\sqrt{x}(-4x^2 - 3 - x)$ B. $-5x\sqrt{x}(-4x^2 + 3 + x)$
   C. $-5x\sqrt{x}(4x^2 - 3 + x)$ D. $-5x\sqrt{x}(-4x^2 + 3 - x)$

28. A retired teacher wants to invest an inheritance worth $160,000 in two separate investments: a municipal bond, paying 3% annual interest, and a corporate bond, paying $4\frac{1}{2}$% annual interest. If she wants an annual income from these investments to total $5775, how much should she invest in the corporate CD?
   A. $65,000   B. $70,000   C. $75,000   D. $80,000

29. $\dfrac{1}{5x^2} =$

   A. $\dfrac{1}{5}x^2$ B. $-\dfrac{1}{5}x^2$ C. $\left(\dfrac{1}{5}x\right)^{-2}$ D. $\dfrac{1}{5}x^{-2}$

30. $-12x^2 - 4x + 20 =$
   A. $-4(3x^2 - x - 5)$ B. $-4(3x^2 + x + 5)$ C. $-4(3x^2 + x - 5)$
   D. $-4(3x^2 - x + 5)$

31. A group of students collected change to help feed animals at a shelter. There was $12.25 of change donated, dimes, nickels, and quarters. There were 4 more quarters than dimes and twice as many dimes as nickels. How many nickels were donated?

    A. 12   **B.** 15   **C.** 18   **D.** 20

32. $\dfrac{3}{4x} + \dfrac{2}{5x-2} =$

    A. 0   **B.** $\dfrac{3}{5x^2 - 2x}$ **C.** $\dfrac{15x - 3}{20x^2 - 8x}$ **D.** $\dfrac{23x - 6}{20x^2 - 8x}$

33. If $2.25x + 1.2 = 1.5x - 0.8$, then

    A. $x = -\dfrac{8}{3}$ **B.** $x = -\dfrac{3}{2}$ **C.** $x = \dfrac{16}{39}$ **D.** $x = \dfrac{8}{15}$

34. A room's length is $1\frac{1}{4}$ times its width. Its perimeter is 90 feet. How wide is the room?

    A. 15 feet   **B.** 20 feet   **C.** 25 feet   **D.** 30 feet

35. $\dfrac{\frac{x^2-4}{9}}{\frac{(x-2)^2}{6}} =$

    A. $\dfrac{2x + 4}{3x - 6}$ **B.** $\dfrac{2}{3}$ **C.** $\dfrac{(x-2)^3(x+2)}{54}$ **D.** -2

36. A large pump can drain a reservoir in 5 hours. If the large pump is used together with a smaller pump, they can drain the reservoir in 3 hours. How long would it take for the smaller pump to empty the reservoir by itself?

    A. 7.5 hours   **B.** 6.5 hours   **C.** 8 hours   **D.** 7 hours

37. $4(6x - 5)^4 + 9x(6x - 5)^3 =$

    A. $(33x - 5)(6x - 5)^3$ **B.** $(15x - 20)(6x - 5)^3$ **C.** $(15x - 5)(6x - 5)^3$
    D. $(33x - 20)(6x - 5)^3$

38. The radius of a circle is decreased by 5 inches, which decreased its area by 165π in². What is the radius of the original circle?

    A. 17 inches   **B.** 18 inches   **C.** 19 inches   **D.** 20 inches

39. What is the solution to $4x + 7 \le 31$?

    A. $x \le \dfrac{19}{2}$ **B.** $x \le 6$ **C.** $x < \dfrac{19}{2}$ **D.** $x \ge 6$

40. $4x^{-2} =$

    A. $\dfrac{1}{16x^2}$ B. $\dfrac{4}{x^2}$ C. $\dfrac{1}{4x^2}$ D. $-\dfrac{1}{4x^2}$

41. $\dfrac{1}{8(3x-10)} =$

    A. $\dfrac{1}{8} \cdot \dfrac{1}{3x-10}$ B. $\dfrac{1}{8}(3x-10)$ C. $8\left(\dfrac{1}{3x-10}\right)$ D. $8(3x-10)$

42. $\dfrac{9}{4x} =$

    A. $9\left(\dfrac{1}{4}x\right)^{-1}$ B. $9\left(\dfrac{4}{x}\right)^{-1}$ C. $9(4x)^{-1}$ D. $\dfrac{9}{4}x$

43. A rectangular box is 8 inches tall. Its length is 1 inch longer than its width. The volume of the box is 240 in^3. How wide is the box?

    A. 5 inches   B. 6 inches   C. 8 inches   D. 10 inches

44. Write $\dfrac{4}{3(2x-1)}$ as a product.

    A. $\dfrac{4}{3}(2x-1)$ B. $\dfrac{4}{3}\left(\dfrac{1}{2x-1}\right)$ C. $4\left(\dfrac{3}{2x-1}\right)$ D. $\dfrac{4}{3(2x-1)}$ cannot be written as a product.

45. If $\dfrac{3x-2}{x+1} = \dfrac{2-x}{5x}$, then

    A. $x = \dfrac{11 \pm \sqrt{249}}{32}$ B. $x = \dfrac{11 \pm \sqrt{233}}{28}$ C. $x = \dfrac{9 \pm \sqrt{209}}{32}$ D. $x = \dfrac{1 \pm \sqrt{3}}{4}$

46. Simplify $\dfrac{10}{6-x}$.

    A. $\dfrac{5}{3-x}$ B. $\dfrac{5}{3} - \dfrac{1}{x}$ C. $\dfrac{5}{3} - \dfrac{10}{x}$ D. $\dfrac{10}{6-x}$ cannot be simplified.

47. A nurse is ordered to administer 10 ml of a solution containing 18% concentrate of a drug. She has vials of 12% concentrate and 20% concentrate. How much of each concentrate should she mix together to produce 10 ml of a 12% concentrate?

    A. 2.5 ml of 12% and 7.5 ml of 20%   B. 3 ml of 12% and 7 ml of 20%
    C. 3.5 ml of 12% and 6.5 ml of 20%   D. 4 ml of 12% and 6 ml of 20%

48. $6x^2 + 5x + 1 =$

   A. $(3x+1)(3x+1)$ B. $(3x+1)(2x+1)$ C. $(5x+1)(x+1)$

   D. $6x^2 + 5x + 1$ cannot be factored.

49. John is $1\frac{1}{3}$ times as old as Jimmy. John is also 4 years older than Robert. The sum of their ages is 62. How old is Robert?

   A. 16 years   B. 18 years   C. 20 years   D. 22 years

50. $\dfrac{1}{x^2 - 1} + \dfrac{3}{x+1} + \dfrac{5}{x^2 + 2x - 3} =$

   A. $\dfrac{3x^2 - 6x + 1}{(x-1)(x+1)(x-3)}$ B. $\dfrac{3x^2 + 6x + 5}{(x-1)(x+1)(x+3)}$

   C. $\dfrac{9x + 15}{(x-1)(x+1)(x+3)}$ D. $\dfrac{3x^2 + 12x - 1}{(x-1)(x+1)(x+3)}$

51. A highway crosses a riding path that runs along a river. At 5:00, a car passes over the riding path, heading west at 60 mph. At the same instant, a bicyclist is riding on the path southward at 11 mph. Assuming they maintain their speed and direction, when will they be 6.1 miles apart?

   A. At 5:04   B. At 5:06   C. At 5:08   D. At 5:10

52. The owner of a shopping village has leased all 80 units, charging monthly rent of $2400. Because of a new shopping center being built, he feels he will lose 5 tenants for every $100 increase in the rent. The owner of the shopping village expects monthly revenue to be $168,000. What rent should he charge so that the shopping village earns enough revenue?

   A. $2500   B. $2600   C. $2700   D. $2800

53. If $x^2 + 6x = 4$, then

   A. $x = -3 \pm \sqrt{13}$ B. $x = -3 \pm 2\sqrt{13}$ C. $x = -3 \pm \sqrt{5}$ D. $x = -3 \pm 2\sqrt{5}$

54. The perimeter of a right triangle is 154 cm. One leg is 23 cm longer than the other. The hypotenuse is 9 cm longer than the longer leg. How long is the hypotenuse?

   A. 56 cm   B. 59 cm   C. 62 cm   D. 65 cm

55. $\dfrac{48xy^3z^2}{25xy^2z} \div \dfrac{42x^3yz^3}{55xyz} =$

A. $\dfrac{88x^2yz^4}{35}$ B. $\dfrac{88x^2}{35yz}$ C. $\dfrac{88y}{35x^2z}$ D. $\dfrac{88yz}{35x^2}$

56. $\dfrac{6}{\sqrt[4]{9}} =$

A. $\dfrac{2\sqrt[4]{9}}{3}$ B. 2 C. $2\sqrt[4]{3}$ D. $2\sqrt[4]{9}$

57. If $-3(x+2) = \dfrac{3}{2}$, then

A. $x = -\dfrac{1}{6}$ B. $x = -\dfrac{1}{3}$ C. $x = \dfrac{1}{6}$ D. $x = -\dfrac{5}{2}$

58. If $(3x+2)^2 = (x+5)(3x+2)$, then

A. $x = \dfrac{3}{2}$ only   B. $x = \dfrac{3}{2}, -\dfrac{2}{3}$ C. $x = -\dfrac{2}{3}, -\dfrac{5}{2}$ D. $x = -5, -\dfrac{2}{3}$

59. Keisha has a 98 homework average in her math class. Her test grades are 84, 85, and 88. If the homework average counts 10%, each tests counts 20%, and the final exam counts 30%, what is the lowest grade she can get on the final to have at least a 90 course average?

A. 92   B. 94   C. 96   D. 98

60. If $2x^2 + 8x + 3 = 0$, then

A. $x = \dfrac{-4 \pm \sqrt{10}}{2}$ B. $x = \dfrac{-2 \pm \sqrt{10}}{2}$ C. $x = -2 \pm 2\sqrt{10}$

D. $x = -8 \pm \dfrac{\sqrt{10}}{2}$

61. $\sqrt[5]{\sqrt{x}} =$

A. $x^{1/10}$ B. $x^{1/7}$ C. $x^{1/3}$ D. $x^{2/5}$

62. $\dfrac{1}{\sqrt{x^2+9}} =$

A. $\dfrac{1}{x+3}$ B. $(x^2+9)^{-1/2}$ C. $\dfrac{\sqrt{x^2-9}}{x^2+9}$ D. $-(x^2+9)^{1/2}$

63. Mary Ann bought a pair of boots that were marked down 25%. If she paid $128.40 after a sales tax was 7% added, what was the original price of the boots?

A. $150.00 B. $151.50 C. $153.60 D. $160.00

64. $\dfrac{8}{x-3} + \dfrac{5}{3-x} =$

A. $\dfrac{3}{3-x}$ B. $\dfrac{-3x+9}{(x-3)^2}$ C. $\dfrac{3}{x-3}$ D. $\dfrac{1}{1-x}$

65. Which interval is the solution for $\dfrac{1}{2} < \dfrac{3x-1}{4} < 5$?

A. $\left(\dfrac{5}{3}, 7\right]$ B. $(-\infty, 1) \cup (7, \infty)$ C. $(1, 7)$ D. $(1, 2)$

66. $\sqrt[3]{54x^{10}y^6} =$

A. $3x^3y^2\sqrt[3]{2x}$ B. $3x^2y\sqrt[3]{2x}$ C. $9x^3y^2\sqrt[3]{2x}$ D. $9x^2y\sqrt[3]{2x}$

67. What is 60 increased by 20%?

A. 68 B. 69 C. 70 D. 72

68. The interval $(-\infty, 6)$ is represented by

69. $\dfrac{x^2+12x+32}{x^2-x-20} =$

A. $\dfrac{x+8}{x+5}$ B. $\dfrac{3x+8}{x+5}$ C. $\dfrac{x+8}{x-5}$ D. $\dfrac{x^2+12x+32}{x^2-x-20}$ cannot be simplified.

70. At 1:50, a news van passed a certain mile marker, heading southward at 60 mph. Ten minutes later, a second news van, going to the same story, passed the same mile marking at 72 mph. Assuming the vans maintain their direction and speed, when will the second news van catch up to the first?

   A. 2:40   B. 2:45   C. 2:50   D. 2:55

71. Completely factor $81x^4 - 16$.

   A. $(9x^2 - 4)^2$ B. $(3x - 2)^2(3x + 2)^2$ C. $(9x^2 - 4)^2$

   D. $(3x - 2)(3x + 2)(9x^2 + 4)$

72. $(6x + 5)^2 =$
   A. $36x^2 + 25$ B. $36x^2 + 60x + 25$ C. $6x^2 + 25$ D. $36x^2 + 30x + 25$

73. Peanuts and a nut mixture containing 60% peanuts will be mixed together to produce 12 pounds of a mixture containing 75% peanuts. How much of the pure peanuts should be used?

   A. 4.5 pounds   B. 4.75 pounds   C. 5 pounds   D. 5.25 pounds

74. Solve for P in the equation $R = \dfrac{kP}{s^2}$.

   A. $P = \dfrac{s^2}{kR}$ B. $P = \dfrac{Rs^2}{k}$ C. $P = \dfrac{kR}{s^2}$ D. $P = \dfrac{k}{Rs^2}$

75. Completely factor $x^4 + 3x^2 - 4$.
   A. $(x - 1)(x + 1)(x^2 + 4)$ B. $(x - 1)(x + 1)(x - 2)(x + 2)$
   C. $(x - 1)^2(x^2 + 4)$ D. $(x - 1)^2(x + 4)^2$

76. Simplify $\dfrac{4x^2y + 20x^3y + 16xy^3}{5x^3y^3 - 10xy + 20xy^2}$.

   A. $\dfrac{4x + 20x^2 + 16y^2}{5x^2y^2 - 10 + 20y}$ B. $\dfrac{4x + x^2 + 16y^2}{5x^2y^2 - 10 + y}$ C. $\dfrac{4 + 20x + 16y^2}{5xy^2 - 10 + 20y}$

   D. $\dfrac{4x + 20x + 16y^2}{5x^2y^2 + 20y}$

77. 160 is what percent of 25?
   A. 640%   B. 156.25%   C. 400%   D. 135%

78. $\left(\dfrac{4x^2 y^{-3}}{3x^{-2} y^3}\right)^{-2} =$

    A. $\dfrac{9}{16}$ B. $\dfrac{9}{16x^8 y^{12}}$ C. $\dfrac{9y^{12}}{16x^8}$ D. $\dfrac{9x^8}{16y^{12}}$

79. If $x^2 - 3x = 54$, then

    A. $x = 54, 57$ B. $x = 9, -6$ C. $x = -9, 6$ D. There is no solution.

80. If $\sqrt{6x - 5} = 7$, then

    A. $x = \dfrac{7}{3}$ B. $x = \dfrac{2\sqrt{2}}{3}$ C. $x = 2$ D. $x = 9$

81. If $-\dfrac{3}{4}x^2 + \dfrac{1}{2}x + 1 = 0$, then

    A. $x = \dfrac{-1 \pm \sqrt{13}}{3}$ B. $x = \dfrac{1 \pm \sqrt{13}}{3}$ C. $x = -1, \dfrac{4}{3}$ D. $x = 1, -\dfrac{4}{3}$

82. Solve for m in the equation $t = \dfrac{4(m + w)}{s}$.

    A. $m = 4st - w$ B. $m = st - \dfrac{1}{4}w$ C. $m = \dfrac{st - w}{4}$ D. $m = \dfrac{st - 4w}{4}$

83. $\dfrac{15}{\sqrt{3}} =$

    A. $\dfrac{5\sqrt{3}}{3}$ B. $\dfrac{5}{\sqrt{3}}$ C. $15\sqrt{3}$ D. $5\sqrt{3}$

84. $(2x^3 y^{-2} z)^3 (3x^2 y^4 z^{-2})^{-2} =$

    A. $\dfrac{8x^2 z^7}{9y^8}$ B. $\dfrac{8x^5 z^7}{9y^{14}}$ C. $\dfrac{8x^6 z^8}{9y}$ D. $\dfrac{8y^{48} z^{12}}{9x^{36}}$

85. A restaurant owner wants to increase the size of his giant pizza from 16 inches to 17 inches in diameter. How much more pizza is there in the giant?

    A. About 3.14 in²   B. About 25.92 in²   C. About 103.67 in²
    D. About 54.26 in²

86. $\dfrac{1}{\sqrt[3]{x^2}} =$

    A. $-x^{2/3}$ B. $x^{-3/2}$ C. $x^{-2/3}$ D. $-x^{3/2}$

87. A spirit club raises money by selling hamburgers at football games. They want to avoid handling change, so they want the hamburger to cost $5 *after* adding $7\frac{1}{2}\%$ sales tax. What should they charge for the hamburger?

    A. $4.65   B. $4.70   C. $4.75   D. $4.80

88. $4x^{-2}y^3(5xy^{-4} + x^3y - 2x^2y^{-3} + 1) =$

    A. $20x^{-1}y^{-1} + 4x^{-6}y^4 - 8x + 4x^{-2}y^3$ B. $20xy + 4xy^4 - 8y + 4x^{-2}y^3$

    C. $20x^{-1}y^{-1} + 4xy^4 - 8 + 4x^{-2}y^3$ D. $20x^3y^7 - 4x^5y^4 + 2x^4y^6 - 4x^2y^3$

89. Fat free milk and whole milk (4%) will be mixed together to produce 5 gallons of 1% milk. How much fat free milk will be used?

    A. 3 gallons   B. 3.25 gallons   C. 3.50 gallons   D. 3.75 gallons

90. $\dfrac{7x^{-4}y^3}{14x^{-2}y^4} =$

    A. $\dfrac{x^2}{2y^2}$ B. $\dfrac{1}{2} \cdot \dfrac{1}{x^2y}$ C. $\dfrac{1}{2}x^2y^{-1}$ D. $\dfrac{1}{2} \cdot \dfrac{1}{x^8y^{12}}$

91. $\dfrac{2}{x^2(x^2 + 2x + 1)} - \dfrac{3}{x(x^2 + 5x + 4)} =$

    A. $\dfrac{-3x^2 - x + 8}{x^2(x + 1)^2(x + 4)}$ B. $\dfrac{-3x^2 + 5x + 8}{x^2(x + 1)^2(x + 4)}$ C. $\dfrac{-x + 5}{x^2(x + 1)^2(x + 4)}$

    D. $\dfrac{-x + 11}{x^2(x + 1)^2(x + 4)}$

92. The diameter of a rectangular room is 29 feet. The room is 1 foot longer than it is wide. How wide is the room?

    A. 20 feet   B. 21 feet   C. 22 feet   D. 23 feet

93. If $5x^2 + 6x - 8 = 0$, then

    A. $x = \dfrac{2}{5}, -4$ B. $x = -\dfrac{2}{5}, 4$ C. $x = -\dfrac{4}{5}, 2$ D. $x = \dfrac{4}{5}, -2$

94. $\dfrac{1}{8x^2} + \dfrac{1}{12x^3} + \dfrac{1}{20} =$

A. $\dfrac{6x^2 + 10 + 15x}{120x^3}$ B. $\dfrac{6x^3 + 10 + 15x}{120x^3}$ C. $\dfrac{18x^3 + 5x + 12}{150x^3}$

D. $\dfrac{18x^3 + 10x + 12}{150x^3}$

95. Working together, Anna and Jean can paint a room in 2 hours, 24 minutes. Working alone, Anna would need 2 hours more than Jean would need to paint the room by herself. How long would Anna need to paint the room by herself?

A. 4 hours   B. 5 hours   C. 6 hours   D. 7 hours

96. $\dfrac{8}{\sqrt{4x^2}} =$

A. $\dfrac{4}{x}$ B. $\dfrac{2}{x}$ C. $-2x$ D. $-4x$

97. A piggybank contains \$3.95 in change. There are 2 more quarters than dimes, and 3 more dimes than nickels. How many nickels are there?

A. 4   B. 6   C. 8   D. 10

98. If $\dfrac{x}{x-5} + \dfrac{6}{x^2 - 4x - 5} = \dfrac{x}{x+1}$, then

A. $x = -\dfrac{6}{5}$ B. $x = -1$ C. $x = -\dfrac{7}{5}$ D. There is no solution.

99. The sum of two consecutive integers is 31. What is their product?

A. 224   B. 210   C. 240   D. The product cannot be determined.

100. A runner and bicyclist pass each other at an intersection. The runner is headed north at the rate of 9 mph. The cyclist is headed west at the rate of 12 mph. How long after they pass each other will they be $4\frac{1}{2}$ miles apart?

A. 16 minutes   B. 18 minutes   C. 20 minutes   D. 22 minutes

Answers to Quizzes and Final Exam

Chapter 1
1. D
2. C
3. C
4. B
5. A
6. A
7. C
8. D
9. C
10. B
11. D
12. D
13. A
14. C
15. A

Chapter 2
1. A
2. D
3. C

4. D
5. B
6. C
7. A
8. B
9. C
10. B
11. B
12. A
13. C
14. A
15. B

Chapter 3
1. C
2. C
3. A
4. A
5. D
6. C
7. B

8. A
9. D
10. B

Chapter 4
1. B
2. B
3. C
4. D
5. A
6. A
7. B
8. B
9. B
10. B

Chapter 5
1. D
2. A
3. C
4. B

5. B
6. A
7. B
8. D
9. C
10. A
11. D
12. C
13. D
14. A
15. D
16. B
17. D
18. C
19. A

Chapter 6
1. C
2. D
3. B
4. A

5. D
6. C
7. A
8. B
9. A
10. C
11. C
12. A
13. C
14. D
15. B
16. A
17. B
18. D
19. C
20. C
21. B
22. A
23. D
24. A

Chapter 7
1. D
2. A
3. D
4. B
5. C
6. D
7. C
8. B
9. B
10. D

Chapter 8
1. B
2. A
3. A

4. C
5. C
6. C
7. A
8. B
9. A
10. C
11. B
12. A
13. C
14. A
15. A

Chapter 9
1. D
2. D
3. A
4. C
5. D
6. C
7. C
8. A
9. C
10. B
11. C
12. A
13. D
14. A
15. B

Chapter 10
1. C
2. C
3. A
4. B
5. B
6. D
7. B

8. A
9. B
10. C

Chapter 11
1. A
2. B
3. D
4. B
5. A
6. D
7. A
8. A
9. A
10. C
11. D

Final Exam
1. A
2. A
3. B
4. D
5. D
6. A
7. B
8. A
9. D
10. C
11. C
12. C
13. B
14. D
15. D
16. B
17. A
18. B
19. D
20. B

21. A
22. B
23. C
24. A
25. C
26. B
27. D
28. A
29. D
30. C
31. B
32. D
33. A
34. B
35. A
36. A
37. D
38. C
39. B
40. B
41. A
42. C
43. A
44. B
45. A
46. D
47. A
48. B
49. C
50. D
51. B
52. D
53. A
54. D
55. C
56. D
57. D

58. B	69. C	80. D	91. A
59. C	70. C	81. B	92. A
60. A	71. D	82. D	93. D
61. A	72. B	83. D	94. B
62. B	73. A	84. B	95. C
63. D	74. B	85. B	96. A
64. C	75. A	86. C	97. B
65. C	76. A	87. A	98. D
66. A	77. A	88. C	99. C
67. D	78. C	89. D	100. B
68. A	79. B	90. B	

a p p e n d i x

Factoring with Prime Numbers

Factoring is a skill that is developed with practice. The only surefire way to factor numbers into their prime factors is by trial and error. There are some number facts that will make our job easier. Some of these facts should be familiar.

- If a number is even, the number is divisible by 2.
- If a number ends in 0 or 5, the number is divisible by 5.
- If a number ends in 0, the number is divisible by 10.
- If the sum of the digits of a number is divisible by 3, then the number is divisible by 3.
- If a number ends in 5 or 0 and the sum of its digits is divisible by 3, then the number is divisible by 15.
- If a number is even and the sum of its digits is divisible by 3, then the number is divisible by 6.
- If the sum of the digits of a number is divisible by 9, then the number is divisible by 9.
- If the sum of the digits of a number is divisible by 9 and the number is even, then the number is divisible by 18.

EXAMPLE _____

126 is even and the sum of its digits is divisible by 9: $1 + 2 + 6 = 9$, so 126 is divisible by 18.

4545 is divisible by 5 and by 9 ($4 + 5 + 4 + 5 = 18$ and 18 is divisible by 9).

To factor a number into its prime factors (those which have no divisors other than themselves and 1), we start with a list of prime numbers (a short list is given at the end of this appendix). We begin with the smallest prime number and keep dividing the prime numbers into the number we want to factor. It might be that a prime number divides a number more than once. We stop dividing when the square of the prime number is larger than the number. The previous list of number facts can help us ignore 2 when the number is not even; 5 when does not end in 5; and 3 when the sum of its digits is not divisible by 3.

EXAMPLE _____

- 120: The prime numbers to check are 2, 3, 5, 7. The list stops at 7 because 120 is smaller than $11^2 = 121$.
- 249: The prime numbers to check are 3, 7, 11, 13. The list does not include 2 and 5 because 249 is not even and does not end in 5. The list stops at 13 because 249 is smaller than the next prime number, 17: $17^2 = 289$.
- 608: The prime numbers to check are 2, 7, 11, 13, 19, 23. The list does not contain 3 because $6 + 0 + 8 = 14$ is not divisible by 3 and does not contain 5 because 608 does not end in 5 or 0. The list stops at 23 because 608 is smaller than $29^2 = 841$.
- 342: The prime numbers to check are 2, 3, 7, 11, 13, 17. The list does not contain 5 because 342 does not end in 5 or 0. The list stops at 17 because 342 is smaller than $19^2 = 361$.

EXAMPLE _____

List the prime numbers to check.

1. 166
2. 401
3. 84
4. 136
5. 465

 **SOLUTIONS** _____

1. 166: The prime numbers to check are 2, 7, 11.

2. 401: The prime numbers to check are 7, 11, 13, 17, 19.

3. 84: The prime numbers to check are 2, 3, 7.

4. 136: The prime numbers to check are 2, 7, 11.

5. 465: The prime numbers to check are 3, 5, 7, 11, 13, 17, 19.

To factor a number into its prime factors, we keep dividing the number by the prime numbers in the list. A prime number might divide a number more than once. For instance, 2 divides 12 twice; $12 = 2 \cdot 2 \cdot 3$.

 EXAMPLE _____

Factor 1224.

The prime factors to check are 2, 3, 7, 11, 13, 17, 19, 23, 29, 31.

$$1224 \div 2 = 612$$
$$612 \div 2 = 306$$
$$306 \div 2 = 153$$
$$153 \div 3 = 51$$
$$51 \div 3 = 17$$
$$1224 = 2 \cdot 2 \cdot 2 \cdot 3 \cdot 3 \cdot 17$$

Factor 300.

The prime factors to check are 2, 3, 5, 7, 11, 13, 17.

$$300 \div 2 = 150$$
$$150 \div 2 = 75$$
$$75 \div 3 = 25$$
$$25 \div 5 = 5$$
$$300 = 2 \cdot 2 \cdot 3 \cdot 5 \cdot 5$$

Factor 1309.

The prime factors to check are 7, 11, 13, 17, 19, 23, 29, 31.

$$1309 \div 7 = 187$$
$$187 \div 11 = 17$$
$$1309 = 7 \cdot 11 \cdot 17$$

Factor 482.

The prime factors to check are 2, 3, 7, 11, 13, 17, 19.

$$482 \div 2 = 241$$
$$482 = 2 \cdot 241$$

 PRACTICE

Factor each number into its prime factorization.

1. 308
2. 136
3. 390
4. 196
5. 667
6. 609
7. 2679
8. 1595
9. 1287
10. 540

SOLUTIONS

1. $308 = 2 \cdot 2 \cdot 7 \cdot 11$
2. $136 = 2 \cdot 2 \cdot 2 \cdot 17$
3. $390 = 2 \cdot 3 \cdot 5 \cdot 13$
4. $196 = 2 \cdot 2 \cdot 7 \cdot 7$
5. $667 = 23 \cdot 29$
6. $609 = 3 \cdot 7 \cdot 29$
7. $2679 = 3 \cdot 19 \cdot 47$

8. $1595 = 5 \cdot 11 \cdot 29$

9. $1287 = 3 \cdot 3 \cdot 11 \cdot 13$

10. $540 = 2 \cdot 2 \cdot 3 \cdot 3 \cdot 3 \cdot 5$

What happens if we need to factor a number such as 3185? Do we really need *all* the primes up to 59? Maybe not. We try the smaller primes first. More than likely, one of them will divide the large number. Because 3185 ends in 5, it is divisible by 5: $3185 \div 5 = 637$. Now all that remains is to find the prime factors of 637, so the list of prime numbers to check stops at 23. The reason this trick works is that the prime factors of $3185 = 5 \cdot 673$ are factors of 5 and 673. Once we divide the large number, the list of prime numbers to check is usually smaller.

The first sixteen prime numbers	
Prime Number	**Square of the Prime Number**
2	4
3	9
5	25
7	49
11	121
13	169
17	289
19	361
23	529
29	841
31	961
37	1369
41	1681
43	1849
47	2209
53	2809

Index